Applications of Cognitive Work Analysis

Edited by

Ann M. Bisantz
Catherine M. Burns

CRC Press
Taylor & Francis Group
Boca Raton London New York

CRC Press is an imprint of the
Taylor & Francis Group, an **informa** business

CRC Press
Taylor & Francis Group
6000 Broken Sound Parkway NW, Suite 300
Boca Raton, FL 33487-2742

International Standard Book Number-13: 978-0-8058-6151-8 (Hardcover)

Library of Congress Cataloging-in-Publication Data

Applications of cognitive work analysis / editors, Ann M. Bisantz and Catherine M. Burns.
    p. cm.
Includes bibliographical references and index.
ISBN 978-0-8058-6151-8 (alk. paper)
    1. Human-machine systems. 2. Human-computer interaction. 3. Work environment. I. Bisantz, Ann M. II. Burns, Catherine M. III. Title.

TA167.A658 2009
620.8'2--dc22

2008025079

Visit the Taylor & Francis Web site at
http://www.taylorandfrancis.com

and the CRC Press Web site at
http://www.crcpress.com

A.M.B.
To Sara

C.M.B.
To my parents

Contents

Acknowledgments

First, we are indebted to the numerous contributors who have provided their time and expertise to produce the chapters for this book, and to critically evaluate and improve its contents.

We would also like to thank our professional mentors, colleagues, friends, and students who have supported our continued efforts in the field of cognitive engineering and CWA, particularly Kim Vicente and Alex Kirlik. In many ways, Kim and Alex have guided us on the paths we follow today.

Finally, we would like to thank our families and friends for their continued support in our lives, especially Albert, Sara, Sara, Jack, Judy, Andrew, Gary, and Jack. They support us in these projects but also balance and enrich our lives in so many other ways.

A.M.B.

C.M.B.

Editors

Ann Bisantz is an associate professor of industrial and systems engineering at the University at Buffalo, State University of New York. Dr. Bisantz received her Ph.D. in 1997 in industrial and systems engineering from the Georgia Institute of Technology, with an area of specialization in human–machine systems and minor in cognitive science. She received B.S. and M.S. degrees in industrial engineering at the University at Buffalo. Her areas of research include communicating uncertainty to decision makers using visual and multimodal displays, methods in cognitive engineering, and modeling dynamic decision-making.

Catherine Burns is an associate professor of systems design engineering at the University of Waterloo, Canada. Dr. Burns received her Ph.D. in 1998 in mechanical and industrial engineering from the University of Toronto, with an area of specialization in human factors engineering. She is a licensed professional engineer in the province of Ontario. She received a B.A.Sc. degree in systems design engineering at the University of Waterloo and a M.A.Sc. in industrial engineering at the University of Toronto. Her areas of research include interface design, displays and visualization, and methods of cognitive work analysis applied to power plant control, healthcare, petrochemical refining, and other domains. In 2004 she coauthored a popular textbook on interface design, *Ecological Interface Design* (also with CRC Press).

Contributors

April Bennett
Psychology Department
Wright State University
Dayton, Ohio
USA

Ann M. Bisantz
Department of Industrial and Systems
Engineering
University at Buffalo
The State University of New York
Amherst, New York
USA

Catherine M. Burns
Systems Design Engineering
University of Waterloo
Waterloo, Ontario
Canada

Bruce A. Chalmers
Maritime Information & Combat
Systems Section
Defence R&D Canada–Atlantic
Dartmouth, Nova Scotia
Canada

Colin G. Drury
Department of Industrial and Systems
Engineering
University at Buffalo
The State University of New York
Amherst, New York
USA

William Elm
Resilient Cognitive Solutions, LLC
Pittsburgh, Pennsylvania
USA

Yukari Enomoto
Toronto, Ontario
Canada

John Flach
Psychology Department
Wright State University
Dayton, Ohio
USA

James Gualtieri
Resilient Cognitive Solutions, LLC
Pittsburgh, Pennsylvania
USA

Tom Hughes
General Dynamics AIS
Dayton, Ohio
USA

Greg A. Jamieson
Department of Mechanical and
Industrial Engineering
University of Toronto
Toronto, Ontario
Canada

Ryan M. Kilgore
Charles River Analytics, Inc.
Cambridge, Massachusetts
USA

Tab M. Lamoureux
Brightdata LP (formerly with
Humansystems Incorporated)
Fergus, Ontario
Canada

Gavan Lintern
General Dynamics–Advanced
Information Systems
Kettering, Ohio
USA

Eric Little
Doctoral Programs
D'Youville College
Buffalo, New York
USA

Natalia Mazaeva
Medtronic
Mounds View, Minnesota
USA

Brian McKenna
Resilient Cognitive Solutions, LLC
Pittsburgh, Pennsylvania
USA

Kathryn Momtahan
The Ottawa Hospital–Civic Campus
Ottawa, Ontario
Canada

Neelam Naikar
Defence Science and Technology
Organisation
Melbourne
Australia

Jonathan D. Pfautz
Charles River Analytics, Inc.
Cambridge, Massachusetts
USA

Stacy L. Pfautz
National Security Innovations, Inc.
Boston, Massachusetts
USA

Scott S. Potter
Charles River Analytics, Inc.
Cambridge, Massachusetts
USA

Emilie M. Roth
Roth Cognitive Engineering
Brookline, Massachusett
USA

Sheldon Russell
Psychology Department
Wright State University
Dayton, Ohio
USA

Olivier St-Cyr
IBM Canada Limited
York University
Toronto, Ontario
Canada

Dan Schwartz
Psychology Department
Wright State University
Dayton, Ohio
USA

Jim Tittle
Resilient Cognitive Solutions, LLC
Pittsburgh, Pennsylvania
USA

Chapter 1

Advances in the Application of Cognitive Work Analysis

Catherine M. Burns and Ann M. Bisantz

Contents

Introduction

After more than three decades of interest in cognitive systems engineering, and in the 10 years since Vicente's (1999) book on Cognitive Work Analysis (CWA) was published, there has been an increasing interest in applying the specific techniques of cognitive work analysis to the analysis and design of complex, human-technology systems. Despite this increased interest, there are few (if any) published accounts other than the analyses in the above-mentioned book that use all of the recommended five phases of cognitive work analysis. Most example applications focus on phases of work domain analysis and control task analysis, which are the phases with the most well-developed and theoretically informed models. There have been fewer examples of CWA methods for strategies analysis, social-organizational analysis, and worker skills and competencies analysis perhaps because the tools and models

1

outlined by Vicente were less developed, or because the goals of these phases were being met with tools and techniques from other theoretical traditions.

CWA, as described by Vicente (1999), and adapted from earlier descriptions of Rasmussen (Rasmussen, Pejtersen, & Goodstein, 1994), consists of a five-phase, iterative analysis focusing successively on constraints inherent in the work environment. Work domain analysis creates a functional description of the objects and relationships in the work environment at different levels of abstraction. Control task analysis describes the fundamental tasks that must occur in terms of information observation, ambiguity resolution, procedures, tasks, and the shortcuts that can occur within these tasks. Strategies analysis looks at various different ways that these tasks can be performed and can highlight the need for strategy shifts based on initial information conditions or workload levels. Worker competency analysis describes the behaviors and skills that individual workers must have to work within the work domain. Social and organizational analysis can provide different ways of examining the organizational or macrocognitive influences on work.

Purposes and Integrating Themes

The purpose of this book is to provide examples of all phases of CWA analyses, across a variety of real-world domains, so that readers can understand how these techniques can be applied in practice. Vicente's (1999) book provides a theoretical basis for CWA, develops the five-phase approach, and provides detailed examples of the approach. Importantly, despite the fact that the initial examples provided by Vicente were based on a microworld process-control simulation, the techniques have proven robust and extendable to real-world, large-scale system design problems in a variety of domains, including nuclear and traditional power generation (Burns, Jamieson, Skraaning, Lau, & Kwok, 2007), petrochemical and metal refining (Jamieson & Vicente, 2001; Burns, Garrison, & Dinadis, 2003), emergency response (Chow & Vicente, 2002); healthcare (Sharp & Helmicki, 1998; Enomoto, Burns, Momtahan, & Caves, 2006; Watson, Russell, & Sanderson, 2000), network management (Burns, Kuo & Ng, 2003), military command and control (Bisantz, Roth, Brickman, Gosbee, Hettinger, & McKinney, 2003; Chin, Sanderson, & Watson, 1999), and aviation (Amelink, van Paasen, Mulder, & Flach, 2003). The chapters in this book provide additional, detailed examples addressing all phases of CWA, derived from a number of real applications. A complete CWA analysis is provided for a more complex simulated environment (Chapter 2), and a three-phase analysis for a real-world health care system (Chapter 7). Detailed applications of work domain, control task, and strategies analysis are provided (Chapters 3–6). Additionally, discussions and examples are provided of techniques drawn from research and design traditions other than CWA for the analysis of strategies, social-organizational factors, and worker skills and competencies (Chapters 6, 8, and 9). These techniques can be used to complement and enrich CWA analyses. Later chapters in the book also emphasize important theoretical and application-oriented advances in CWA, related to the integration of CWA within a larger systems-design process (Chapters 10 and 11) and the integration of ontological theories to more rigorously define the links and relationships identified through a work domain analysis (Chapter 12). The final chapter takes a larger view of the progress of CWA as a cognitive engineering technique and outlines both the theoretical underpinnings of CWA and a path for the future of this approach.

Additionally, throughout the book, examples of CWA analyses and extensions to the method demonstrate important themes that have arisen through the development and application of CWA techniques. The first is the importance of carefully defining the system boundary as a key component of the analysis. The choice to include or exclude aspects of the work domain has important implications for the analysis outcome, which are explored by authors throughout the book, particularly in Chapters 2, 3, and 7.

A second important theme is the concept of identifying, using, and documenting connections among the phases of CWA, and between CWA and resulting design concepts. Because many previously published examples focused on only one phase of analysis, it has been hard to identify how the earlier phases of analysis inform later phases. These connections are made explicit in several of the chapters (including Chapters 2, 5, and 10), as well as connections from CWA to the larger systems-design process and to explicit design recommendations (Chapters 5, 6, 10, and 11).

Third, the chapters illustrate the variety of methods that analysts are using to elicit information about the domain and those from expert practitioners. Methods such as interviews, observations, and document analysis are common across CWA analyses and consistent with cognitive systems and task analyses in general (see Chapter 5, 6, and 7, and also Bisantz & Roth, 2008, for an extensive overview). Chapter 8 suggests a variety of new, sociological methods that can be added to the repertoire of knowledge-gathering and analysis methods.

Fourth is the diversity of approaches taken by different analysts in applying the methods of CWA. Even within the work domain and control task analysis phases, which are guided by well-documented and explicit theoretical structures (e.g., the abstraction hierarchy and decision ladder), the examples provided in this book illustrate a diversity in system characteristics, analysis purposes, analysis components, and methodologies, which demonstrate the flexibility with which these methods can, and should, be tailored to meet the analyst's purposes and goals. Analysts are increasingly developing knowledge elicitation techniques specific to CWA phases (see Chapter 7 for examples) as well as submethods (Chapters 2, 4, and 5) and micro-artifacts (Chapters 2, 4, 5, and 7) across CWA phases for representing CWA products. Lintern emphasizes the "power of a theoretically motivated and well-organized set of representations for assimilating, building, archiving, and transferring knowledge" (p. 323). Representational forms from other traditions are also applicable (e.g., the social network and link models described in Chapter 8, or hierarchical task models from Chapter 9). Performing CWA that is useful and informative for design purposes is more than following a set of steps or filling in boxes: it requires deep analytical understanding of aspects of the work domain and practitioner behaviors in order to provide real insights and design guidance. The structured outputs are representations and the means of communicating that understanding, rather than ends in themselves. As noted by Kilgore, St-Cyr, and Jamieson in Chapter 2, "it is the job of the cognitive work analyst to critically leverage the outputs of knowledge elicitation activities to support construction of formative models of the work domain" (p. 44). Later chapters of the book provide examples of the diverse ways CWA fits within the systems design process (Chapters 10 and 11), and the variety of theoretical foundations relevant to CWA (Chapters 12 and 13).

Chapter Contributions: Applications and Themes

The degree to which these themes are represented throughout the book, as well as the particular analysis phases or extensions provided by each chapter, is reviewed here.

Chapter 2 by Kilgore, St-Cyr, and Jamieson provides an example of a complete CWA of a limited scope system: an air traffic control simulation. In this chapter, the authors demonstrate canonical applications of the more commonly found CWA phases (work domain analysis using an abstraction hierarchy, and control task analysis using a decision ladder) and demonstrate the use of an "information flow map" for documenting strategies analysis and the allocation of activities across different human and computerized actors in the domain (social-organizational analysis). Although this tool has been suggested previously (Vicente, 1999) few examples exist in the literature. Finally, the authors apply a novel tool (an SRK inventory) for the worker competency phase, to document display design suggestions and worker competencies related to skill-, rule-, and knowledge-based behaviors associated with the strategies.

In addition to providing a complete, multiphased example of the analysis, the authors provide a useful illustration of how outputs and insights from the different phases of analysis can be linked, and also how the determination of system boundaries necessarily determines contents and aspects of the analysis. For instance, the generalized functions identified in the WDA modeling phase were used to guide identification of control tasks in the second phase. Information-processing activities and knowledge states were linked to AH components. Strategies were expressed as further decompositions of aspects of the corresponding decision-ladder components (e.g., a formulated procedure). Additionally, the authors describe how control tasks (e.g., "rerouting") can occur in different contexts, which trigger different subcategories of tasks and potential strategies. The social-organizational analysis showed how strategy components could be allocated across actors or agents in the domain. Later chapters (particularly, Chapters 10 and 11) provide further insights regarding how different analysis phases as well as analysis outputs and design artifacts can be linked, particularly in very large, long-term projects.

Chapter 3 by Bisantz and Mazaeva provides two detailed examples of work domain analysis. These examples illustrate the diversity in the application of work domain analysis across different types of systems for different purposes of analysis and at different levels of detail. Specifically, the chapter contrasts case studies of two systems: one with primarily physical constraints (an automated camera), and one with important intentional constraints (an emergency response system), done for different purposes (modeling to support an understanding of system-automation interactions, and modeling to support identification of information needs) at very different levels of detail. An important component of the modeling in both cases was the determination of system boundaries, which explicitly included automation (in the case of the camera) and the physical environment and civilian population (in the case of the emergency response system) in order to illustrate constraints and information needs that result from interactions among these systems. Bisantz and Mazaeva discuss the importance of considering automation within the system to be analyzed, noting that "having appropriate models of automation functioning is critical to safe system operation—in a way that is not fundamentally different from the knowledge and understanding required of other (nonautomation) system components" (pp. 52).

In **Chapter 4,** Naikar focuses on the first two phases of CWA—work domain analysis and control task analysis. Whereas much of the previous CWA work is directed towards interface design, Naikar extends the application of CWA successfully to the areas of evaluating design proposals, designing teams, examining training needs, and defining training requirements, stating, "Although cognitive work analysis has become widely known as a framework for designing ecological interfaces, the relevance of this approach to other applications is not as well recognized" (p. 70). Case studies are presented in which new techniques and ways to apply and interpret CWA methods are described. Thus, Chapter 4 not only emphasizes the diversity of purpose with which CWA phases can be applied but provides examples of diversity in analysis methodologies and new representations, such as the use of an abstraction-hierarchy framework for representing potential training needs.

Chapter 5 by Lamoureux and Chalmers provides multiple applications of control task analysis to describe complexities of military command-and-control tasks. Similar to the analysis in Chapter 2, the authors indicate how the control tasks modeled can be related to locations within an AH model. In addition to providing detailed examples of how tasks can be represented with a decision ladder, and how analysts and SMEs can collaborate to complete the analysis, the chapter focuses on design traceability and demonstrates the diversity in CWA approaches and methods. For instance, the authors seek to provide a "traceable design thread that directly links knowledge elicitation and work analysis outputs to specific design hypotheses for supporting operator work demands" (p. 96) through the creation of design "seeds" that "represent some specific and relatively independent design concept to support some specific aspect of the work" (p. 98). These seeds can be evaluated through SME feedback and through iteratively developed and tested prototypes. Design outputs based on these seeds were related to classes of activities (e.g., seeds related to supporting perception/understanding; external representations of important constructs; support for resolution of ambiguity/goal conflicts; or the support of response development, selection, and execution). In many cases, opportunities for automation of response were identified, thus, in some sense, performing a social-organizational analysis as well.

Additionally, the authors demonstrate two diverse approaches to control task analysis. Although both approaches use the decision ladder model, one approach relied on SME task descriptions to populate the model, whereas the other approach had analysts and SMEs work together to produce an inventory that attempted to capture all possible activities, processes, and knowledge states associated with a higher-level activity. For instance, they catalogued all the situationally dependent instances and triggering information for process "activation." Lamoureux and Chalmers suggest a hybrid approach as well, where one would create the exhaustive inventory of states and inputs and then specify paths (or ladders) corresponding to different circumstances, each utilizing different aspects of the inventory. This has similarities to the strategies analysis described in Chapter 2, where multiple strategies were identified for subcategories of tasks, based on different contexts.

As noted by the authors, "These results plainly show several worthwhile paths toward generating useful design seeds. As such, CTA [cognitive task analysis], and CWA more generally, comes well recommended for use in future research concerning complex, dynamic, cognitive work domains. In addition, the fact that neophytes to this analysis, SMEs and analysts alike, could grasp CWA and its outputs as quickly and easily as they did lends further support to the validity of this relatively new cognitive research approach" (p. 126).

Chapter 6 by Roth provides a detailed description of how to identify strategies used by practitioners (through case examples from process control and railroad maintenance operations), and the role that strategy analysis serves compared to other phases of CWA, stating that "empirical approaches to strategies analysis examine the actual strategies used by domain practitioners. They provide an efficient way to uncover knowledge and skills that underlie expert performance; strategies that have evolved to compensate for limitations in the current environment (e.g., workarounds); and 'buggy knowledge' and suboptimal strategies that characterize less skilled performance that can lead to inefficiencies and error. The results can point to opportunities for improvements through new training and support systems" (p. 145).

Roth notes the diversity of approaches to strategy identification and analysis. In a formative approach, strategies are based on a work domain analysis (in some sense, like the inventory analysis suggested in Chapter 5, the formative analysis provides a "space of possible strategies"). Such analyses are "useful early in design, particularly for first-of-a-kind systems, as a way to establish requirements for displays, training, and/or automation" (p. 145). The descriptive approach focuses on identifying strategies that are descriptive of practitioner expertise. These are necessarily cognitively feasible and, additionally, reveal the knowledge and skills required to accomplish the task (and in this sense, contribute to the last CWA phase). Additionally, descriptive strategies can reveal faulty strategies (e.g., those used by more novice controllers that could lead to problems). Unlike the previous chapters, Roth does not utilize a specific representational form for presenting strategies (i.e., the information flow maps from Chapter 2). Although Roth does not represent strategies in terms of CWA constructs, it is possible to see the mapping from the strategies she describes to more typical CWA representations. For instance, the description of the strategy railroad dispatchers use to maintain awareness of the location of railroad roadway workers provides a specific set of activities that could be mapped to the alert, observation, and interpretation segments of the decision ladder.

Chapter 7 by Burns, Enomoto, and Momtahan describes a project that integrated three phases of cognitive work analysis: work domain analysis, control task analysis, and strategies analysis, for a cardiac care tele-triage environment. This chapter provides an example of a project that used multiple CWA phases and specifically notes the need for analysts to acknowledge the types of information (e.g., interactions with hospital management and economic goals; skills and competencies of nurses) that would not be provided in the analysis, thus making analysis boundaries explicit. Burns et al. provide detailed examples regarding their information-gathering methods, such as sample interview questions, and, similar to having SMEs generating decision ladders as in Chapter 5, asking nurses to generate typical call sequences in order to understand triage processes and questioning strategies. As in Chapters 2 and 3, there is a discussion of how and why the system boundary was defined for analysis. In this case, the boundary was determined based on what the nurses could "control," including some aspects of the patient and some care resources; additionally, medications were included because these exert control over patient systems. This is analogous to the inclusion of automation in the WDA analysis described in Chapter 3, where the automation was used to control aspects of the camera system. Chapter 7 also provides a useful example of how to model patient health in terms of physiological processes.

Like the control task analysis in Chapters 2 and 5, which include the task context as an important consideration, the control tasks analysis of the tele-health nurses introduces the notion of patient "modes" (e.g., pre- or postsurgery), which lead to different contexts for the control tasks. Strategies were identified that added detail to components of the decision ladder (DL) analyses (e.g., methods by which nurses acquired "observations"). Burns et al. introduce a new formalism (a question sequence chart) for explicating strategies, as well as provide insights regarding methods for enumerating and distinguishing among strategies: "A key clue to differentiating strategies from other aspects of cognitive work is that strategies differ between workers and, in some cases, with work context. ... In contrast, work domain constraints hold across all work, and control tasks identify models of operations that are also common to all workers. ... Strategies can be thought of as an energy state—the worker, given his or her background, experiences, and context, will work in that state until some perturbation arises that changes the state of work. At this time, the worker may shift into a different strategic pattern" (pp. 171).

Chapter 8 by Pfautz and Pfautz provides new insights for an important but less documented phase of CWA—social-organizational analysis. They note that this analysis phase is critical because

- Work is situated not only in a physical environment but also in a social environment.
- Work can include communication of data, information, and knowledge.
- Work is often performed collaboratively and cooperatively.
- Social interaction self-evolves and adapts through work. (p. 177)

Examples of social-organizational analysis in this book, as well as other studies, have tended to use CWA representations and formalisms to show the mapping of different functions and phases of activities to different actors (e.g., in Chapter 2, showing what actors/agents are active during different phases of activity). Other researchers have used interconnections among decision ladders to show interaction across different timescales and organizational levels (e.g., Mazaeva & Bisantz, 2007, showed how control tasks are allocated across time, and across the designer, automated system, and camera user).

Chapter 8 seeks to expand social-organizational analyses beyond the understanding or design of function allocation to the analysis of social and organizational relationships and constraints, using formalisms and theories from disciplines such as organizational psychology, management science, and social science. As the authors note, social and organizational analysis is "an important aspect of CWA for which there are many possible approaches but little guidance on their applicability" (p. 176). Pfautz and Pfautz discuss in a detailed way how techniques for analyzing social organization and cooperation can inform CWA in terms of design decisions regarding function allocation, automation schemes, and technology support and design, in part because they help in identifying relevant system entities. The focus is on identifying the types of relationships between entities such as communication, or coordination/consensus relationships, and their implications for system design decisions. The authors describe and contrast methods for capturing data about these relationships (e.g., through observations, self reports, or simulation), methods for processing data (e.g., automatic text processing, data mining), and analytic frameworks such as social network and link analysis for understanding patterns in the data.

In **Chapter 9**, Drury discusses the historical background of CWA and task analysis (particularly HTA). In particular, he describes how task analysis has evolved from systems design, whereas CWA has evolved from a sociotechnical systems perspective. Whereas task analysis evolved from the breakdown of tasks, CWA has evolved from the perspective of breaking down functions.

Despite these historical differences, Drury shows how both approaches meet the common goal of the analysis of human work to identify demands and improve design. He demonstrates how the two approaches, when pursued carefully and thoroughly, can elicit many of the same aspects of human work. Drury emphasizes that the breakdown of tasks, often referred as task analysis, is really only a starting point for a deeper analysis of human work. Through the example of aircraft inspection, he shows how a thorough task analysis can identify rule and knowledge-based work, worker strategies, and social organizational factors. HTA techniques can complement CWA techniques in the first, third, and fourth phases of CWA, as well as in the worker competencies phase of CWA. As Drury notes, "using the Plans function of HTA, a broader interpretation is possible. As shown, there is clearly a role for HTA as currently used in Phase 3 of CWA, although Phase 5 is still the obvious place to use HTA within a CWA context" (p. 246). The net result, assuredly, is that by becoming more deeply versed in each of these techniques, whether the starting point is analyzing tasks, or functions, the human factors analyst gains new and fruitful dimensions for analyzing human work.

In **Chapter 10**, Elm, Gualtieri, Tittle, Potter, and McKenna present a concrete way to work from CWA analyses to design ideas, while maintaining design rationale. This approach takes advantage of Woods' (1991) Mapping Principle to extract design insights and create links between analysis and design. Maintaining these strong connections is an important aspect of integrating CWA in a systems-engineering process. In the words of these authors, "It is insufficient to have a hard break in the analysis-to-design effort or to have vague information requirements from the CWA serve as the basis for design. Rather, it is essential to have specific CWA artifacts that explicitly link to both specific content of the information spaces and specific properties of the representation" (p. 255).

Within this chapter the mapping principle and its implications for design based on CWA are broken down. A critical contribution of this chapter is the emphasis that the external world modeled by CWA must, through the design, be mapped in a compatible way with the operator's internal mental models and cognitive structures. This gap is at the core of the design problem. How well the designer can span this gap, while at the same time representing the true nature of the external world (i.e., "inaccurate" operator models should not serve as a design template), will be reflected in how well the operator can use the functional information of CWA.

To span this design gap, Elm et al. have proposed a series of intermediate artifacts that work like stepping stones between the CWA and the eventual design. These artifacts demonstrate both how information developed in earlier phases of analysis in used in further analyses, and how analysis outputs are linked to design (similar to the production of "design seed" in Chapter 5, or the technology implications based on the strategies identified in Chapter 6). Elm et al. highlight how strong interface design practices fit within an applied CWA framework and outline a structure for decision-centered testing, in essence closing the system engineering loop and effectively showing how CWA can stretch across the entire systems-engineering life cycle.

In **Chapter 11**, Flach, Schwartz, Bennett, Russell, and Hughes also describe the role of CWA in the system life cycle. Rather than characterizing CWA as a static analysis completed before a design, they view it as a living and evolving analysis that evolves with system design and changes over time. In their words, "work analysis will never be complete. In other words, the horizon of any work analysis will always be finite/limited—such that most work systems will outlive any specification based on finite work analysis—regardless of the framework used for that analysis. For a system to be truly adaptive, it will be required that work analysis be integrated as part of the system life cycle" (p. 280). This broad and dynamic view describes not what CWA as an approach currently is but rather what it could be.

To support this concept, these authors provide the example of a tool to allow a CWA to evolve and change over time. Largely based on the idea of a dynamically changing database, the tool is both a reminder of what analytical goals have been met and a future resource for new analytical work. This tool broadens the notion of links among analysis phases to include links across time and analysts (supporting organizational knowledge and reuse), and to links from raw data (e.g., interview snippets, primary source documents) and analysis results to sources of information and to different intersections of system goals, functions, objects, and constraints. They demonstrate how the tool works in the context of a military targeting system.

Chapter 12 by Little diverges from previous chapters in that it addresses a fundamental characteristics of one CWA component: the nature of the relationships and entities described through work domain modeling. Little discusses Work Domain Analysis (WDA) from an ontological perspective. He analyses the structure of objects and links that we commonly see in WDA and compares WDA with Task Analysis (TA), another frequently used cognitive engineering artifact. His analysis shows clearly that these two approaches generate structures that are fundamentally different. This provides strong support to argue that WDA and TA can independently complement a design process. In his words, "Although WDA and TA certainly share relationships to one another, it is important to distinguish between them because, as will be discussed later in this piece, the formal structure of a work domain is metaphysically distinct from the formal structure of tasks and behaviors carried out within that domain" (p. 304).

Little describes how the formal relationships found in ontological models can be mapped to the relationships included in an abstraction hierarchy model, and how conceptual distinctions in ontology (e.g., between ontologies that describe reality at a point in time and those that describe processes that occur over time) are expressed within CWA. Little explores the ontology of WDA further through example analyses of a home and emergency management. His ontological perspective provides a theoretical grounding for the relationships typically included in CWA, particularly within the abstraction hierarchy, and also provides directions for further development of these ideas. In particular, by considering other relationships typical in an ontological analysis, there may be opportunity for work structures to be expanded to model more relationships, or for more explicit use of inheritance relationship to be used in generating outputs and design concepts based on work domain modeling. For instance, nodes at the level of physical form might inherit properties (e.g., location, status) based on their nature (e.g., people, vehicles, physical structures) that might need to be displayed to system operators.

Chapter 13 by Lintern provides a conclusion full of challenges for CWA and indeed any cognitive engineering approach that is trying to understand human work more fully. Lintern begins his chapter by outlining the current challenges to CWA. He argues that CWA is often misunderstood, and that this misunderstanding may result from an inadequate explanation of the roots of CWA. As an approach, CWA is formulated for solving larger and more complex problems than other human factors engineering techniques. It has had a strong focus on knowledge representation over knowledge elicitation. The notion of functional structure as a determinant of human behavior is also a challenge.

To address these challenges, Lintern provides a thorough foundation of theories that reflect into the concepts of CWA. He explores CWA's roots in ecological psychology, situated cognition, and distributed cognition. In all of these approaches, the notion exists that human behavior adapts to the complexity of its surroundings. Continuing with this theme, Lintern also explores systems theory and Ashby's Law of Requisite Variety, and Self-Organizing Systems. He convincingly shows that ecological interface design, rather than reducing complexity, instead controls complexity, and in this way is compatible with Ashby's law.

Lintern argues that there is structure in human work and that this is a fundamental principle of CWA. He proceeds to demonstrate through the five levels of CWA how this structure can be explored at each and every level. Finally Lintern leaves the reader and the CWA community with important challenges for the future. Can problem-solving protocols be as effectively mapped on to other knowledge representations as on to the abstraction decomposition space? Can CWA take advantage of its knowledge representations to contribute back to the literature on reasoning and problem solving? Can the role or importance of key information outlined in a work domain analysis be validated or explained further through deeper experimental work that explores the sensitivity of ecological designs to various information requirements? Can CWA scale to larger problems or cast insights into questions of human work at various scales of magnitude? Above all, Lintern emphasizes the importance of an individual exploration of CWA, as this promising approach continues to develop.

Conclusions

To conclude, CWA as an approach has matured in the last 10 years. The phases are more defined, and there are many more examples in the literature. Researchers and practitioners in the field continue to enrich and expand the theoretical and applied components of CWA. The most notable developments have been adding a richer set of knowledge elicitation techniques, intermediate modeling artifacts, and real-world examples demonstrating how the techniques can be, and are, applied in practice. The overall impact of these new developments has been to shorten the analysis time required when using CWA, improve its learnability, and decrease the gap between analysis and design. There are now numerous examples of how CWA can be used effectively in a timely manner to generate useful design insights, providing examples for both students and practitioners of the approach, and providing evidence that CWA is a mature technique that should be a part of every cognitive engineer's toolbox.

References

Amelink, M. H. J., van Paassen, M. M., Mulder, M., & Flach, J. M. (2003) Applying the abstraction hierarchy to the aircraft manual control task. *Proceedings of the 12th annual symposium on aviation psychology.* April 14–17, 42–47.

Bisantz, A. M., & Roth, E. M. (2008). Analysis of Cognitive Work. D. Boehm-Davis (Ed.) Volume 3 of the *Annual Review of Human Factors and Ergonomics*, Human Factors and Ergonomics Society, Santa Monica, CA.

Bisantz, A. M., Roth, E., Brickman, B., Gosbee, L. L., Hettinger, L., & McKinney, J. (2003). Integrating Cognitive Analyses into a Large Scale System Design Process. *International Journal of Human-Computer Studies*, 58, 177–206.

Burns, C. M., Garrison, L., & Dinadis, N. (2003). WDA for the petrochemical industry. *Proceedings of the 47th Annual Meeting of the Human Factors and Ergonomics Society*, 258–262.

Burns, C. M., Jamieson, G. A., Skraaning, G., Lau, N., & Kwok, J. (2007). Supporting situation awareness through ecological interface design. *Proceedings of the 51st Annual Meeting of the Human Factors and Ergonomics Society*, 205–209.

Burns, C. M., Kuo, J., & Ng, S. (2003). Ecological interface design: A new approach for visualizing network management. *Computer Networks*, 43, 369–388.

Chin, M., Sanderson, P., & Watson, M. (1999). Cognitive work analysis of the command and control work domain. *Proceedings of the 1999 Command and Control Research and Technology Symposium,* United States Naval War College, Newport, RI, June 29–July 1.

Chow, R., & Vicente, K. J. (2002). A field study of emergency ambulance dispatching: implications for decision support. Paper presented at the *Human Factors and Ergonomics Society Conference,* 30 September–4 October 2002, Baltimore, MD, 313–317.

Enomoto, Y., Burns, C. M., Momtahan, K., & Caves, W. (2006). Effects of Visualization Tools on Cardiac Telephone Consultation Processes. *50th Annual Meeting of the Human Factors and Ergonomics Society*, San Francisco, CA, 1044–1048.

Jamieson, G. A., & Vicente, K. J. (2001) Ecological interface design for petrochemical applications: Supporting operator adaptation, continuous learning, and distributed, collaborative work. *Computers and Chemical Engineering*, 25: 1055–1074.

Mazaeva, N., & Bisantz, A. M. (2007) On the Representation of Automation Using a Work Domain Analysis. *Theoretical Issues in Ergonomics Science*, i-print (online preview) 1–22.

Rasmussen, J., Pejtersen, A. M., & Goodstein, L. P. (1994). *Cognitive Systems Engineering*. New York: John Wiley.

Sharp, T. D., & Helmicki, A. J. (1998) The application of the ecological interface design approach to neonatal intensive care medicine. *Proceedings of the 42nd Annual Meeting of the Human Factors and Ergonomics Society*, 350–354.

Vicente, K. J. (1999). *Cognitive Work Analysis*. Mahwah, NJ: Lawrence Erlbaum Associates.

Watson, M., Russell, W. J., & Sanderson, P. M. (2000). Anesthesia monitoring, alarm proliferation, and ecological interface design. *Australian Journal of Information Systems*, 7, 109–114.

Woods, D. (1991). The cognitive engineering of problem representations. In G. R. S. Weir and J. L. Alty (Eds.), *Human–Computer Interaction and Complex Systems*. Academic Press, London.

Chapter 2

From Work Domains to Worker Competencies: A Five-Phase CWA

Ryan M. Kilgore, Olivier St-Cyr, and Greg A. Jamieson

Contents

Overview

In this chapter we describe the process of applying all five phases of a Cognitive Work Analysis (CWA) to a single domain that reasonably approximates the complexity of a real-world sociotechnical system. For each of these phases, we provide an overview of analytical activities performed by the CWA practitioner and explicitly place the inputs and outputs of these activities within the context of preceding and subsequent phases. Where concrete descriptions of analytical methodologies are lacking in prior texts—in particular for the Phase 5 worker competencies analysis—we introduce tools and techniques developed through our own experiences in applying the framework. Through this process, we establish a basic template for performing a CWA that may be extended by practitioners to larger, more complex sociotechnical systems. It is our hope that this chapter will help students and practitioners alike to develop a better understanding of the interrelationships between the five phases of CWA and the methods used to fully implement the CWA framework.

Introduction

Vicente's (1999) seminal text on Cognitive Work Analysis provided a pedagogical account of a framework for the analysis and design of work in complex sociotechnical systems. This account was necessary because many practitioners and scientists had encountered difficulty learning and applying CWA based on the literature available at the time (e.g., Rasmussen, Pejtersen, & Goodstein, 1994). The text addressed these difficulties by encapsulating CWA in a coherent theoretical framework and couching the tools developed by Rasmussen and his colleagues within that framework. The surge in applications of CWA since the book's publication is compelling evidence that it was effective in meeting its pedagogical goal.

Although use of CWA is growing, the spread of the framework has followed more of an apprenticeship model than a classroom or self-learning model (Naikar, Hopcroft, & Moylan, 2005). A likely reason for this is that Vicente's text primarily addresses the *what* and *why* questions of CWA, leaving the methodological *how* question only partially answered. For this reason, learning how to apply the concepts and tools of CWA is still largely facilitated by having an experienced mentor. Our hope is that this current book will better enable selflearning of the framework by addressing three key limitations of Vicente's (1999) text that have not yet been fully addressed in the CWA literature:

- First, the case study of the DURESS II microworld provided by Vicente is incomplete, lacking a Social Organization Analysis. In fact, there are very few published accounts of "complete" CWA available in the literature, with practitioners rarely going beyond the first two phases of the five-phase framework (although, see Cummings & Guerlain, 2003; and Chapter 7 in this book). This paucity of fully worked examples gives skeptical students a potent objection to producing a complete analysis themselves and may also stand as a barrier to industry practitioners interested in adopting CWA techniques.
- Second, although the text does an excellent job of relating the five analytical phases to an organizing theory, it provides little practical guidance on integrating across them. Students usually grasp the concept that each phase of analysis reveals new constraints that are nested within those identified in the preceding phase, but there is no detailed demonstration of this principle in practice. Instead, the output of each analytical phase appear to be unrelated to each other. Again, the broader CWA literature does little to alleviate this shortcoming (although, see Hajdukiewicz & Vicente, 2004; Jamieson, Miller, Ho, & Vicente, 2007).
- Third, the DURESS II microworld is, at best, a marginally complex system. Although it is sufficient to exemplify most of the phases of analysis and to demonstrate the separate analytical tools in action, it pales in comparison to the complex systems for which use of CWA is advocated. Students often express unease in scaling the application of CWA to the analysis of real-world systems.

Our purpose in applying all five phases of CWA to a single system was pedagogical in nature. The first two authors came together as an analytical team within the context of a graduate CWA course. Responding to a challenge to identify a work system of manageable complexity, the students suggested the air traffic control microworld introduced in the following text as an object of analysis. To manage their workload, the analysts selected a challenging and interesting subset of the domain, at the completion of each phase, to develop further in the succeeding phases. The result is a CWA that traverses all five analytical phases but only for an increasingly narrow slice of the work domain. In other words, the CWA is complete in *depth*, but not in *breadth*.

It is worth emphasizing that the goal of this analysis was not to develop a "correct" and complete CWA for any existing air traffic control system. Rather, the goal was to deepen the analysts' understanding of the framework, associated tools, and interrelationships of the individual analytical phases. As a course-based application of the framework, the modeling effort was severely limited in terms of time, funding, and access to domain experts. Although these limitations may have impacted the overall robustness of the resultant domain model, we believe that this exercise still demonstrates the potential design value of CWA in real-world domains.

In the remainder of this chapter, we describe our application of the five phases of the CWA framework to a simulated Terminal Radar Approach Control (TRA-CON) aviation microworld. For each phase, we first revisit the purpose of the analysis and then provide a description of the conceptual tool used to support its execution. Following this, we discuss our methodology for implementing this tool within the context of the respective analytical phase as well as that of the greater CWA as a whole. In doing so, we highlight the key activities associated with completing each analytical phase, as well as how these activities are fed by the products of prior phases within the CWA framework.

Domain Description

TRACON for *Windows* (TRACON, 1991) is a microworld simulation of an air traffic control environment. The TRACON domain consists of an airspace surrounding a major airport and its regional satellite airports. Within the domain, a radar system continuously scans the skies. Aircraft, along with their relevant flight parameters (i.e., flight number, altitude, destination), are presented on the radar display screen as they enter the operator's boundary of control (Figure 2.1). Other pieces of contextual information, including airports, airways, radio beacons, intersection fixes, instrument landing systems, and significant ground markings, are

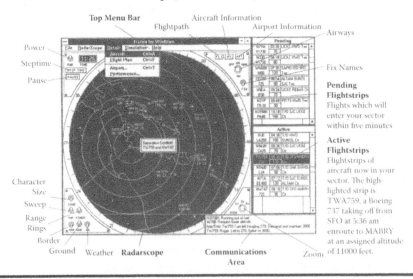

Figure 2.1 Visual interface of the TRACON simulator, from *TRACON for Windows,* **1991.**

also depicted in the TRACON display. Using information from the radar system, as well as automated flightstrip information and text-based communications with aircraft and other neighboring control zones, operators must handle all aircraft in their designated zone. This involves keeping the aircraft on course, vectoring them in and out of airports, and handing them off to adjacent controllers. The fidelity of the TRACON microworld is representative of many existing air traffic control systems, although at a significantly reduced scale.

Phase 1: Work Domain Analysis

Our CWA of the TRACON microworld began with a detailed investigation of the work domain. The purpose of this first analytical phase was to identify the underlying functional structure of the TRACON microworld, as well as the physical and intentional constraints that must be respected to achieve the system's intended purposes. This analysis provided a foundation for subsequent phases by establishing a representation of the functional ecology of the TRACON simulator.

Phase 1 Analytical Tool: The Abstraction Hierarchy

The abstraction hierarchy (AH; Rasmussen, 1979, 1985) served as the primary tool in supporting the work domain analysis (WDA). The AH consists of five different levels, each containing critical information regarding the work domain to be modeled: *functional purpose, abstract function, generalized function, physical function,* and *physical form.* Links between the levels of the AH are means–ends or how–why relationships. The AH is also decomposed across the part–whole dimension, typically using three levels of resolution: *whole system, subsystems,* and *components.* The AH is the most commonly applied tool of the CWA process. Numerous well-documented examples are presented in the literature (e.g., Bisantz & Vicente, 1994; Ham & Yoon, 2001; Jamieson et al., 2007; Reising & Sanderson, 2002), although there are ongoing discussions regarding the techniques and outputs of work domain analyses (e.g., Miller, 2004; Naikar et al., 2005). As a tool for performing the first phase of CWA, the AH provides a structured approach for identifying both the means–ends and part–whole relationships of the TRACON domain. This allows for a full depiction of the necessary constraints to achieve the work domain's purposes, *independent* of any actions and actors. It also aids in identifying the underlying ecology and limitations of the domain (i.e., the space in which operator control *must* occur to satisfy purposes). The outputs of this analysis are a series of well-defined diagrams.

Phase 1 Methodology

The abstraction hierarchy served not only as a tool for capturing the outputs of the work domain analysis but also as a means of structuring the Phase 1 analytical activities themselves. We approached our analysis of the domain simultaneously from both ends of the abstraction hierarchy. First, we identified the Functional Purposes of the domain and enumerated the individual physical components comprising the TRACON simulator ecology (Physical Functions). Following this, we progressed inwards from the top and bottom of the abstraction hierarchy, populating the Abstract and Generalized Function levels by asking how—why or means—ends relationship questions while transitioning between levels. This approach was informed by existing examples of WDA in the literature, including those of Vicente (1999) and Burns and Hadjukiewicz (2004).

Although it is an idealized microworld, the TRACON simulator still represents a very complex work domain. To accomplish the WDA and produce deliverables within aggressive time limitations, we paid careful attention to developing appropriate domain boundaries, which play a critical role in shaping the overall CWA effort (Burns et al., 2004). Generally speaking, components relevant to directing aircraft both within and across regional airspaces were included in this analysis. Aspects of the system that we determined to be outside the domain boundary included the activities of area control centers that handle high- versus low-altitude traffic, as well as physical subcomponents of aircraft (e.g., engines, control surfaces, landing gear). Although these omitted subcomponents are part of the overall air traffic control domain, they fall outside the responsibility of individual TRACON centers. In addition, the TRACON simulator environment used in this analysis is itself limited to single-operator air traffic control cells and does not address the shared, collaborative monitoring of cells by multiple human operators. It is important to note that our decisions to limit the scope of the WDA in these ways impacted all subsequent phases of our cognitive work analyses, particularly the Phase 4 social organization analysis.

As the Phase 1 analysis progressed, our understanding of the domain deepened. Through several iterations, we refined the contents of the abstraction hierarchy table while focusing on maintaining a common language within each level of abstraction. We also simultaneously developed additional part–whole decomposition tables to accommodate the organization of domain components across multiple levels of detail. Throughout this process, the evolving domain depiction was continually assessed by explicitly questioning the nature and validity of each relationship represented within the abstraction hierarchy and part–whole decomposition documents. The accuracy and completeness of the means–ends links between each node of the hierarchies and decompositions were probed repeatedly by asking how–why questions while traversing these links across the four levels of domain abstraction that we investigated.

Several information sources supported our WDA activities. Primary among these were the operating manuals for the domain. These references were particularly useful for identifying the physical equipment modeled within the simulator, as well as the goals, functions, and laws governing the domain ecology. Unfortunately, project constraints made it infeasible to observe domain experts engaged in air traffic control activities using the simulator. However, we were fortunate to have the opportunity to discuss the TRACON domain with a subject matter expert who was knowledgeable of both the simulator and real-world air traffic control systems. We used these discussions to both confirm and augment our developing domain representation.

Phase 1 Products

Our team captured concrete products of the Phase I analytical activities at various levels of detail through a series of abstraction hierarchy and part–whole decomposition diagrams. An example of one such abstraction hierarchy diagram, in this case providing a coarse-grained overview of the entire domain, is shown in Figure 2.2.

Through our initial analysis we identified two functional purposes of the work domain: routing aircraft safely and routing aircraft efficiently. Six abstract functions (Figure 2.2, level 2) serve as the means of satisfying these two functional purposes. They were:

- Aircraft responsibility: each and every aircraft is the responsibility of exactly one air traffic controller at all times.
- Pilot situation awareness: the understanding of all relevant air traffic, terrain, and weather constraints.
- Maintenance of a field of safe travel: the minimal distances required between aircraft at all times.
- Performance abilities of individual aircraft: the limits on parameters to keep a given aircraft safely traveling in the air, including minimum speed, landing approach speed, turning radius, etc.
- Passenger comfort parameters: the flight parameters needed to keep passengers comfortable, including limits on rates of climb and descent, climb/descent cycles, turning radius, etc.
- Scheduling demands: landing, takeoff, and handoff times specified by existing airline or airport schedules.

We also identified four subsystems of generalized functions within the domain (Figure 2.2, level 3). These subsystems comprised of processes and attributes involving negotiating with neighboring regional ATC centers, establishing and updating aircraft flight paths, locomotion of aircraft through the sector, and the transitioning between the zones of responsibility of neighboring regional ATC centers.

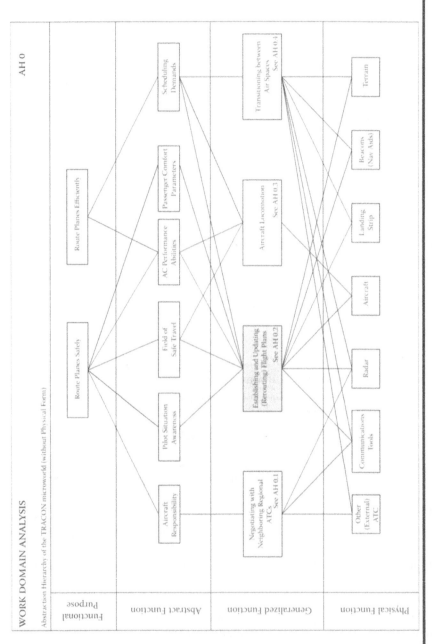

Figure 2.2 Main overview of the abstraction hierarchy of the TRACON domain.

We recorded the individual components contributing to these four subsystems using separate part–whole decomposition diagrams. For example, the generalized function "establishing and updating flight plans" (Figure 2.2) is comprised of seven unique components (Figure 2.3), which can themselves be further decomposed.

Finally, we identified seven primary components at the domain's physical function level (Figure 2.2, level 4). These components served as means to achieving the ends represented in the generalized function level of the abstraction hierarchy. They included external regional ATC centers, communication tools, radar systems, the aircraft themselves, landing strips, navigational beacons (NAV-AIDS), and the terrain encompassing the controller's zone of responsibility.

We did not decompose these physical functions into their constituent components (e.g., the individual pieces of equipment making up the communication tools), nor did we investigate the physical form level of abstraction investigated in this analysis. Detailed knowledge of the physical characteristics of individual components modeled in the domain would not have significantly benefited our understanding of the software-based simulator. Examples of work domain analyses that include this final level of abstraction are readily available in the literature (e.g., Burns & Hajdukiewicz, 2004).

Throughout the entire CWA process, we designated each resultant design document with a unique label. These labels were chosen in a preliminary attempt to trace vertical streams of the overall CWA across multiple phases and to facilitate discussion by allowing for explicit referencing of analytical by-products. The labels consist of two parts: (1) an abbreviation indicating the type, or structure, of the design document itself; and (2) a numeric string indicating its relationship to the system as a whole. For example, the coarse-grained overview of the entire work domain (shown previously in Figure 2.2) was designated AH 0, as the primary abstraction hierarchy depiction of the entire work domain. Additional abstraction hierarchies, which were used to describe the four subsystems of generalized functions in greater detail (and are not reproduced here for reasons of space), were labeled AH 0.1, AH 0.2, AH 0.3, and AH 0.4, respectively.

Phase 2: Control Task Analysis

Following the WDA, we completed a control task analysis for the TRACON microworld. While the Phase 1 analysis identified the underlying physical and intentional constraints of the domain ecology, the purpose of this Phase 2 analysis was to better understand the actions involved in completing the recurring classes of control activities required to realize the domain's functional purposes. Through this control task analysis we examined the cognitive aspects of work within the air traffic control microworld, including the information processing activities and resultant knowledge states required to complete component tasks. The outputs of this analytical phase described in detail *what* tasks must

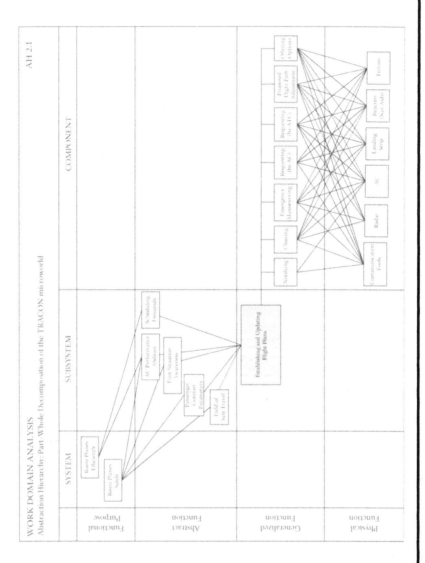

Figure 2.3 Part–whole decomposition showing component-level detail of the "Establishing and Updating Flight Plans" subsystem.

be effectively performed within the domain, but not *how* these tasks should be performed, nor *who* should perform them. Our understanding of component tasks within the TRACON microworld fed directly into the subsequent three phases of the CWA.

Phase 2 Analytical Tool: The Decision Ladder

We used the decision ladder (DL; Rasmussen, 1974, 1976) to support our Control Task Analysis (CTA) of the TRACON microworld. The DL is a tool for capturing the logical cascade of information-processing activities and resultant knowledge states comprising a single control task, from an initial perception that a control action may be necessary to execution of an appropriately formulated control response. The DL describes *what* tasks must be done to achieve the functional purposes of the work domain identified through the first phase of CWA. Although they are applied less commonly than the abstraction hierarchy (AH), examples of DL and methodological perspectives on their use in supporting control task analyses are available in the literature (e.g., Jamieson et al., 2007; Lamoureux et al., 2006; Moray et al., 1992; Naikar et. al., 2006; Vicente, 1999).

Decision ladders are comprised of *information-processing activities* (depicted as boxes) and *states of knowledge* (depicted as circles). Information-processing activities are the cognitive or computational activities actors must engage in to complete a task, whereas states of knowledge are the products of these activities. These components are arranged in alternating order, prescribing a serial progression of the task. Actors may begin a control task from different entry points in the DL. While novice actors may follow the linear sequence between the different information-processing activities and states of knowledge, expert actors are likely to take shortcuts. Two types of shortcuts are defined, *shunts* and *leaps*. Shunts connect information-processing activities to nonsequential states of knowledge within the DL structure, whereas leaps represent a link between two states of knowledge through direct association.

Phase 2 Methodology

We used DL templates to model the information-processing constraints of the domain's control tasks. We first defined the control tasks themselves, with reference to the generalized functions identified through the WDA. To do this, we combined insight from the abstraction hierarchy and part–whole decomposition with our own experiences operating the TRACON simulator and with the domain knowledge we garnered from documentation and subject matter experts. Given our limited access to ATC or TRACON domain experts, we relied heavily on introspection and discussion of our own experiences with the simulation to populate these decision ladders.

Although we enumerated over a dozen general tasks during this phase, we chose to focus the scope of our analysis by examining three tasks in great detail: *approaching*, *receiving*, and *rerouting*. The approaching task refers to obtaining a flight path for landing at a given airport. This path must have an appropriate heading, altitude, and speed, as determined by the physical nature of the landing strip and the flight characteristics of the aircraft. The receiving task involves an aircraft transitioning between zones of air traffic control. Before an aircraft can enter a new airspace, the pilot must request permission from the controller who has jurisdiction over that airspace. The controller then chooses whether or not to accept the aircraft based on whether the safety of his or her flight zone can be maintained. Finally, the rerouting task involves changing the proposed flight path of one or more aircraft. This control task may become necessary when there is a potential violation of the field of safe travel, which was defined as an abstract function of the domain.

We used DLs to record the knowledge states and information-processing activities needed to complete each of the three representative tasks safely and efficiently. Based on these DLs, we created a formal description of the superset of cognitive activities an operator would follow to complete this task. We developed detailed descriptions of the information-processing activities and resultant knowledge states—including indications of corresponding levels of the abstraction hierarchy (AH)—and recorded these in a separate table for each of the three control tasks. Following this, we developed several unique instantiations of each DL to identify leaps and shunts across the informational pathway that could be made by operators with sufficient skill or experience. We completed several iterations, and numerous revisions, of this analysis before deciding upon final DLs and expert pathways for each of the three tasks.

Phase 2 Products

The products of our Phase 2 control task analysis included complete DLs for the *approaching*, *receiving*, and *rerouting* tasks. A sample DL showing one expert pathway for the rerouting task is shown in Figure 2.4. It is important to note that, as pointed out by Vicente (1999), these DLs do not provide any information as to *how* the information processing steps are achieved for a given task, nor *who* is responsible for processing information. These issues are addressed by subsequent phases of the CWA.

We identified four distinct series of expert leaps and shunts for the receiving task, two for the approaching task, and three for the rerouting task. The specific expert pathway for the rerouting task illustrated in Figure 2.4 is described here in greater detail.

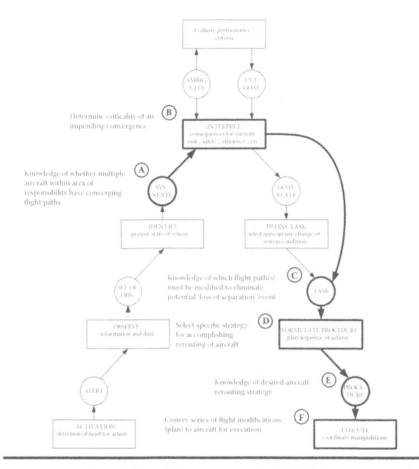

Figure 2.4 **Decision ladder of the** *rerouting* **task, highlighting one potential expert pathway. In this diagram, each rectangle represents a separate information-processing activity, and each circle represents the resultant knowledge state of that activity.**

A. System State

Based on visual interpretations of the TRACON simulator display, an experienced controller perceives the existence of a potential violation in aircraft safety constraints (for example, the minimum separation mandated by air safety regulations).

B. Interpret

Using the existing system state knowledge, the controller determines the potential for, as well as the criticality of, a pending problem—in this case a loss of separation between two aircraft.

C. Task

Following the *interpret* activity, the expert makes a leap to the knowledge of what task must be completed to cope with the pending loss of separation—in this case rerouting one or more of the aircraft. This leap is supported by the operator's previous experience that all potential losses of separation must be dealt with by changing the flight paths of one or more of the concerned aircraft.

D. Formulate Procedure

Once the controller has arrived at the knowledge that rerouting is necessary, he or she must then derive a plan of action for realizing this goal. This step involves planning a sequence of actions to accomplish the rerouting of relevant aircraft, based on the state of individual aircraft relative to each other.

E. Procedure

The *formulate procedure* activity leads to the operator's knowledge of a particular plan for remedying the potential loss of separation. Thus, a series of temporal flight modifications have been selected to realize the desired rerouting of flight paths. At this point, only the controller has knowledge of which procedure will be implemented for rerouting one or more of the concerned aircraft.

F. Execute

To complete the rerouting task, the operator must convey the series of flight modifications to relevant aircraft for execution.

We described each of the expert pathways identified for every one of the control tasks in a separate DL, for a total of 12 ladders between the three control tasks. We also created individual tables to summarize the pathways shown in each of these DLs and linked them to the information contained within specific abstraction hierarchy documents of the previous WDA. An example of one such table for the rerouting task is shown in Figure 2.5.

As with the earlier Phase I products, each of the individual DL design documents generated during the Phase II analysis was given a unique identifying label (in this case beginning with the abbreviation DL to designate "Decision Ladder"). The labels also included a string of numbers indicating the parent generalized function of the control task (using the same conventions as the previous AH documents), the unique tasks within each function, and the unique expert pathways identified within each unique task. Using this convention, the ladder shown in Figure 2.4 was labeled DL 2.7.1. This indicates that the ladder was the first expert pathway identified for the seventh task (rerouting) of the second generalized func-

CONTROL TASK ANALYSIS Decision Ladder Summary (Expert A)			DLSUM 2.7.1	

Rerouting Task (Control Task 2.7)
Related Documents
    Abstraction Hierarchies: AH0, AH0.2, AH2.1
    Decision Ladders: DL2.7.1, DL2.7.2, DL2.7.3				

Step	**Description**	**Type**	**Ladder Code**	**Abstraction Level**
A	Knowledge of whether multiple aircraft within area of responsibility have converging flight paths	Knowledge State	System State	Functional Purpose
B	Determine criticality of a pending convergence	Information Processing Activity	Interpret	Abstract Function
C	Knowledge of which flight path(s) must be modified to eliminate potential 'loss of separation' event	Knowledge State	Task	Generalized Function
D	Select specific strategy for accomplishing rerouting of aircraft	Information Processing Activity	Formulate Procedure	Generalized Function
E	Knowledge of desired aircraft rerouting strategy	Knowledge State	Procedure	Physical Function
F	Convey series of flight modifications (plan) to aircraft for execution	Activity	Execute	Physical Form

Figure 2.5   Example of one expert pathway through the rerouting task, with links to relevant artifacts of the previous work domain analysis (WDA).

tion subcomponent of the work domain (establishing and updating flight plans, shown earlier in Figure 2.2). The corresponding summary description of the expert pathway (Figure 2.5) was designated DLSUM 2.7.1

Phase 3: Strategies Analysis

After completing the control task analysis, we transitioned to the third phase of the CWA strategies analysis. The purpose of this phase was to investigate the different ways in which each of the control tasks could be accomplished, regardless of the actor performing them. Thus, the Phase 3 strategies analysis uncovered insight regarding *how* these tasks may be performed. Emphasis was placed on decomposing the formative, constraint-based descriptions of the DLs into process descriptions illustrating the multiple ways by which execution of the control tasks could be approached.

Phase 3 Analytical Tool: The Information Flow Map

Rasmussen (1981) describes a strategy as "a category of cognitive task procedures that transform an initial state of knowledge into a final state of knowledge" (Vicente,

1999, p. 220). In that sense, strategies can be viewed as a collection of generic control responses, each having the potential to transform different states of knowledge. Rasmussen (1980) points out that the strategy adopted by an actor at any time will be the product of a multitude of factors in a manner that is highly situation- and actor-dependant. For this reason, it becomes very difficult to identify each particular exemplar response to a given control task. The strategy analysis phase of CWA allows designers to identify potential categories of generic strategies. These strategy descriptions are intended to aid in the subsequent design of computer-based information systems that will support the actor in strategy selection. The ultimate goal of this approach is to develop interfaces that will effectively support operator control regardless of which strategy is adopted in a given set of circumstances.

Information flow maps (IFMs) were used to conduct the strategies analysis. IFMs are graphical representations of the information-processing activities and knowledge states of particular strategies. Although IFMs have not reached the same level of maturity as the DL or the AH, the tool still provides a method for formally describing the information-processing activities and states of knowledge used in the selection of a given strategy. Moreover, IFMs visually map each of the strategies identified for a given task. This creates a formal representation of each strategy that can be described and compared with representations of other strategies.

Phase 3 Methodology

We fed the formal descriptions of the three classes of control tasks—approaching, receiving, and rerouting—into our strategies analysis. Because of time constraints, we further narrowed our analytical focus at this point to include only a single task. However, it is important to note that a complete strategies analysis would involve exploration of *all* of the tasks identified by the decision ladders of a *full* CTA.

Based on our own experiences, we identified the rerouting task as a cognitively challenging control task with many degrees of freedom. The ways in which each analyst completed the rerouting task were highly varied. Our methods also appeared to be very sensitive to individual preferences, as well as numerous environmental factors, such as time pressures or the cognitive burden of dealing with other air traffic. Additionally, the appropriate selection of a rerouting strategy appeared to be highly dependent upon domain expertise. For all of these reasons, the rerouting task was selected for detailed examination in this and all subsequent phases of the CWA.

As defined through the control task analysis, rerouting is not a single task, but rather a class of tasks, all of which are concerned with changing the proposed flight paths of one or more aircraft. Distinct forms of rerouting may be invoked for different reasons, for example, when a single aircraft requests a new flight path to save time in reaching an airport, or when multiple aircraft are determined to be converging in a manner that threatens loss of separation. Each type of rerouting task can be accomplished with its own unique set of strategies, depending on the motivation of the air traffic controllers and their relative workload (e.g., Sperandio, 1978). During this third phase we further confined the focus of our CWA by exclusively examining strategies for rerouting two aircraft on a convergent path. This situation threatens to violate the domain's "field of safe travel" abstract function constraint.

We began the strategies analysis by first establishing a list of the generic strategies that an operator could invoke to reroute two convergent aircraft. We then mapped each of the generic strategies in this list to a separate IFM. This effort was largely supported through group discussions of our own operating experiences. Rerouting convergent air traffic is a common task within the simulator, which meant that all members of the analytical team had numerous experiences in completing this task (to varying degrees of success). Through these discussions, we attempted to describe and classify the rerouting strategies we had employed by using the IFM graphical structure, paying particular attention to how the individual strategies varied with the overall workload. In this manner, three generic strategies were identified, each of which differed in terms of their overall efficiency and the relative amount of additional cognitive workload they placed on the operator. These strategies were identified as (1) *hold one aircraft*, (2) *reroute one aircraft*, and (3) *"tweak" one aircraft*.

Phase 3 Products

The products of the Phase 3 strategies analysis included written descriptions of each of the three strategies identified (Figure 2.6) and IFMs illustrating the information-processing activities (rectangles) and knowledge states (circles) required to execute each strategy (Figure 2.7). Within the context of the previously described control task analysis, these IFMs represent a more detailed decomposition of the "formulate procedure" information-processing activity of the rerouting task DL (shown previously in Figure 2.4). It is important to note that the strategies identified through this decomposition represent only a subset of those available to operators engaged in rerouting tasks. A complete understanding of all of these strategies would require a greater level of knowledge elicitation from domain experts than was feasible during the limited duration of this exercise. Also, although this phase was useful in identifying potential strategies for completing control tasks, it was not intended to result in a prioritization of these strategies, nor in the selection of a single "optimal" strategy to support through interface design. Instead, the outputs of this phase provide a concrete description of the superset of activities that a control interface should accommodate to allow for the operator's own opportunistic, context-driven strategy selection.

STRATEGIES ANALYSIS		IFSUM 2.7.1
Information Flow Map Summary		

Rerouting Two Convergent Aircraft (Control Task 2.7)
Related Documents
    Abstraction Hierarchies: AH0, AH0.2, AH2.1
    Decision Ladders: DL2.7.1, DL2.7.2, DL2.7.3

ID	Strategy Name	Description
1	Hold One Aircraft	One aircraft is placed in a holding pattern (e.g., "AC1, hold at VICTOR until further notice"), while the other aircraft continues on its original flight path, unaffected. Once it has been determined that sufficient time has passed for the two aircraft to no longer be on intersecting courses, the held aircraft is released and allowed to continue on its original flight path (e.g., "AC1, you are cleared to continue to TANGO"). This strategy will likely produce the least amount of cognitive overhead, as it allows operators to reduce workload by temporarily removing one of the two aircraft from the operational picture by effectively 'pausing' that aircraft at a fixed location. This strategy offloads the need to continually monitor or redirect this aircraft and allows the operator to quickly focus attention exclusively on the second aircraft's needs, for example preparation for a runway approach. However, this strategy is fairly inefficient, as the holding aircraft makes no forward progress towards its destination while the second aircraft is being dealt with. For this reason, the Hold One Aircraft strategy would likely be used only when operators are under the greatest of cognitive demands in dealing with the second aircraft.
2	Reroute One Aircraft	One aircraft is redirected to a new navigational beacon, or NAV-AID, (e.g., "AC1, redirect to TANGO, via BRAVO, at current speed and altitude"), while the other aircraft continues on its original flight path. In this case, the reroute that is proposed is selected so that it eliminates the convergence of the two aircraft paths. Alternatively, the reroute could be selected to increase the time differential between when the two aircraft reach the convergence point of their flight paths, thus eliminating the potential loss of separation. This strategy is more efficient than the previous Hold One aircraft strategy because the first aircraft continues to make progress towards its destination, albeit in a less-direct manner. However, this strategy will also place greater cognitive demands on the operator, as they must go through the process of selecting a new route for the first aircraft, making it suitable only if the operator has some amount of spare attentional resources to focus on the task of redirecting.
3	'Tweak' One Aircraft	A series of commands are given to adjust one aircraft's flight path in a manner that eliminates a potential loss of separation with the other aircraft, while minimizing the magnitude of change to its original flight path (for example: "AC1, increase alt to 10,000 feet"; "AC1, hold altitude and turn left to heading 1-7-0"; "AC1, descend to 5,000 feet

Figure 2.6 Formal verbal descriptions of three expert strategies for coping with the task of rerouting two convergent aircraft.

Remaining consistent with the previously described numbering conventions, design product documents were given the abbreviations IF and IFSUM, for "Information Flow Map" and "Information Flow Map Summary," respectively) and labeled IF 2.7.1 and IFSUM 2.7.1 to indicate their direct relationship to the DLs developed during the previous analytical phase.

Phase 4: Social Organization and Cooperation Analysis

After completing the strategies analysis of the rerouting task, we progressed to a social organizational analysis of the same task. The purpose of this Phase 4 analysis was to identify potential schemes for allocating responsibility for the domain's individual information-processing activities and knowledge states—as identified through the Phase 2 and Phase 3 analyses—across a team of actors. This allocation could be comprised of any number of human operators or pieces of computerized

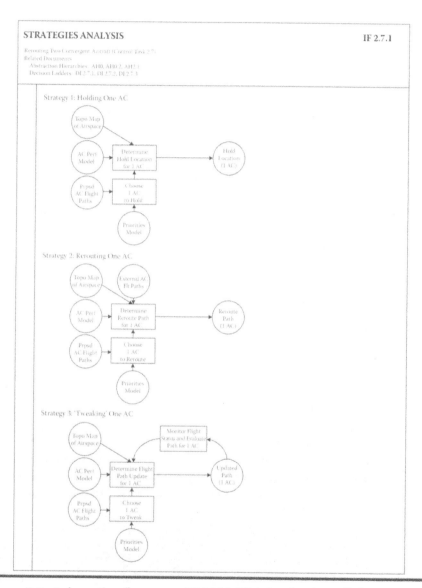

Figure 2.7 Information flow maps (IFMs) highlighting the three distinct strategies for coping with the task of rerouting two convergent aircraft. In this diagram each rectangle represents a separate information-processing activity, and each circle represents the resultant knowledge state of that activity.

automation, as constrained by the work domain. To execute this analytical phase, we applied our understanding of the potential strategies for completing the rerouting task to an evaluation of several potential social–organizational structures for the TRACON domain.

Phase 4 Analytical Tool: The Information Flow Map (IFM)

As with the previous phase, we used IFMs to structure the activities of our social organization analysis. Unlike the previous strategies analysis, however, we augmented these IFMs to reflect the potential allocation of individual task components across a team of actors. Although we chose to use IFMs for this phase of modeling, Vicente (1999, pp. 256–272) describes how other modeling tools introduced in earlier CWA phases may also be applied to this end. For the purpose of the current analysis, however, we believed that IFMs would provide an appropriate level of detail within our narrow, targeted slice of the work domain to map responsibility for control strategy components to a small team of actors.

Phase 4 Methodology

We evaluated the demands of the strategies used to complete TRACON control tasks with respect to the relative strengths and weaknesses of the actors in our organizational team—in this case a single human operator working with computerized automation. This team configuration was informed by the earlier decision to limit the boundaries of the WDA to *individual*, single-operator air traffic control cells (as simulated by the TRACON microworld). This precluded any analysis of how tasks and responsibilities might be allocated between neighboring regional air traffic control cells or even between multiple human actors controlling a single cell. In analyses of larger domains, however, this fourth analytical phase would also serve to determine how activities could be best distributed across multiple human and automated agents, informing the design of mixed-initiative collaborative teams.

As in previous phases, we chose to limit the scope of analytical activities by focusing exclusively on the three strategies for rerouting aircraft that were described in Phase 3. We began our social organization analysis of these strategies by directly copying the IFM products of the previous analytical phase. Following this, we critically evaluated how each of the information-processing activities and knowledge states could be best allocated to our team of actors. To do this, we examined the component tasks with respect to known strengths and weaknesses of human and computer actors. We attempted to allocate information-processing activities and responsibility for maintaining knowledge states to each potential actor in a manner that maximized the potential effectiveness of the team. However, we also paid attention to keeping the human operator actively engaged in meaningful control activities and decision making. Using a color-coding scheme, we annotated the IFMs for each strategy to visually reflect several iterations of potential task-allocation schemes.

During this entire Phase 4 analysis, we attempted to simultaneously consider the effectiveness of our proposed task allocations across all three of the rerouting strategies. This encouraged the development of a single organization that would adequately support operator actions regardless of which of the three strategies they chose to invoke. Such an organization would afford the operator with the context-conditioned adaptability necessary to effectively "finish the design" of the control system by selecting whichever strategy they felt to be most appropriate to complete the rerouting task at any given time.

Phase 4 Product

The products of the Phase 4 social organization analysis were augmented IFMs for each of the strategies (Figure 2.8). These augmented maps depicted a design strategy for distributing the information-processing activities and knowledge states of the three rerouting strategies across a mixed-initiative team consisting of one human operator and a computerized automation system. As seen in Figure 2.8, the automation system is tasked with maintaining detailed models of the flight characteristics of individual aircraft, as well as up-to-date knowledge of the structure of the airspace (e.g., the geospatial location of local aircraft at any given time) and future predictions of aircraft locations (based on current flight parameters and proposed flight paths). All of these tasks are data- and computation-intensive, two attributes for which computerized systems are well adapted but which would place large cognitive loads on human operators' working memory. In contrast, the human operator is allocated judgment tasks that draw upon these sources of information, such as selecting which of the two convergent aircraft to hold, reroute, or "tweak." Finally, some components of the rerouting strategies were designated as "shared" between the human operator and the computer system. This indicates that multiple levels of automation may be invoked by the operator to complete these task components. For example, in the *Holding One Aircraft* strategy, an operator may choose to select which navigation aid to use as an aircraft's hold location, or he or she may allow the system's automation to suggest—or even to select—an optimal navigation aid for this purpose.

The label selected for this design product, IF 2.7.1.1, reflects that the document is a direct extension of the IFM developed previously under the Phase 3 analysis (IF 2.7.1, shown in Figure 2.7).

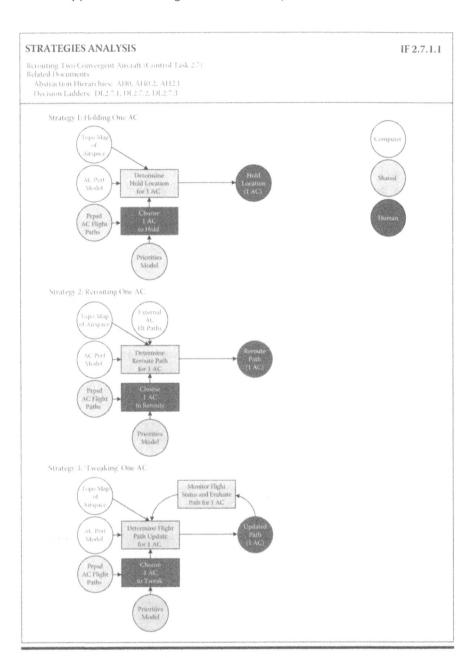

Figure 2.8   IFMs showing the mixed-initiative allocation of information-processing activities and knowledge states between a computer and a human actor. Each rectangle represents a separate information-processing activity, and each circle represents required or resultant knowledge states.

Phase 5: Worker Competencies Analysis

Following the social organization analysis, we began a worker competencies analysis, the fifth and final phase of the CWA framework. The purpose of this analysis was to identify the competencies required by members of the operational team to effectively accomplish the domain's numerous control tasks, with the ultimate goal of identifying additional psychological constraints applicable to systems design. As the final phase of CWA, the worker competencies analysis inherited all of the constraints identified through the previous four analytical phases. However, it is important to note that the output of this analysis was *not* a finished design for a system interface. Instead, the analytical products of *all five phases* of the CWA are used in the development of information requirements to feed subsequent interface design activities. As such, it is critical for this final phase of analysis to effectively organize and represent all the nested constraints of the system domain.

The Skill, Rule, and Knowledge taxonomy (SRK) forms the backbone of this final analytical phase. This taxonomy outlines basic distinctions between three main psychological processes: Skill-Based Behavior (SBB), Rule-Based Behavior (RBB), and Knowledge-Based Behavior (KBB) (Rasmussen, 1983). The outputs of the worker competencies analysis are intended to inform the design of information displays that appropriately accommodate these three levels of system control. The goal of this approach is to motivate the development of interfaces that allow operators to perform control tasks using efficient skill- and rule-based behavior when possible, but that also support operators' knowledge-based behavior whenever necessary, for example during troubleshooting or in coping with abnormal situations.

Phase 5 Analytical Tool: The SRK Inventory

In contrast to the other four phases of the CWA framework, Vicente (1999) does not recommend a specific methodology for completing the worker competencies analysis. To address this gap, we developed the SRK inventory (Kilgore & St-Cyr, 2006) as a means of supporting the Worker Competencies Analysis. The goal of the SRK inventory is to facilitate the process of identifying informational design concepts that support skill-, rule-, and knowledge-based behaviors throughout the control activities of a given work domain. Particular emphasis was placed on creating a methodology that would both structure the analytical process and capture the results of this analysis in a manner that supported subsequent display design initiatives.

The SRK inventory is comprised of a series of tables describing analysts' conceptualizations of behaviors that operators may invoke to complete information-processing activities. Separate SRK inventory tables are created for each of the domain's control tasks. Each row within a given table describes a single information-processing step and resultant knowledge state for a particular control task.

Within these rows, three separate columns describe how an operator may execute this information-processing step using skill-, rule-, or knowledge-based behavior. Taken as a whole, the behavioral concepts captured within the completed SRK inventory matrix motivate subsequent display design activities by aiding in the development of information requirements. The tables accomplish this by providing an explicit checklist of behaviors that must be accommodated to ensure that an interface allows for skill- and rule-based behaviors whenever possible and supports knowledge-based behavior whenever necessary (see Vicente & Rasmussen, 1992). Additionally, the completed SRK inventory can also be used to generate profiles of competencies that operators must possess to adequately perform control tasks across all three of the behavior types. In this manner, the SRK inventory can be used to inform worker selection and training.

Phase 5 Methodology

We performed a worker competencies analysis of the domain's rerouting control task. This task was selected for its relative complexity and because it had been analyzed across all four of the previous CWA phases. Again, it is important to note that a full worker competencies analysis would include analyses of every control task across the entire domain. However, as with the preceding three phases, this was prohibited by a limited project scope.

We began the worker competencies analysis by creating an SRK inventory template for the rerouting task. This table included references to all the individual design products (abstraction hierarchies, decision ladders, and information flow maps) that were developed in the prior four phases of the CWA and fell within the context of our increasingly narrow slice of the work domain. We populated the cells within the first two columns of the SRK inventory table by directly importing each of the information-processing steps and resultant knowledge states from the rerouting task DL descriptions (DL SUM 2.7, not shown here). Cells in the remaining three columns of the SRK inventory served to motivate a series of design-related brainstorming activities. The analytical team worked its way through the entire table, focusing its attention on each cell in turn. For every cell, the team attempted to conceptualize ways in which operators might interact with an information display to perform the associated information-processing activity using the particular type of skill-, rule-, or knowledge-based behavior. Descriptions of these potential behaviors were recorded directly within the respective cells of the SRK inventory. Empty cells within the table served as reminders of areas where specific styles of control behavior had not yet been addressed for a particular information-processing activity within the rerouting task.

Phase 5 Products

The product of the worker competencies analysis was a completed SRK inventory for the rerouting control task (Figure 2.9). As a whole, the table provides a structured representation of potential interaction design concepts for each level of the SRK taxonomy.

WORKER COMPETENCIES ANALYSIS				SRK 2.7
SRK Inventory				

Rerouting Task (Control Task 2.7)
Related Documents
  Abstraction Hierarchies: AH0, AH0.2, AH2.1
  Decision Ladders: DL2.7.1, DL2.7.2, DL2.7.3, DLSUM2.7.1, DLSUM 2.7.2, DLSUM 2.7.3
  Information Flow Maps: IF2.7.1., IF2.7.1.1, IF2.7.2, IF2.7.2.1, IF2.7.3, IF2.7.3.1				

Information Processing Step	Resultant Knowledge State	Skill-Based Behavior	Rule-Based Behavior	Knowledge-Based Behavior
1. Scan for aircraft presence in area of responsibility	2. Whether multiple aircraft are within area of responsibility	Monitoring of time-based spatial representation of aircraft in area of responsibility	Perceive explicit indication multiple aircraft are currently within area of responsibility	Reason, based on proposed flight plans, that multiple aircraft may be present in area of responsibility within similar time frames
3. Determine future flight vector for each aircraft	4. Whether multiple aircraft within area of responsibility have intersecting flight paths	Perceive headings of related aircraft as convergent, divergent, or parallel	Use heuristics to determine whether flight paths are intersecting (e.g., flight paths A→B and C→D are convergent paths; flight paths A→C and B→D are divergent)	Reason, based on geospatial knowledge of to/from points for each flight, that aircraft are on convergent or divergent paths
5. Predict future, time-based location states for aircraft on convergent paths	6. Whether converging aircraft will arrive at point of convergence within a similar time frame	Perceive time-to-collision (tau) of each aircraft with the convergence point, based on spatial representations of heading and speed	Use heuristics to estimate whether aircraft will arrive at convergence point within a similar time frame (e.g., if distance A distance B AND airspeed A airspeed B, aircraft will arrive at approximately same time)	Calculate, using airspeed, heading, and location of each aircraft, the time at which each aircraft will arrive at the convergence point
7. Determine the criticality of a pending convergence	8. Whether future distances between converging aircraft will constitute a 'loss of separation' event	Perceive whether the zones of safe travel surrounding each aircraft will overlap at or near their closest point	Use heuristics to determine proximity as being greater or less than the minimum required envelope of separation	Calculate distance between each aircraft at their closest future states and compare with the minimum value of separation required for safe travel.
9. Choose to modify aircraft flight path(s) to address future problem	10. Which aircraft flight path(s) must be modified to eliminate potential 'loss of separation' event	Directly perceive that one or more aircraft must be redirected	Apply doctrine: (e.g., If loss of separation will occur, MUST reroute one or more aircraft)	Reason from knowledge of proposed flight paths, current locations, and expected future behavior that aircraft must be rerouted
11. Select specific strategy for accomplishing rerouting of aircraft	12. Desired aircraft rerouting strategy	Respond automatically to perception of loss of separation by directly manipulating a representation of aircraft flight path(s)	Classify loss of separation within a set of generalized scenarios and select appropriate stereotypical control rule	Develop new, optimized flight path(s) based on weighted criteria including urgency, flight priority, passenger convenience, efficiency etc.
13. Convey flight modifications to aircraft for execution	14. Aircraft's awareness of new flight path(s)	Direct, simultaneous interaction with communication equipment through control interface through input of rerouting information	Apply stereotypical control rules to select method/sequencing for conveying proposed flight path (e.g., if one aircraft involved, contact that aircraft; if two, contact both)	Reason using knowledge of aircraft systems, priorities, urgency, etc., the best means and order for contacting each aircraft to convey proposed flight path(s)

Figure 2.9   Completed SRK inventory for the rerouting control task within the TRACON simulator domain.

The following design insights were captured through the completed SRK inventory for the *rerouting* task.

Skill-Based Behavior

SBB can be supported by providing signals to operators in the form of time-based spatial representations of aircraft flight. Any operator engaging in SBB must be able to perceive aircraft headings, time-to-collision, potential overlap zones, and boundaries of minimal separation. It is crucial that the interface provide adequate support for such competencies in light of known perceptual limitations. For example, to predict the future locations for converging aircraft (Figure 2.9, Step 5) using SBB, operators should directly perceive a potential a loss of separation between two aircraft—a product of their individual flight headings, altitude, and traveling speeds—as a continuous signal. Slow visual update and poor resolution of the traditional TRACON simulator interface make this task difficult. SBB may be better supported by augmenting the interface with the ability to provide accelerated, hi-resolution animation of aircraft locomotion through the airspace.

Rule-Based Behavior

To support RBB, signs should be presented to operators via isomorphic mapping between display geometries and system states (Vicente & Rasmussen, 1990). Operators must have the necessary competencies to use these signs to trigger stereotypical control rules, as outlined in Figure 2.9. Such signs include indications of the presence of multiple aircraft, the convergence/divergence of particular aircraft routes, and whether future proximities of converging aircraft may necessitate intervention. Additionally, control concepts captured through the SRK inventory suggest that operators should be trained to recognize and address some predefined set of generic loss-of-separation scenarios.

Knowledge-Based Behavior

To support KBB, the interface must facilitate operators' integration of multiple pieces of information to predict future system states and develop rerouting strategies. Figure 2.9 highlights information requirements for effective problem solving during rerouting, including geospatial knowledge of aircraft origins and destinations, as well as performance characteristics, airspeeds, headings, altitudes, and scheduling criteria, among others. To support problem-solving across multiple levels of system abstraction, these informational requirements must be conveyed in a manner that reflects the underlying relationships and constraints uncovered through Phase 1 WDA activities. Finally, the SRK inventory highlights necessary worker competencies to engage in effective control, including the ability to integrate relevant flight parameters into a calculation of future aircraft locations for determining suitable rerouting procedures.

Again, following previously described labeling conventions, this design document was given the abbreviation SRK, for "SRK Inventory," and was labeled SRK 2.7 to indicate its incorporation of insight from all of the component DL and IFM documents developed for the rerouting task (DL 2.7.X, DL SUM 2.7.X, IF 2.7.X, and IFSUM 2.7.X).

The insight captured through the SRK inventory facilitates the generation of information requirements and aids in subsequent interface design activities in two key ways: First, the inventory tables provide a formalized checklist for the evaluation of potential display forms during preliminary design activities. For a display concept to be considered viable, it should accommodate all of the skill-, rule-, and knowledge-based behaviors identified for relevant tasks. Multiple design alternatives can be compared with respect to how successfully they address the control activities described in particular SRK inventory tables. Second, the SRK inventory provides the basis for developing explicit links between display designs and the findings of a CWA. During the design process, graphical forms can be annotated as supporting one or more cells within referenced SRK inventory tables. This provides interface developers with a tangible justification of design decisions and supports traceability of the design process across large and complicated work domains.

Discussion and Conclusions

In this chapter, we have described our experiences in applying all five phases of the CWA framework to an air traffic control microworld system. Such an account is unusual in the CWA literature. It is therefore instructive to briefly review what we have learned from the process and highlight opportunities for continued development of the framework.

As both students and practitioners, one of the greatest benefits of this process was our increased appreciation of the relationships between the domain knowledge elucidated through each phase of the CWA. It is one thing to be aware of the intended nesting of the five analytical phases of the CWA in theory but quite another to perform all of these analyses first hand for a single complex work system. Managing analytical activities across the individual phases and coordinating their resultant domain descriptions proved challenging. We provide an overview of our entire analytical process in Figure 2.10, which lists the goal of each analytical phase, the specific tools used to accomplish these goals, and descriptions of how our activities in performing each component analysis were supported by the products of previous phases.

Analytical Phase	Goal	Tools	Relationship to Previous Phases
1 Work Domain	Identify the underlying functional structure, or ecology, of the system of interest, as well as the physical and intentional constraints that must be respected to achieve the system's intended purposes (why?)	Abstraction Hierarchy Part-Whole Decomposition	Iteratively updated, expanded, or contracted, through insight gleaned from the activities of all subsequent analytical phases
2 Control Task	Identify the information processing activities and knowledge states required to complete the recurring classes of activities, or tasks, that must be performed to achieve the system's functional purposes (what?)	Decision Ladder	Unique classes of recurring tasks were established, in part, through reference to the generalized functions identified in the Phase I work domain analysis as captured by the abstraction hierarchies and part-whole decompositions For a full-breadth CWA, each control task of the entire work domain described in Phase I would be analyzed in detail
3 Strategies	Identify the different ways in which each of the system's control tasks can be accomplished, regardless of the actor performing them (how?)	Information Flow Map	Instances of recurring task classes that were identified through the Phase II control task analysis established the granularity of the strategies analysis focus For a full-breadth CWA, a separate strategies analysis would be performed for each control task identified in Phase II
4 Social Organization	Identify potential design concepts for allocating responsibility for the system's individual information processing activities and knowledge states across a team of human and automation-based actors (by whom?)	Information Flow Map	Information flow maps were directly copied from Phase III products and then augmented to visually identify responsibility for information processing activities and resultant knowledge states across actors For a full-breadth CWA, this would be done for every strategy identified in Phase III (and thus for every control task identified in Phase II)
5 Worker Competencies	Identify techniques—and required competencies of members of the operational team—for effectively accomplishing the information processing steps of the system's control tasks through skill-, rule-, and knowledge-based behavior (by what means?)	SRK Inventory	The first two columns of the SRK Inventory table ('Information Processing Steps' and 'Resultant Knowledge States') were directly imported from the information processing and knowledge state components of the decision ladders generated during the Phase II analysis For a full-breadth CWA, this would be done for every control task established in Phase II

Figure 2.10   Description of the CWA process and relationships between the five analytical phases.

With respect to analytical scope, it is important to remember that the analysis described in this chapter represents only a subset of a complete CWA undertaking. Our chosen path through the analytical space increasingly focuses on a limited slice of the work domain (Figure 2.11). For example, in Phase 1, we selectively defined the scope of our WDA to exclude neighboring zones of regional air traffic control. We also established a limit of one single air traffic controller within the human/automation team and chose not to model the domain at the physical form level of abstraction. In Phase 2, we subjected only three of the domain's dozen or so major operational tasks to a detailed control task analysis. In Phase 3, we performed a strategies analysis for only one of these three control tasks: rerouting. Finally, we limited our Phase 4 social organization analysis and our Phase 5 worker competencies analysis by considering only the rerouting control task.

Despite the limited breadth of our analysis, we generated over 20 unique design documents. This work was completed over several weeks by a team of three analysts working at a part-time level (approximately 60 person-hours in total). A full-fledged analysis at a similar level of detail would reasonably generate 200 or more of such design documents, and this for a *small* microworld domain with limited complexity. What does this say about the feasibility of applying CWA to very large very complex domains? Teams of analysts would obviously be necessary for such undertakings, suggesting that formalization (or, at least, standardization) of the entire CWA process is critical. A selection of robust tools suitable for each phase would facilitate tailoring the analysis to a particular work system. Also, more consistent interrelationship between the tools used and the documents generated for each phase would aid in integrating a collaborative team's analytical work.

Combined, our experiences across the five phases of our CWA led to additional insight in the following areas:

- **Subject matter experts:** It was evident to us throughout the process that access to subject matter experts is extremely valuable (and probably necessary for any real-world system). In particular, such experts would have enabled more detailed and exhaustive control task and strategies analyses. Some caution is called for in employing subject matter experts as they are likely to provide descriptive, as opposed to formative, accounts of the work. However, it is the job of the cognitive work analyst to critically leverage the outputs of knowledge elicitation activities to support construction of formative models of the work domain.

- **Experience:** Jamieson et al. (2007) discussed how the abstraction–decomposition space acts as a scaffold for learning about a work domain. Our experience here extends that observation across the subsequent phases of CWA. Each phase of analysis provided us with a deeper understanding of the intrinsic constraints of the TRACON system. Moreover, the structure of the CWA facilitated a structuring of that new understanding in a coherent and logical manner. One indicator of this structuring is that newly discovered constraints

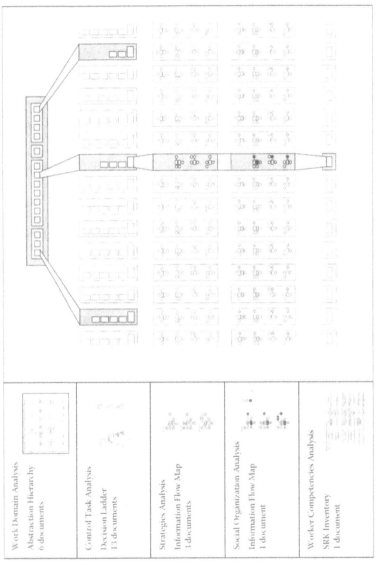

Figure 2.11 Comparison of design products of the current analytical undertaking, within the context of a CWA that is complete with respect to both depth *and* breadth.

fit intuitively into one of the five phases of analysis. Similarly, Sanderson's (2004) observation that work domain analysis "prepares your mind to recognize the profitable design insight" also extends across the remaining four phases of the CWA. At each phase we identified important considerations for subsequent design efforts.

■ **Maturity of Tools:** Although some of the CWA phases were supported by clearly defined tools and methodologies, others were not. For example, the WDA was conducted using the abstraction hierarchy, for which Vicente (1999) provides thorough examples. In contrast, we found the worker competencies analysis to be a difficult undertaking, as little information was provided on the nature of the tools used to carry out the analysis. In this case, it was necessary to develop our own methodology to complete the analysis. Whether this tool serves the needs of other practitioners remains to be seen. However, its introduction highlights the continuing evolution of the CWA framework and invites further innovation.

We believe that this example illustrates the potential design value of performing a CWA for real-world domains across all five analytical phases even if time or resource constraints limit the overall breadth of the analysis.

References

Bisantz, A. M., & Vicente, K. J. (1994). Making the abstraction hierarchy concrete. *International Journal of Human–Computer Studies, 40*, 83–117.

Burns, C. M., Bisantz, A. M., & Roth, E. M. (2004). Lessons from a comparison of work domain models: Representational choices and their implications. *Human Factors, 46*(4), 711–727.

Burns, C. M., & Hajdukiewicz, J. R. (2004). *Ecological Interface Design.* Boca Raton, FL: CRC Press.

Cummings, M.L., & Guerlain, S. (2003). The tactical Tomahawk conundrum: Designing decision support systems for revolutionary domains. In *Proceedings of the IEEE International Conference on Systems, Man and Cybernetics*, 1583–1588. Piscataway, NJ: IEEE.

Hajdukiewicz, J. R., & Vicente, K. J. (2004). A theoretical note on the relationship between work domain analysis and task analysis. *Theoretical Issues in Ergonomics Science, 5*, 527–538.

Ham, D. H., & Yoon, W. C. (2001). Design of information content and layout for process control based on goal-means domain analysis. *Cognition, Technology, and Work, 3*, 205–223.

Jamieson, G. A., Miller, C. M., Ho, W. H., & Vicente, K. J. (in press). Integrating Task- and Work Domain-based Work Analyses in Ecological Interface Design: A Process Control Case Study. *IEEE Transactions on Systems, Man, and Cybernetics 37*, 887–905.

Kilgore, R.M., & St-Cyr, O. (2006). The SRK Inventory: A tool for structuring and capturing the execution of a Worker Competencies Analysis, *Proceedings of the 50th Annual Meeting of the Human Factors and Ergonomics Society*, 506–509. San Francisco, CA: HFES.

Lamoureux, T., Bos, J., Rehak, L., & Chalmers, B. (2006). Control Task Analysis in Applied Settings, *Proceedings of the 50th Annual Meeting of the Human Factors and Ergonomics Society*, 391–395. San Francisco, CA: HFES.

Miller, A. (2004). A work domain analysis framework for modeling intensive care unit patients. *Cognition, Technology, and Work, 6*, 207–222.

Moray, N., Sanderson, P. M., & Vicente, K. J. (1992). Cognitive task analysis of a complex work domain: A case study. *Reliability Engineering and System Safety, 36*, 207–216.

Naikar, N., Hopcroft, R., & Moylan, A. (2005). *Work domain analysis: Theoretical concepts and methodology* (DSTO-TR-1665). Air Operations Division, Defence Science and Technology Organisation, Australia.

Naikar, N., Moylan, A., & Pearce, B. (2006). Analysing activity in complex systems with cognitive work analysis: Concepts, guidelines and case study for control task analysis. *Theoretical Issues in Ergonomics Science, 7(4), 371–394.*

Rasmussen, J. (1974). *The human data processor as a system component: Bits and pieces of a model* (Report No. Risø-M-1722). Roskilde, Denmark: Danish Atomic Energy Commission.

Rasmussen, J. (1976). Outlines of a hybrid model of the process plant operator. In T. B. Sheridan & G. Johannsen (Eds.), *Monitoring behaviour and supervisory control*, 371–383. New York: Plenum.

Rasmussen, J. (1979). *On the structure of knowledge: A morphology of mental models in a man-machine system context.* (Report No. Risø-M-2192). Roskilde, Denmark: Danish Atomic Energy Commission.

Rasmussen, J. (1980). The human as a systems component. In H. T. Smith & T. R. G. Green (Eds.), *Human interaction with computers*, London: Academic Press.

Rasmussen, J. (1981). Models of mental strategies in process plant diagnosis. In J. Rasmussen & W. B. Rouse (Eds.), *Human detection and diagnosis of system failures* (pp. 241–258). New York: Plenum.

Rasmussen, J. (1983). Skills, Rules, and Knowledge; Signals, Signs, and Symbols, and Other Distinctions in Human Performance Models. *IEEE Transactions on Systems, Man, and Cybernetics, 13*, pp. 257–266.

Rasmussen, J. (1985). The role of hierarchical knowledge representation in decision making and system management. *IEEE Transactions on Systems, Man, and Cybernetics, 15*, pp. 234–243.

Rasmussen, J., Pejtersen, A. M., & Goodstein, L. P. (1994). *Cognitive System Engineering.* New York: John Wiley.

Reising, D. V. C., & Sanderson, P. M. (2002). Work domain analysis and sensors II: Pasteurizer II case study. *International Journal of Human–Computer Studies, 56*, 597 637.

Sanderson, P.M. (2004). Tactics for conveying CWA concepts to undergraduate engineering vs. psychology students. Presented at the *Workshop on Cognitive Work Analysis*, Seattle, WA. November 2004.

Sperandio, J. C. (1978). The regulation of working methods as a function of work-load among air traffic controllers. *Ergonomics, 21*, 195–202.

TFW, TRACON for Windows (1991). Manual for the TRACON for Windows Multi-Player Air Traffic Control Simulator. Wasson International.

Vicente, K. J. (1999). *Cognitive work analysis: Towards safe, productive, and healthy computer-based work*. Mahwah, NJ: Lawrence Erlbaum Associates.

Vicente K. J., & Rasmussen, J. (1990). The ecological of human-machine system II: Mediating "direct perception" in complex work domains. *Ecological Psychology*, 2(3), 207–250.

Vicente, K. J., & Rasmussen, J. (1992). Ecological interface design: Theoretical foundations. *IEEE Transactions on Systems, Man, and Cybernetics*, 22(4), 589–606.

Chapter 3

Work Domain Analysis Using the Abstraction Hierarchy: Two Contrasting Cases

Ann M. Bisantz and Natalia Mazaeva

Contents

Overview

Work Domain Analysis (WDA) has been applied extensively within cognitive engineering as an analytic framework for the evaluation of complex sociotechnical systems in support of design. The analysis provides valuable information about the structure and the constraints operating in a work domain that define the normal functioning of a system, and limit the actions of either human or automated agents within the domain.

This chapter presents two case examples of WDA that demonstrate the application of the technique across different types of systems, at different levels of detail, and for different purposes. The first case presents a highly detailed analysis of a relatively small, well-defined physical system—a camera. The purpose of this analysis was to explore the utility of modeling automated components in tandem with the system itself. The second case presents an analysis of a very large, intentional system—an emergency response system. The purpose of this analysis was to identify information needs required for decision makers in the emergency response environment.

Background

WDA has been applied extensively within cognitive engineering as an analytic framework for design and evaluation of complex sociotechnical systems (Bisantz et al., 2003; Burns, Barsalou, Handler, Kuo, & Harrigan, 2000; Burns, Bryant, & Chalmers, 2000; Naikar & Sanderson, 2001; Reising & Sanderson, 2002a, 2002b). The analysis provides valuable information about the structure and the constraints operating in a work domain that define the normal functioning of a system, and limit the actions of either human or automated agents within the domain. This framework is useful in analyzing complex systems because it presents a description of the work domain or the system to be controlled separately from users' tasks, goals, events, or automation. It thus provides an event-independent representation of the system and its performance boundaries, presenting information about functional constraints on whatever actions users select, such as available equipment and functions (Vicente, 1999).

WDA has been used to support different design-related purposes. For instance, work domain analyses are a key element in the design of computer interfaces (Burns & Hajdukiewicz, 2004; Elm, Potter, Gualtieri, Easter, & Roth, 2003; Vicente, 2002). Naikar and Sanderson (2001) employed WDA to evaluate design proposals for a military defense system, focusing the evaluation on whether proposed designs of physical devices for the future system met the purposes, priorities, and requirements of that system. Naikar, Pearce, Drumm, and Sanderson (2003) used WDA to identify requirements for the structure of teams that would operate a new military system. WDA has further been applied for the design of sensors, specifically for placement of sensors, requirements for information to be measured by sensors, design of instrumentation, and representation of information needs for interface design (Riesing & Sanderson, 2002a,b). WDA has also been used in order to develop design recommendations for naval systems, such as staffing, functionality, and capabilities of automation (Bisantz et al., 2003; Burns, Bryant, & Chalmers, 2000). Burns, Bisantz, and Roth (2004) evaluated methods in work domain analysis by comparing two separate modeling efforts of similar naval combat vessels. The differences in treatment of the environmental components (e.g., including sensor systems and environmental constraints in the model as opposed to modeling the two separately) revealed several issues in modeling work domains, such as questions of defining system boundaries, representing the sensing systems, and incorporating the environment in which the vessels operated.

The Abstraction Hierarchy (AH) (Rasmussen, 1986; Rasmussen, Pejtersen, & Goodstein, 1994), a modeling technique consisting of multiple levels of abstraction, has been used to represent the structure of a work domain for WDA (Rasmussen et al., 1994; Vicente, 1999). The higher levels of the hierarchy describe the overall purpose of the work domain, and the lower levels describe how objects in the system are used to achieve the higher-level purposes of the system. Moving from higher to lower levels is equivalent to moving from functional to physical representation of system properties where each level of abstraction is a separate representation of the work domain (Vicente, 1999). The levels are linked structurally through the means–ends connections indicating why the device, function, or purpose exists when going bottom-up (ends), and how the device, function, or purpose is implemented when moving top-down (means) (Rasmussen et al., 1994). The means–end hierarchy is typically supplemented with a part–whole decomposition to allow the description of system components at multiple levels of decomposition (Bisantz & Vicente, 1994).

Case Study 1: WDA of a Camera

Motivation

Systems with advanced automated capabilities have become standard in many domains. Researchers in human–machine systems have clearly identified the role of humans in the management and control over automated systems and potential difficulties that arise during interaction with these systems (Endsley, 1996; Parasuraman, 1987; Parasuraman, Mouloua, & Molloy, 1996) For example, problems can arise if operators fail to adequately monitor and understand the performance of the system and automation, or if the actions of the system and automation are not clearly displayed (Sarter & Woods, 2000).

Operators need to have an adequate knowledge of the structure and functioning of the automated system in order to appropriately manage it and to reduce the likelihood of "automation surprises" (Sarter & Woods, 2000), or situations in which the behavior of the automation is unexpected. It is equally important to provide users with models of automation that are instrumental in coping with unanticipated situations. Thus, having appropriate models of automation functioning is critical to safe system operation in a way that is not fundamentally different from the knowledge and understanding required of other (nonautomation) system components. Rasmussen et al. (1994) further emphasize the importance of providing operators with information about the behavior of automated control systems, stating that understanding the functions and behavior of automation requires that information about reasons for specific system designs be given to operators such that they can interact with the system effectively. For instance, operators need information about the purposes for which automated systems were implemented, the functions that the automation can carry out, and the physical systems that comprise, and that are controlled by, the automation. Such information needs are congruent with the type of information provided through a WDA, and therefore, it is valuable to explore how representations of automation components can be included as part of the WDA.

Despite these requirements for operator support, WDAs described in the literature have generally not included a representation of automated system components despite their application to systems in which automation plays a key role (Burns, Bryant, & Chalmers, 2002; Nadimian, Griffiths, & Burns, 2002; Sanderson & Naikar, 2000; Sanderson, Naikar, Lintern, & Goss, 1999). Ho and Burns' (2003) research on automated aircraft collision avoidance systems is an exception. The dearth of WDA applications to automation modeling may be explained by the fact that, within the CWA framework, automation design is usually addressed as part of a different component of CWA once control tasks and their requirements have been identified. Automation is considered in terms of its ability to accomplish control tasks within the constraints set out in the WDA *rather than part of the work domain to be controlled*. In this way, automation receives a similar treatment to that of sensors within the WDA framework, in that it is not represented within the work domain models.

Burns et al. (2004) have argued that when sensor systems (particularly, sensors that sense the state of the environment external to the system being controlled) also need to be controlled and interact with important system goals, they should be included explicitly in a WDA. Likewise, in this case study, automation was treated as a system that must be managed and understood, and therefore, that was modeled as part of the work domain. Critically, this representation was comprised not of the tasks that automation is accomplishing but of the purposes, balances, functions, and physical systems that make up the automated system. Interconnected AH models of both the camera and its automation were developed and examined to identify implications for automation design and operator information needs with respect to the automation.

Method

A Canon AE-1P single lens reflex automatic camera was modeled. This camera incorporates a central processing unit that regulates electronic control of exposure, display of signals, and computation of information necessary for control of its functions. The camera is equipped with an exposure meter that uses a silicon photocell to measure the light as it passes through the lens and as it is reflected from the object being photographed. The camera supports both fully manual and fully automatic settings of the aperture and the shutter speed, as well as shutter-priority photography that allows for manual setting of the shutter speed and automatic setting of the aperture. The primary sources of information used to perform the analysis were the camera operating manual, design specifications, the general repair guide of the Canon AE-1P, and the *Manual of Photography* (Horder, 1971). These sources provided information about the physical structure of the camera, specific functions of various parts of the camera, as well as the functions of its automated components. In addition, an expert in camera operation was interviewed in order to confirm the physical and functional decomposition of the camera, AH models of the camera, and automation.

Work Domain Models

Separate Camera and Automation Models

Separate AH models of the camera and automation were constructed (Figures 3.1 and 3.2). The AH of the camera system consisted of five means–end levels and three part–whole levels of decomposition, which described the combination of parts in the camera to form subsystems. The functional purpose (or highest level of abstraction) of the camera was to take pictures (e.g., record an image onto film). Taking a picture involves the recording of light onto film (i.e., exposure), which is possible when a shutter in the camera is opened to allow the light to pass through a lens and form an image on a light-sensitive film. The abstract function level included balances of energy related to kinetic energy, corresponding to subject motion and film motion, and light. The generalized function level included processes such as focusing light and film movement, whereas the physical function level included the physical processes required to achieve these functions (e.g., lens system, viewfinder system, and film movement system). The physical form level of the AH represents states, operational conditions, and settings of systems and components affecting systems at the physical function level.

The means–end and part–whole links connecting the nodes allowed the interdependencies and constraints of the camera work domain to be represented. For instance, the quality of the recorded image depends on the intensity of the light incoming through the lens and the time for which the light is recorded, which are controlled by the camera exposure or the combination of lens aperture and shutter

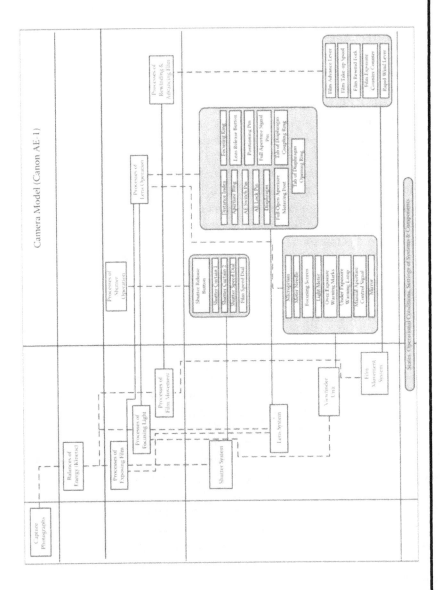

Figure 3.1   AH model of the camera system.

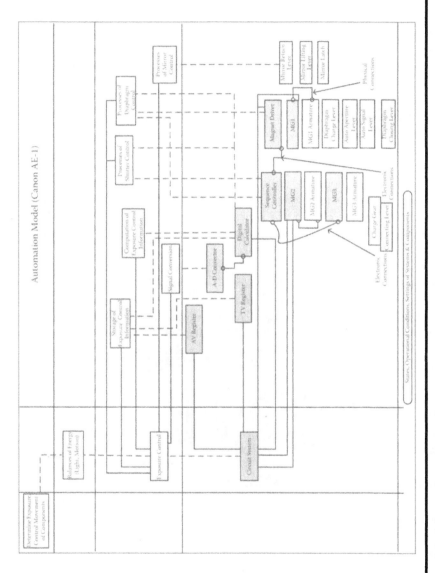

Figure 3.2 AH model of the camera automation.

speed in use. Hence, the function of exposing film (generalized function level), which was decomposed to the processes of shutter operation and lens operation at the component level, is achieved by the shutter system and the lens system, which were further decomposed to parts such as shutter curtains, shutter speed dial, and shutter release button in the shutter system. As evident from the connections in the model, both systems affect exposure, and changes in the functioning of either will have an impact on the processes of exposing film. Similarly, the viewfinder system (at the physical function level), which shows the limits of the field of view of the lens and permits the user to compose the picture, serves as a means to accomplish the function of focusing light. Components of the viewfinder system, in addition to components of the lens system, affect the processes of lens operation at the component level of part–whole decomposition. For example, at the instance of exposure, the mirror (part of the viewfinder system), which is used for focusing and viewing in its stationary position, flips up to permit the light to hit the film, whereas the diaphragm (part of the lens system) closes down to the predetermined aperture.

In contrast to the camera, the functional purpose represented in the AH of the automation was to determine the exposure and control automated movement of components in the camera. The function is automated in order to allow for an optimum combination of shutter speed and aperture setting through automatic operation and setting of both components. This purpose is achieved by exposure control (a generalized function) with constituent processes of shutter and diaphragm control, as well as storage of exposure control information, signal conversion, and computation of exposure information. These functions are carried out through a variety of physical systems, including circuitry as well as some mechanical systems. As in the camera system model, the physical form level of the abstraction described states, operational conditions, and settings of systems and components affecting systems at the physical function level.

As in the camera model, the AH of the automation allows the automation components to be understood in terms of their higher levels, functions, and purposes. For instance, electromechanical control of both the diaphragm and the shutter functions (part of the camera model) is required in order to control exposure. Because correct exposure requires a combination of aperture and shutter speed, the control system automatically computes the exposure and controls mechanical components (represented in the camera model) that complete implementation of the two functions. Specifically, processes of exposure computation include reference to a fixed program of shutter speeds and apertures. Processes of information storage involve storage of digital signals corresponding to subject brightness, maximum aperture, and speed, as well as storage of shutter speed and aperture values. The circuit system and its components, represented at the physical function level, provide the means to carry out the digital control functions. For example, the exposure control information described earlier is stored in storage registers—components of the circuit system. The digital calculator serves several functions as evident from the means–ends connections in the model. In particular, it accomplishes the processes of computation of exposure information, diaphragm and shutter control, as well as processes of information storage by computing the exposure for both the shutter and the diaphragm and storing it in the registers.

The sequence controller is used to achieve the functions of diaphragm and shutter control; specifically, it shuts off the current to the diaphragm control magnet (MG1), which induces the magnet to release its armature, causing movement of the diaphragm. Although MG1 armature is not a part of the circuit system, it is used for diaphragm control and is physically connected to the MG1. Similar physical connections between other components that are not parts of the circuit system but are essential in the electromechanical control of camera functions (i.e., processes of shutter, diaphragm, and mirror control) are represented in the model. In addition, electrical connections between components, which signify the flow of control information within the circuit system, are shown at this level. For instance, the connections between the sequence controller, the magnet driver, and shutter control magnets (MG2 and MG3) illustrate that the sequence controller signals the magnet driver when to shut off current through MG2 and MG3.

Automation and Camera with Interconnections

To make explicit what functions, subsystems, and components in the camera were controlled by the automation and how they were controlled, the functions, subsystems, and components of the automation were connected to those of the camera system (see Figure 3.3). At the generalized function level, connections between the two models demonstrate the relationship between processes of exposure control in the automation model and processes of exposing film at the subsystem level, as well as processes of shutter and diaphragm control and shutter and lens operation in the camera model. At the physical function level, the armature of the MG3 is connected to the second shutter curtain because the release of its armature causes the movement of the second curtain. The link between MG2 armature and the mirror indicates that MG2 armature releases the mirror. At the same time, components of the automation that control the movement of the mirror, such as the mirror-lifting lever, and components that control the shutter operation, such as the first curtain-disengaging lever and the lever latch, control the movement of the first shutter curtain. The mirror is the part of the camera system; however, the mechanical components that instantiate its control and movement are shown in the automation model. The mirror-lifting lever that moves the mirror also strikes the first curtain, as indicated by the connections between the first curtain and the mirror-lifting lever. The control and movement of the diaphragm is accomplished by the combined action of the MG1 armature and the levers in the automation system, which control the movement of the diaphragm. For example, the MG1, the diaphragm-control lever, the diaphragm-closing lever in the camera, and the aperture signal lever, the automatic aperture lever on the lens allow the camera to set the desired aperture.

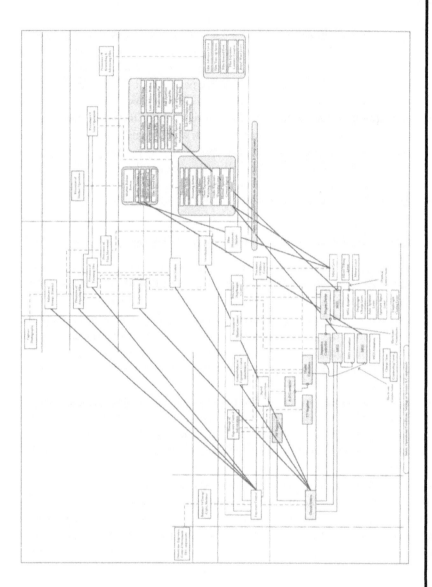

Figure 3.3   AH model showing links between the camera and automation.

Modeling Outcomes

The camera and automation models demonstrate the following applications of WDA accomplished using an AH model. First, the camera model alone illustrates an AH of a well-defined system governed by physical constraints. Second, the case study showed how modeling the automated portion of a system can explicitly provide information relevant for fault diagnosis, training, or real-time aiding with respect to the operation of automated systems.

For instance, the AH of the automation revealed that the sequence controller impacts multiple processes (e.g., both shutter and diaphragm control), which could assist in diagnosing a fault involving both processes. Knowledge of the causal relationships provided with the AH of the automation, as well as the links between the automation and the camera itself, can improve users' ability to effectively take photographs. Although all the subsystems within the camera are controlled by the automation, not all are controlled fully. Such relationships may also be mode dependent (e.g., a camera might have a manual, semiautomated, or fully automated model). The AH representation can represent which parts of the system are being controlled by the automation and to what degree. To illustrate the degree of control, consider single components controlled within the camera subsystems. For instance, considering the viewfinder system, it is obvious that only the mirror is directly controlled by the automation. Such representations are instrumental in showing to the user the extent of automation and the degree of control, and may be helpful in systems with multiple modes and variable degrees of control (e.g., automated systems with multiple modes of operation with different control mechanisms and control over different parts of the system). Given the complexity of the structure of these systems and the complexity of the automated control, the AH allows identification of mode-specific representations of the mechanisms and the degree of control. Moreover, the advantages of having a representation of automatic control connections and separating the automated components become clear when considered in the context of mirror and shutter control. There are causal relationships between mechanical mirror control and control of the first curtain (i.e., movement of components that control the mirror also causes the first curtain movement). In addition, this example makes explicit the structure of the control of the shutter system or what automated components are specifically implicated in computing the shutter speed and moving the curtains.

Case Study 2: WDA of an Emergency Management System

Motivation

The purpose of this case study was to evaluate the potential role of WDA in informing the development of information fusion algorithms to support operations in emergency management and response. Information fusion uses techniques from signal processing, statistics, numerical methods, and artificial intelligence to formally combine information (numeric data) from numerous, disparate, and uncertain sources (often, physical sensors), in order to provide information for further processing and/or human decision makers (Hall & Llinas, 2001). Information processed through data fusion algorithms typically provides an improved estimate (i.e., one with less statistical uncertainty) than can be ascertained from a single sensor or source. The WDA modeling effort was part of a large, multiyear project that used post-earthquake emergency management as a testbed to support information fusion research. WDA was used as a tool to identify relevant information needs with a large, complex, sociotechnical system comprising multiple governmental and nongovernmental organizations and individuals. Additionally, in contrast to the previous case, emergency management and response is a system in which the execution of system purposes are constrained not only by the laws of nature but also by laws, policies, and various altruistic and risk-assessing motivations.

Because of the scale and complexity of the system, nodes in the WDA represent system elements at a more aggregated level of detail than in the previous case (e.g., fire engines, rather than components of the engine). Additionally, this model illustrates the use of multiple models spanning several domains of interest. Researchers (Burns et al., 2004; Burns, Bryant et al., 2000) have suggested that this approach is beneficial because, for instance, it allows the sensing or information needs of one domain with respect to another to be clearly identified. Emergency management can be characterized as spanning multiple, interacting domains: the controlled system consisting of emergency management and response entities, the physical world in which the emergency management system operates, as well as other actors (e.g., the civilian population) who have resources and capabilities that are independent of, but interact with, the emergency management system.

Method

System and Sources of Information

In general, the process of emergency management and response is one in which teams of operators in the field (e.g., firefighters, medical personnel) act to resolve problems (e.g., care for and transport injured civilians, restrict access to unsafe buildings), and provide status information as well as resource requests to higher-level operational response centers (e.g., city emergency operations centers). The role of these higher-level centers is to monitor the current state of the situation, predict possible consequences of the current state, coordinate and provide requested resources, and plan for future resource needs and overall action plans. In the event that local (e.g., city) resources are or will likely be unable to meet the needs of those in the field, requests are made (based on mutual aid agreements and other disaster management agreements and plans) to other municipalities (e.g., other city fire departments, the county). Likewise, if county resources are overcommitted, requests are made to the state, and so on (Bisantz, Rogova, & Little, 2004).

The primary sources of information used to model the emergency management system were government documents related to the structure of emergency management and response systems, research and published accounts of emergency response following earthquakes; discussions and interviews with subject matter experts and practitioners in emergency management (including officials at different jurisdictional levels) in California (state, county, city) as well as an official with the Federal Emergency Management Agency; and observations of a full-scale emergency management exercise at the city level (conducted over 4 hours in a fully staffed, city emergency operations center) (Bisantz et al., 2004).

Work Domain Models

Figure 3.4 shows a condensed version of the model created to represent the emergency management environment. The model explicitly represents the system under control of the decision makers (e.g., the broad array of fixed, mobile, and human resources that can be mobilized for emergency response, and their associated functions and processes), as well as the physical environment and the civilian population. Purposes are represented for the emergency management (one purpose, casualty management, was selected for focus) and civilian population domains but not the physical environment. In contrast to the first case study, many of the abstract functions represent nonphysical constraints, such as those involving economic resources, risk, and authority (for the emergency management system) or altruism (for the civilian population). Balances of physical resources are also of critical importance. General processes included financial, planning, logistics, and operations processes; most of the modeling effort focused on operations processes (defined in this context as involving the deployment and use of resources for the support of and by emergency personnel) because of the purpose of the project. Note that multiple operations processes (e.g., fire and rescue, law enforcement) may rely on the same physical resources (e.g., police officers, 911 call centers). This is similar to the first case, in which multiple processes of the camera relied on the same physical subsystems. Additionally, some civilians may perform operations (e.g., first aid, transport to hospitals) that fulfill emergency management goals; however, these resources may be under the loose of or under no control by the emergency management system. The physical environment includes aspects of the built as well as natural environment.

Modeling Outcomes

The primary outcome from the modeling effort was the identification of information needs, some of which could be potentially supplied by an information fusion system, and/or serve to guide the design of algorithms and information-sensing

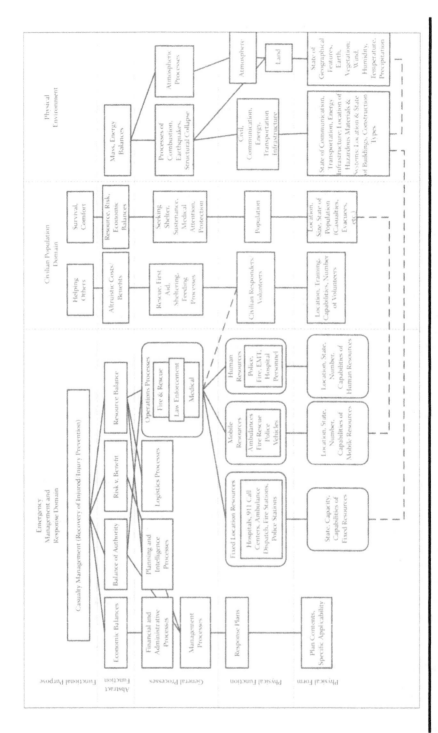

Figure 3.4   AH model of the emergency management system.

functions for such a system. At higher levels, information needs included over-all assessment of the safety of the population (the purpose of the system) as well as actual and predicted resource demand versus availability. The need to manage different jurisdictional and governmental levels of authority implied the need for intelligent access to a diverse set of response plans, laws, and mutual aid agreements. Information needs associated with the physical form node "mobile human resources" correspond to the locations, assignments, and instance-specific capabilities of the resources (e.g., the location, response status, and ability to perform advanced medical interventions for an emergency medical technician).

The complete model noted important interactions among locations and states of different aspects of the physical environment, and the physical environment, civilian population, and emergency resources, which led to the identification of specific information needs (some of these links are shown in Figure 3.4 as dashed lines). For instance, it is important to understand the relationship between hazardous materials sites and utility systems because broken gas lines could lead to explosions or fires near chemical storage facilities. The state of utility systems can impact the capacity and capabilities of hospitals to care for preexisting patients as well as those injured in the event. Dry vegetation, high winds, and damaged water lines could result in fire spread. The state of the ground (e.g., shifted, liquefied) postearthquake has important consequences for predicting the state of damage of structures of particular construction types; likely population density in the structure (which may be dependent on time of day) and likely damage to the structure will have implications for the extent and type of injury as well as rescue resource needs. Such information could be generated from a combination of a priori modeling results fused with real-time sensor and information reports.

Discussion

The two cases presented in this chapter illustrate the use of the work domain analysis phase of CWA for different purposes, across different systems types and levels of detail. Each include abstraction hierarchy models showing nodes at five hierarchical levels linked by means–end relationships: system purposes, abstract functions and constraints, general processes, physical functions, and physical form. The models in the first case also include explicit part–whole levels.

The contrast between the cases illustrates the breadth of systems that can be described using this technique, as well as issues related to the application of abstraction hierarchy models for work domain analysis. First, the cases demonstrate how the modeling technique can be applied to systems that are guided by intentional constraints, such as laws, human motivations, or policies, as well as systems that are primarily governed by physical constraints. Second, the cases illustrate two types of information that can be provided by a WDA: information related to understanding how and why a system functions (in the camera case), which can be critical to training or fault diagnosis; and information regarding system state at multiple levels for control and decision-making purposes (in the emergency management case). Third, both cases illustrate the importance and implications of modeling choices involving system boundaries.

Burns, Bisantz, and Roth (2004) describe trade-offs of different boundary choices in their comparison of work domain models of two navy ships. One issue discussed in that paper is the level of model integration: whether to include aspects of the environment (or other actors) within a system model, or to model these domains separately. Choosing the former approach allows interactions among the domains (in their case, the ship, the sea environment, and other friendly or threatening contacts) to be explicitly represented, whereas choosing the latter approach makes it clear that the environment and contact domains have different purposes, constraints, processes, and functions that are not under the direct control of the ship (though it may be possible to control contacts, for instance, through different system actions—e.g., use of weapons systems).

A second issue involved the decision to represent sensing systems explicitly in the models or to remain silent on the sensing processes. Choosing the former approach emphasized the fact that sensors, like other components of the ship, are essential to achieving ship purposes, and may have important interactions with other functions and high-level goals. Choosing the latter, combined with electing to use multiple domain models, emphasized the need to design sensors for the controlled system (the ship), which captured essential elements of the two other domains. Burns et al. (2004) suggested that similar issues could be raised in the representation of automation: whether automation should be considered during the WDA as part of the system of interest, or whether it should be seen as input to the design of automated functions. Miller (2004) presents an analysis framework built upon work domain analysis, applied to model human physiologic systems in the context of an intensive care medical setting. The approach emphasized the need to model the internal regulatory and control processes of the patient as a key component of the work domain (rather than as something to be designed based on an analysis) because these processes are an inherent component of the patient and must be understood by clinicians in order to diagnose and impact the processes causing illness.

The cases provided in this chapter illustrate choices regarding both of these issues. In the camera example, we elected to explicitly represent automation, rather than consider automation only in task analysis or strategy analysis phases of CWA. Here, the camera model illustrates the value of explicitly including automation in the WDA (thus, recognizing its role in achieving the purposes of the camera) but, at the same time, separating the automated functions so that their separate structure, and relationship to the camera components, could be understood. Also, in both the camera and emergency management cases, domains were represented separately. In the camera case, dual models of the camera and automation were created. In the emergency management case, models of the emergency management system, civilian population, and physical environment were created. Making the choice to create separate models emphasized the different purposes, constraints, processes, and components across the domains. However, in both cases, explicit links were created to show relationships and interactions among the domains. For instance, links represented the manner in which the automation exerted control over aspects of the camera. Links in the emergency management case indicated how aspects of the civilian population could serve emergency management system purposes, and also how interactions among the physical environment (e.g., utilities) and emergency response resources (e.g., hospitals) could affect system purposes.

Conclusions

In summary, this chapter presents two work domain analyses, providing detailed examples of the approach as well as illustrating the flexibility of WDA in treating systems of different size, scope, and nature and for different purposes. Both cases also illustrated important modeling choices regarding the definitions of system boundaries with respect to the representation of multiple domains, the explicit representation of automation, and the benefit of representing interactions among domain models within the work domain analysis.

Acknowledgments

Portions of this chapter, including Figures 3.1 to 3.3 were excerpted and adapted with permission of Taylor and Francis, from Mazaeva, N. and Bisantz, A. M. (2007). "On the representation of automation in a work domain analysis," *Theoretical Issues in Ergonomics Science*. Work on the first case study was supported by the National Science Foundation under grant number IIS9984079 to the first author for support of this work. Work on the second case study was supported by AFOSR Grant F49620-01-1-0371, "Information Fusion for Natural and Man-Made Disasters," to investigators at the Center for Multisource Information Fusion at the University at Buffalo.

References

Bisantz, A. M., Rogova, G., & Little, E. (2004). On the integration of cognitive work analysis within a multisource information fusion development methodology. *Proceedings of the 2004 Human Factors and Ergonomics Society Meeting*, September 20–24, New Orleans, LA.

Bisantz, A. M., Roth, E. M., Brickman, B., Gosbee, L., Hettinger, L., & McKinney, J. (2003). Integrating cognitive analyses in a large scale system design process. *International Journal of Human–Computer Studies, 58*, 177–206.

Bisantz, A. M., & Vicente, K. J. (1994). Making the abstraction hierarchy concrete. *International Journal of Human–computer Studies, 40,* 83–117.

Burns, C. M., Barsalou, E., Handler, C., Kuo, J., & Harrigan, K. (2000). A work domain analysis for network management. *Proceedings of the IEA 2000/HFES 2000 Congress,* 469–472.

Burns, C. M., Bisantz, A. M., & Roth, E. M. (2004). Lessons from a comparison of work domain models: Representational choices and their implications. *Human Factors, 46*(4), 711–727.

Burns, C. M., Bryant, D., & Chalmers, D. (2000). A work domain model to support naval command and control. In *Proceedings of the 2000 IEEE International Conference on Systems, Man, and Cybernetics* (pp. 2228–2233).

Burns, C. M., Bryant, D., & Chalmers, D. (2002). Assessment of the TADMUS DSS with Work domain analysis. In *Proceedings of the Human Factors and Ergonomics Society 46th Annual Meeting* (pp. 453–455). Santa Monica, CA: HFES.

Burns, C. M., & Hajdukiewicz, J. (2004). *Ecological Interface Design.* Boca Raton FL: CRC Press.

Elm, W. C., Potter, S. S., Gualtieri, J. W., Easter, J. R., & Roth, E. M. (2003). Applied cognitive work analysis: A pragmatic methodology for designing revolutionary cognitive affordances. In E. Hollnagel (Ed.), *Handbook of Cognitive Task Design* (pp. 357–382). Mahwah, NJ: Lawrence Erlbaum Associates.

Endsley, M. (1996). Automation and situation awareness. In R. Parasuraman & M. Mouloua (Eds.), *Automation and human performance: Theory and applications* (pp. 163–182). Mahwah, NJ: Lawrence Erlbaum Associates.

Hall, D., & Llinas, J. (2001). Multisensor Data Fusion. In D. Hall & J. Llinas (Eds.), *Handbook of multisensor data fusion* (pp. 1.1–1.10). Boca Raton, FL: CRC Press.

Ho, D., & Burns, C. M. (2003). Ecological interface design in aviation domain: Work domain analysis of automated collision detection and avoidance. In *Proceedings of the Human Factors and Ergonomics Society 47th Annual Meeting* (pp. 119–123). Santa Monica, CA: HFES.

Horder, A. (1971). *The Manual of Photography.* New York: Focal Press.

Miller, A. (2004). A work domain analysis framework for modelling intensive care unit patients. *Cognition Technology and Work, 6,* 207–222.

Nadimian, R. M., Griffiths, S., & Burns, C. M. (2002). Ecological interface design in aviation domains: Work domain analysis and instrumentation availability of the Harvard aircraft. In *Proceedings of the Human Factors and Ergonomics Society 46th Annual Meeting* (pp. 116–120). Santa Monica, CA: HFES.

Naikar, N., Pearce, B., Drumm, D., & Sanderson, P. (2003). Designing teams for first-of-a-kind complex systems using the initial phases of cognitive work analysis: A case study. *Human Factors, 42*(2), 202–217.

Naikar, N., & Sanderson, P. (2001). Evaluating design proposals for complex systems with work domain analysis. *Human Factors, 43,* 529– 542.

Parasuraman, R. (1987). Human–computer monitoring. *Human Factors, 29,* 695–706.

Parasuraman, R., Mouloua, M., & Molloy, R. (1996). Effects of adaptive task allocation on monitoring of automated systems. *Human Factors, 38,* 665–679.

Rasmussen, J. (1986). *Information Processing and Human–Machine Interaction: An Approach to Cognitive Engineering.* New York: North-Holland.

Rasmussen, J., Pejtersen, A. M., & Goodstein, L. P. (1994). *Cognitive Systems Engineering.* New York: John Wiley.

Reising, D. V. C., & Sanderson, P. (2002a). Ecological interface design for Pasteurizer II: A process description of semantic mapping. *Human Factors, 44*(2), 222–247.

Reising, D. V. C., & Sanderson, P. (2002b). Work domain analysis and sensors II: Pasteurization II case study. *International Journal of Human–computer Studies, 56*, 597–637.

Sanderson, P., & Naikar, N. (2000). Temporal coordination control task analysis for analyzing human–system integration. In *Proceedings of the IEA2000/HFES2000 Congress* (pp. 206–209). San Diego, CA.

Sanderson, P., Naikar, N., Lintern, G., & Goss, S. (1999). Use of cognitive work analysis across the system life cycle: From requirements to decommissioning. In *Proceedings of the Human Factors and Ergonomics Society 43rd Annual Meeting* (pp. 318–322). Santa Monica, CA: HFES.

Sarter, N. B., & Woods, D. D. (2000). Team play with a powerful and independent agent: A full mission simulation study. *Human Factors, 42*, 390–402.

Vicente, K. J. (1999). *Cognitive Work Analysis.* Mahwah, NJ: Lawrence Erlbaum Associates.

Vicente, K. J. (2002). Ecological interface design: Progress and challenges. *Human Factors, 44*, 62–78.

Chapter 4

Beyond the Design of Ecological Interfaces: Applications of Work Domain Analysis and Control Task Analysis to the Evaluation of Design Proposals, Team Design, and Training

Neelam Naikar

Contents

69

Overview

Cognitive work analysis (Rasmussen, Pejtersen, & Goodstein, 1994; Vicente, 1999) is well established as a framework for ecological interface design. Vicente (2002) reviews a number of studies in which cognitive work analysis was used to create ecological interfaces for a variety of systems. These systems included process control, aviation, computer network management, software engineering, medicine, command and control, and information retrieval. Vicente's review showed that ecological interface design was consistently associated with better performance compared with current design approaches in industry. Moreover, in several of the studies, cognitive work analysis identified new information requirements for displays. In addition to Vicente's paper, Burns and Hajdukiewicz (2004) review a range of studies in which cognitive work analysis was used for ecological interface design.

Although cognitive work analysis has become widely known as a framework for designing ecological interfaces, the relevance of this approach to other applications is less well recognized. The aim of this chapter is to show that work domain analysis and control task analysis, the first two phases of cognitive work analysis, can be applied to a variety of other problems in industrial settings. Specifically, the chapter presents four case studies of projects in which work domain analysis or control task analysis were used to (1) evaluate design proposals; (2) design teams for future, first-of-a-kind systems; (3) examine training needs and training-system requirements; and (4) define training requirements for error management. All of these projects were undertaken for the Australian Defence Organisation.

The four projects have been reported individually (Naikar, Pearce, Drumm, & Sanderson, 2003; Naikar & Sanderson, 1999, 2001; Naikar & Saunders, 2003) and reviewed collectively (Naikar, 2006) in the past. This chapter builds on the earlier work by examining explicitly the usefulness and feasibility of applying work domain analysis or control task analysis to each project. *Usefulness* is assessed in terms of both impact and uniqueness. Impact is evaluated in terms of whether or not work domain analysis or control task analysis influenced practice on each project, while uniqueness is evaluated in terms of whether or not the analysis made a unique contribution to each project relative to standard techniques that are used in industry. *Feasibility*, on the other hand, is assessed in terms of whether or not the analysis was implemented within the schedule, staff limitations, and financial budget of each project.

The amount of detail that can be conveyed about all four projects in this chapter is limited. Further information on the projects can be obtained by consulting the original papers (Naikar et al., 2003; Naikar & Sanderson, 1999, 2001; Naikar & Saunders, 2003) and a recent publication by Naikar (forthcoming). The last offers the most detailed treatment of the first three case studies in this chapter.

Case Studies

Evaluation of Design Proposals

Background

The aim of the first case study is to show that work domain analysis provides a useful and feasible approach for evaluating design proposals for complex sociotechnical systems. The context for this case study is a stage of the military acquisition cycle for large-scale systems that is referred to as tender evaluation in the United Kingdom and Australia, and source selection in the United States. At this stage of the acquisition cycle, the organization intending to acquire a system (e.g., a military aircraft) may evaluate several proposals for its design, submitted by competing manufacturers. This is a critical stage of acquisition because, following the selection of a particular proposal, the procuring organization becomes "locked into" a specific design concept. Any subsequent evaluations, later in the acquisition cycle, are limited to modifying or refining the chosen design.

Complex sociotechnical systems have special characteristics that place unique requirements on the evaluation of designs. First, complex sociotechnical systems are comprised of an interdependent set of physical devices or components that are necessary for fulfilling the system's work demands. This means that a framework for evaluation must be concerned, not only with the functional capabilities and limitations of each component, but also with the interactions between multiple components. In addition, a framework for evaluation must take into account how well the interactions between multiple components will satisfy the system's work demands.

Second, the work demands of complex sociotechnical systems cannot be described simply in terms of stable sets of tasks or procedures (Meister, 1996; Rasmussen, et al., 1994; Vicente, 1999). This is because stable sets of tasks or procedures can only be specified for routine or anticipated events. In complex sociotechnical systems, it is novel or unanticipated events that pose the greatest threats to performance and safety (Perrow, 1984; Reason, 1990; Vicente, 1999). Thus, a framework for evaluation must offer descriptions of the work demands of a system that can support judgments of how well proposed designs will perform across many different situations, including those that are novel or unanticipated.

Work domain analysis provides a framework for evaluation that is well suited to the unique requirements of complex sociotechnical systems. To illustrate, Figure 4.1 depicts a generic abstraction hierarchy with five levels of abstraction, namely, functional purposes, value and priority measures, purpose-related functions, object-related processes, and physical objects. Using this framework, the physical devices (i.e., individual components) of a proposed design and their functional capabilities and limitations of those physical devices can be mapped onto the respective levels of abstraction: physical objects and object-related processes. This means that the lowest two levels of abstraction will represent the technical solutions of a proposed design. These technical solutions can then be evaluated in terms of how well they satisfy the work demands of a desired system, which are described in terms of its purpose-related functions, value and priority measures, and functional purposes. In Figure 4.1, Physical device B and functional capability and limitation B can be evaluated in terms of their impact on purpose-related functions Y and Z. The effect on these purpose-related functions can then be evaluated against value and priority measures V and W, and the functional purpose of the desired system.

This framework can also be used to evaluate the interactions between the technical solutions of a proposed design, particularly with respect to how well these interactions support the work demands of a desired system. For example, this framework may highlight that although physical devices B and C may each be satisfactory in terms of functional capabilities and limitations B and C, respectively, the interactions between these technical solutions may compromise the purpose-related function Z and the value and priority measure W. Another possibility is that functional capabilities and limitations B and C may both have positive effects on the purpose-related

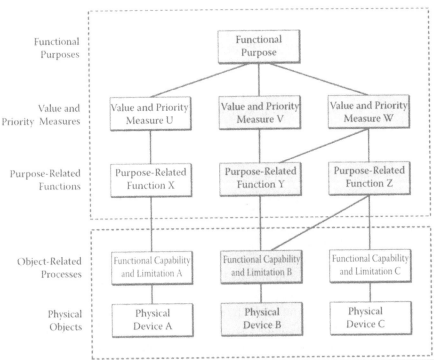

Figure 4.1 **The technical solutions (physical devices and functional capabilities and limitations of those devices) of a proposed design can be mapped onto the levels of abstraction (physical objects and object-related processes) and, subsequently, evaluated in terms of how well they satisfy the purpose-related functions, value and priority measures, and functional purposes of a desired system.**

function Z, which may result in an enhanced effect on value and priority measure W. Alternatively, functional capability and limitation B may have a positive effect on purpose-related function Z, whereas functional capability and limitation C may have a negative impact on this function. The positive effect of functional capability and limitation B on purpose-related function Z may cancel the negative effect of functional capability and limitation C on this function so that overall there is no effect on value and priority measure W.

The preceding discussion illustrates how work domain analysis may be used to evaluate whether or not an interdependent set of physical devices and their functional capabilities and limitations are suitable for satisfying the work demands of a desired system. The work demands defined by work domain analysis represent the fundamental constraints on a system's operation, regardless of situation (Rasmussen, et al., 1994; Vicente, 1999). Thus, by evaluating whether or not proposed designs will satisfy these work demands or constraints, work domain analysis can promote an understanding of how well the proposed designs will perform across a large range of situations, including those that have not occurred or been anticipated previously.

Application

In the first case study, work domain analysis was used to evaluate three competing design proposals for the provision of an Airborne Early Warning and Control (AEW&C) system to the Australian Defence Organisation. The AEW&C system includes a set of six aircraft, each of which will be crewed by a pilot and a copilot in the cockpit and a team of people in the cabin of the aircraft. This crew will be responsible for developing a tactical picture and coordinating the activities of other defense assets in an allocated area of operations. Each AEW&C aircraft will be equipped with a suite of physical devices including sensors, satellite intelligence links, and voice- and data-communication subsystems. The AEW&C system may be likened to the Airborne Warning and Control System (AWACS) of the United States Air Force and the E2C system of the United States Navy.

Naikar (forthcoming) provides a comprehensive account of how the work domain analysis of AEW&C was performed. Furthermore, Naikar offers a detailed description of how this analysis was used to evaluate the three competing design proposals for AEW&C.

In addition to work domain analysis, two other approaches for evaluation were implemented on this project—a technical evaluation and an operational evaluation. For the technical evaluation, a team of evaluators judged whether the technical solutions of the proposed designs complied with, exceeded, or were deficient with respect to pre-specified functional requirements. For the operational evaluation, a team of operations analysts developed computer-based models of aspects of the technical solutions of the proposed designs and then used Monte Carlo simulation to test the performance of the proposed designs in six mission scenarios.

Outcomes

Impact. There were two major demonstrations that work domain analysis influenced practice on this project. The first was a decision by the AEW&C System Program Office, the main defense organization responsible for the acquisition of AEW&C, to adopt work domain analysis as a third approach for evaluation, having initially planned to employ only the technical and operational approaches. Naikar (forthcoming) explains the rationale for the inclusion of work domain analysis on this project. The second demonstration was that in awarding an acquisition contract to one of the manufacturers, the Australian Defence Organisation took into account that work domain analysis had identified that manufacturer's proposal as offering the best design for AEW&C. It is important to note that the operational evaluation identified the same proposal as work domain analysis as offering the best design for AEW&C.

Uniqueness. The technical and operational approaches to evaluation may be regarded as standard techniques since the basic elements of these approaches are commonly used for military acquisition throughout the world (Charlton & O'Brien, 1996; Department of Defence, 1992, 1999; Gabb & Henderson, 1995, 1996; Malone, 1996; O'Brien, 1996). On the AEW&C project, work domain analysis complemented the technical and operational approaches and made a unique contribution to the evaluation of the competing design proposals. The technical evaluation resulted primarily in an appreciation of the functional capabilities and limitations of the individual physical devices of AEW&C. On the other hand, work domain analysis promoted an understanding of how well the functional capabilities and limitations of each physical device and the interactions between them would support the work demands of AEW&C. The operational evaluation led to an appreciation of how well the technical solutions of the proposed designs would perform in six likely or typical mission scenarios, which were specified in terms of stable sequences of events and tasks anticipated by subject matter experts. In contrast, work domain analysis promoted an understanding of how well the technical solutions of the proposed designs would perform across a broad range of situations, including those that are novel or unanticipated.

Feasibility. The application of work domain analysis to evaluate competing design proposals for AEW&C was achieved within the schedule, staff limitations, and financial budget of the project. Naikar (forthcoming) provides estimates of the amounts of time and money and the number of personnel that were required for this application of work domain analysis.

Team Design

Background

The aim of the second case study is to show that work domain analysis and control task analysis provide a useful and feasible approach for designing teams for future, first-of-a-kind systems during the early stages of development. First-of-a-kind systems are those without comparable existing systems, often because significant advances in technology have made it possible to create systems with vast improvements in functionality relative to older systems. A significant benefit of designing a team for a future, first-of-a-kind system in the initial stages of its development (when the system is still only a concept on paper) is that the team design can be tailored to the fundamental work demands of the proposed system. As the development of the system proceeds, its technical solutions are more likely to become "locked in," and the team design is more likely to be determined by the characteristics of the technical system. Such a team design may not be well suited to fulfilling the work demands of the proposed system.

Designing teams for future, first-of-a-kind systems in the initial stages of development poses special challenges for work analysis. First, as these systems are still at a conceptual stage, observable information about workers' activities is not readily available. Second, as these systems have no comparable existing systems, the work demands of such systems cannot be inferred entirely from studying the work demands of existing systems. Third, it is difficult to specify the full range of workers' tasks or procedures for such systems because many of the details of their technical solutions are still undefined, workers will often develop novel ways of working as they gain experience with a new system, and workers may be required to improvise new forms of behavior to deal with novel or unanticipated situations.

Cognitive work analysis offers an approach for dealing with the special challenges of designing teams for such systems. This framework recognizes that workers in complex sociotechnical systems can generate a large variety of work patterns for dealing with a range of situations, including those that are novel or unanticipated, and that not all of these work patterns can be observed or specified in advance. Therefore, rather than analyzing the work demands of a system in terms of how work is done or how work should be done under particular conditions, cognitive work analysis focuses on analyzing the fundamental constraints on workers. These constraints can accommodate many different work patterns, including tasks or procedures that are difficult to predict a priori. Furthermore, the analysis of constraints can take place in the absence of an existing or functioning system.

Four main steps characterize the process of using cognitive work analysis to design a team for a future, first-of-a-kind system during the early stages of development. The first step is to perform a work domain analysis. The second is to perform a control task analysis. The third is to conduct a table-top analysis (Kirwan & Ainsworth, 1992) in which work domain analysis and control task analysis are used to examine different team concepts for the proposed system and to generate requirements for a new team design. The final step is to create a new team design that fulfills the requirements generated from the table-top analysis. Naikar (forthcoming) offers a detailed explanation of the rationale for each of these steps.

Application

In the second case study, cognitive work analysis was used to design a team for the Australian Defence Organisation's AEW&C system. Naikar (forthcoming) offers a detailed account of how cognitive work analysis was used to design a team for this system.

Outcomes

Impact. There were two main demonstrations that cognitive work analysis influenced practice on this project. The first was that the team design proposed for AEW&C was adopted by the AEW&C System Program Office. Given that this team design could not be easily evaluated empirically, a number of strategies were put in place by the AEW&C System Program Office to manage the risk that this team design may not be suitable for AEW&C (Naikar, forthcoming). The second demonstration was that modifications were made to the AEW&C technical-system specification on the basis of the analysis that had been performed. These modifications were made before the contract for the AEW&C system was signed by the selected manufacturer and thus at no cost to the Australian Defence Organisation.

Uniqueness. Standard techniques for team design span a variety of disciplines such as engineering (Davis & Wacker, 1982, 1987; Mundel, 1985; Niebel, 1988), social and organizational psychology (Medsker & Campion, 1997; Hackman & Oldham, 1980; Sundstrom, De Meuse, & Futrell, 1990), and engineering psychology (Dieterly, 1988; Dubrovsky & Piscoppel, 1991; Lehner, 1991). These techniques cannot easily be applied to future, first-of-a-kind systems during the initial stages of development. Some of the techniques rely on descriptive methods of work analysis, which seek to establish how work is done in a system (Vicente, 1999). Other techniques rely on normative methods of work analysis, which seek to specify how work should be done in a system in terms of stable sets of tasks or procedures (Vicente, 1999). As highlighted previously, such information is difficult to ascertain for future, first-of-a-kind systems during the early stages of development. Cognitive work analysis, in contrast, is a formative approach that seeks to establish how work can be done in a system within a set of boundaries or constraints. On the AEW&C project, cognitive work analysis made it possible to design a team for this future, first-of-a-kind system during the early stages of its development.

Feasibility. The application of cognitive work analysis to design a team for AEW&C was implemented within the schedule, staff limitations, and financial budget of the project. As above, estimates of the resources necessary for this application of cognitive work analysis are offered by Naikar (forthcoming).

Training Needs and Training-System Requirements

Background

The aim of the third case study is to show that work domain analysis provides a useful and feasible approach for examining training needs and training-system requirements for complex sociotechnical systems. This application of work domain analysis is presented in the context of the acquisition of training systems for military organizations. During the preliminary stages of procurement, the organization intending to acquire a training system may issue an invitation to training-system manufacturers to register their interest in the project. The *invitation to register interest* is typically accompanied with a broad statement of requirement that provides an idea of the scale and nature of the training system that is sought. Subsequently, the procuring organization may issue a *request for proposal* or a *request for tender*, or both. The request for proposal, which calls for manufacturers to present an initial description of the training-system solution that they can offer, is usually also accompanied with a broad statement of requirement. The request for tender, which calls for manufacturers to submit an extensive plan for the analysis, design, development, and delivery of the desired training system, is normally accompanied with a well-defined and detailed statement of requirement.

To facilitate the procurement of effective training systems, the statement of requirement for a desired training system must be based on a systematic examination of the training needs of the relevant organization. In this chapter, the term *training needs* is used to refer to those work demands of a system for which workers require training. A work analysis is necessary to define a system's work demands and, subsequently, to examine its training needs and training-system requirements.

Identifying training needs and training-system requirements for complex sociotechnical systems poses special challenges for work analysis. One reason for this is that complex sociotechnical systems are subject to disturbances (Vicente, 1999). This means that although many of the events that workers encounter will be routine or familiar to them, they will also have to deal with novel or unanticipated events, which pose a considerable threat to a system's performance and safety (Perrow, 1984; Reason, 1990; Vicente, 1999). Workers generally cannot rely on preplanned tasks or procedures for dealing with novel or unanticipated events. Instead, flexible or adaptive and innovative problem-solving behavior is usually necessary. The training needs of workers, therefore, cannot be established purely in light of work demands described as stable sets of tasks or procedures for dealing with known or anticipated events. Furthermore, it is no longer sufficient for training-system requirements to be formulated entirely on the basis of such descriptions of work demands.

Work domain analysis is well suited to examining training needs and training-system requirements for complex sociotechnical systems. This framework recognizes that complex sociotechnical systems are subject to disturbances and that flexible and innovative problem-solving behavior is usually necessary for dealing with these events. Hence, work domain analysis does not specify work demands in terms of stable sets of tasks or procedures for dealing with known or anticipated events. Instead, it specifies work demands in terms of the fundamental constraints that workers must be capable of satisfying, irrespective of situation. The training needs of workers and the requirements for training systems can then be examined in terms of these fundamental constraints. As these constraints are event independent, the training needs and training-system requirements highlighted by work domain analysis will be relevant to a broad range of situations, including those that are unusual or unanticipated. Moreover, as workers still have multiple options for action within these constraints, the training needs and training-system requirements highlighted by work domain analysis will support possibilities for flexible and innovative behavior.

Table 4.1 portrays how work domain analysis can be used to examine training needs and training-system requirements for complex sociotechnical systems. The first column of the table lists labels for the five levels of an abstraction hierarchy. The second column of the table provides a brief description of the types of constraints represented at each level of abstraction. The third and fourth columns of the table show how the constraints at each level of abstraction highlight qualitatively different potential training needs and training-system requirements. The term *potential* is used to indicate that not all of the constraints that are identified at each level of abstraction may pose training needs or training-system requirements. For instance, workers may already have the ability to fulfill particular work demands or constraints or it may be determined that the real or actual system is best for training workers to fulfill some of the work demands or constraints.

Application

In the third case study, work domain analysis was used to examine training needs and training-system requirements for the acquisition of a new training system for F/A-18 fighter aircraft by the Australian Defence Organisation. The F/A-18 aircraft, which is usually manned by a single pilot, is designed for attacking both air and ground targets. Its suite of physical devices includes a multimode radar, global positioning and inertial navigation systems, a "head-up" cockpit display, multifunction displays, and a range of weapons. The acquisition of a new training system

Table 4.1 Work domain analysis provides a framework for examining potential training needs and training-system requirements in terms of the fundamental constraints on workers

Levels of abstraction	Constraints	Potential training needs	Potential training-system requirements
Functional purposes	The ultimate purposes or objectives of a system and its external constraints	The ultimate purposes or objectives and external constraints that workers must be capable of fulfilling	The ultimate purposes or objectives and external constraints that a training system must have the capability to train workers to fulfill
Value and priority measures	The criteria that must be met for a system to achieve its functional purposes	The definitive criteria that workers must be capable of satisfying	The definitive criteria that a training system must have the capability to train workers to satisfy
Purpose-related functions	The functions that are necessary for a system to achieve its functional purposes	The fundamental set of functions that workers must be capable of managing	The fundamental set of functions that a training system must have the capability to train workers to manage
Object-related processes	The functional capabilities and limitations of physical objects and/or the conditions created by physical objects	The functional capabilities and limitations of physical objects and/or the conditions created by physical objects that workers must be capable of exploiting	The functional capabilities and limitations of physical objects and/or the conditions created by physical objects that a training system must have the capability to train workers to exploit
Physical objects	The natural and artificial objects in the physical environment	The functionally-relevant properties of natural and artificial physical objects that workers must be capable of recognizing	The functionally-relevant properties of natural and artificial physical objects that a training system must have the capability to train workers to recognize

for the F/A-18, which was expected to include a number of high-fidelity simulators, was necessitated by a planned upgrade to the F/A-18 aircraft that would result in a significant enhancement to its capability. In addition to work domain analysis, a training task analysis was performed to support the acquisition of a new training system for the F/A-18 (Wallace, Walkden, & Lintern, 1999).

Naikar (forthcoming) provides a comprehensive description of how the work domain analysis of the F/A-18 system was performed and how this analysis was used to examine the training needs and training-system requirements of the F/A-18 system.

Outcomes

Impact. There were two main demonstrations that work domain analysis influenced practice on the F/A-18 project. The first was a decision by the Capability Development Group and the F/A-18 System Program Office, the two main organizations concerned with the acquisition of the F/A-18 training system, to adopt work domain analysis as an additional approach for examining training needs and training-system requirements. Initially, only a training task analysis (Wallace et al., 1999) was to be employed. The second demonstration was that the F/A-18 work domain analysis was used to examine training needs and training-system requirements and, subsequently, to inform the development of a broad statement of requirement for the F/A-18 training system. This statement of requirement formed part of the "invitation to register interest" documentation that was issued to training-system manufacturers. It is important to note that the training task analysis was also used to inform the statement of requirement for the F/A-18 training system.

Uniqueness. A common approach to the development of training systems, particularly in the military domain, is the instructional systems development process (Childs & Bell, 2002; Department of the Air Force, 1993; Dick & Carey, 1990; Dick et al., 2005; Gagne et al., 2005; Hays & Singer, 1989). This process may be portrayed as consisting of four main phases, namely, analysis, design, development, and delivery (Childs & Bell, 2002). A critical component of the first phase (the analysis phase) is a task analysis. The aim of the task analysis is essentially to define a system's work demands, which provides a basis for examining training needs and training-system requirements (Hays & Singer, 1989).

The training task analysis performed on the F/A-18 project (Wallace et al., 1999) was based on a standard approach to task analysis. On the F/A-18 project, work domain analysis complemented the training task analysis and made a unique contribution to the examination of training needs and training-system requirements. Whereas the training task analysis identified the work demands of the F/A-18 system in terms of stable sets of tasks for dealing with known or anticipated events, work domain analysis identified the work demands of the system in terms of the fundamental constraints that must be satisfied by workers, regardless of the situation. By focusing on these fundamental constraints, work domain analysis provided a framework for examining training needs and training-system requirements that was relevant to a broad range of situations, including those that were novel or unanticipated, and that accommodated possibilities for flexible and innovative behavior.

Two other points are noteworthy. First, standard techniques for task analysis may be concerned with some of the same information as work domain analysis, such as the functionality of physical devices and the criteria or measures for evaluating performance. However, unlike work domain analysis, standard techniques for task analysis focus on analyzing this information in light of stable sets of tasks or procedures for dealing with known or anticipated events. Second, it may be argued that techniques such as hierarchical task analysis, cognitive task analysis, and functional decomposition may be used to uncover the same information as work domain analysis for examining training needs and training-system requirements. Irrespective of whether or not this claim is true, none of these techniques have a theoretical foundation that suggests that such information is important to consider in the development of training systems.

Feasibility. The use of work domain analysis to examine training needs and training-system requirements for the F/A-18 system was accomplished within the schedule, staff limitations, and financial budget of the project. Naikar (forthcoming) presents some approximations of the requirements of this project in terms of time, money, and number of personnel.

Training Requirements for Error Management

Background

The aim of the fourth case study is to show that control task analysis provides a useful and feasible approach for identifying training requirements for managing human error. In the aviation domain, initial efforts to improve safety were based on the view that human error is avoidable. Thus, safety interventions focused on preventing human error. Over the last decade, there has been increasing recognition that errors are not only impossible to eliminate but that they are a consequence of the same cognitive mechanisms that allow humans to operate flexibly in dynamic conditions. Furthermore, although humans often make errors, they also provide a critical line of defense in averting the adverse effects of errors because of their ability to adapt to changing situations. The focus of safety programs has therefore shifted from error prevention to error management, or the creation of systems that are better able to tolerate the occurrence of errors and contain their damaging effects.

The concept of error management may be visualized in light of a characterization of systems as having boundaries of safe operation (Flach & Rasmussen, 2000; Rasmussen et al., 1994).

A system is in a safe or desired state when it is operating within these boundaries (Figure 4.2). When a system crosses these boundaries, it is in an unsafe or

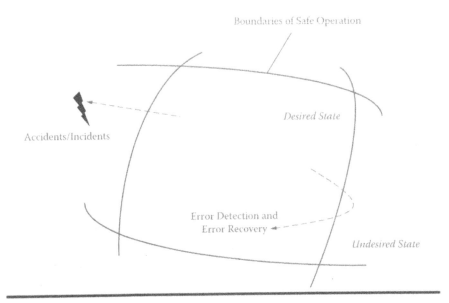

Figure 4.2   A visualization of the concept of error management.

undesired state and an accident or incident can result. Crossing the boundaries of safe operation is inevitable in complex sociotechnical systems in which the pressure to achieve objectives within tight resource and operating constraints leads to systematic migration towards the boundaries. Thus, safety initiatives should focus not only on preventing errors to keep systems within these boundaries but also on managing errors by making it possible for systems to return to a desired state after these boundaries have been crossed.

This chapter describes a novel approach to error management that involves training workers in detecting and recovering from errors. Explained in terms of the characterization supplied above, the approach involves presenting workers with the opportunity to cross the boundaries of safe operation in a training environment, so that they can practice detecting the cues that the system is in an undesired state and recovering the system to a desired state. The assumption is that if workers then cross these boundaries inadvertently in real situations, they will be more likely to detect and recover from their errors, thereby averting an accident or incident. Some empirical support for this assumption may be gleaned from studies in the areas of human-computer interaction (Dormann & Frese, 1994; Frese & Altmann, 1989; Frese, Brodbeck, Heinbokel, Mooser, Schleiffenbaum, & Thiemann, 1991) and expert decision making in naturalistic settings (Klein, 1997; Klein, Calderwood, & McGregor, 1989).

To develop training programs for error detection and error recovery, it is necessary to analyze the potential errors by workers that may place a system in an undesired state, the problem-solving difficulties they may experience in detecting and recovering from those errors, and the problem-solving strategies they may adopt to detect and recover from those errors. This analysis makes it possible to identify training requirements for error management in terms of the undesired states that workers should experience during training and the problem-solving strategies they should practice to enable error detection and error recovery. Training for error management will generally require some kind of training device, such as a simulator, because crossing the boundaries of safe operation is likely to be dangerous when operating the actual system.

Control task analysis offers an approach for analyzing workers' errors and their error-detection and error-recovery processes. The decision ladder (Figure 4.3), one of the main modeling tools of control task analysis, departs from traditional models of information processing in several respects. First, the decision ladder need not be followed in a linear sequence. Shortcuts are possible from one part of the decision ladder to another. Second, the decision ladder accommodates various start and end points. Activity need not begin in the *activation* node of the decision ladder and finish at the *execute* node, but can begin, for example, with an understanding of the *target state* to be achieved. Third, the flow of activity in the decision ladder need not be in the left-to-right sequence, but can occur from right to left. The decision ladder therefore accommodates the opportunistic form of cognitive activity that Rasmussen (1974) found was the norm in expert behavior in complex sociotechnical systems.

To illustrate the use of the decision ladder to identify training requirements for error management, imagine a situation in which the pilot of an aircraft executes a maneuver manually without first disengaging the autopilot. The pilot has difficulty performing the maneuver because the autopilot is trying to retain control of the aircraft. However, the pilot perseveres with completing the maneuver. The autopilot has a bank-angle limit of 45°. When the aircraft reaches this bank angle as a result of the pilot's actions, the autopilot produces a fail tone and then disengages from controlling the aircraft. Because the autopilot disengages while the pilot is applying excessive force to the controls of the aircraft in order to complete the maneuver, the aircraft is "thrown" into a hazardous attitude and subsequently hits the ground.

Three main steps are necessary to identify training requirements for error management using the decision ladder. The first is to identify the critical events in an accident or incident. Critical events refer to points at which workers' actions placed the system in an undesired state, workers had the opportunity to detect that the system was in an undesired state, or workers had the opportunity to recover the system to a desired state. In the situation described earlier, the first critical event occurred when the pilot executed a maneuver manually without disengaging the autopilot. The second critical event occurred when the pilot did not respond appropriately to the fail tone that was produced when the autopilot disengaged.

The second step is to use the decision ladder to examine workers' errors and their error-detection and error-recovery processes during each critical event. Table 4.2 provides a sample of prompts relating to the different nodes of the decision ladder. By using these prompts to review accident or incident data, analysts can explore

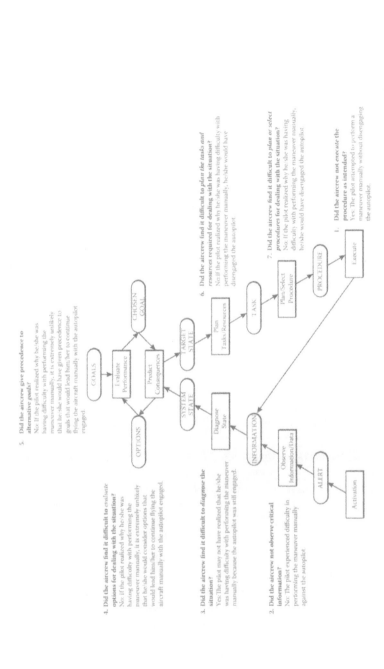

Figure 4.3 The decision ladder and an illustration of its use to examine workers' errors and their error-detection and error-recovery processes. The annotations to the decision ladder restate the prompts from the second column of Table 4.2 and describe the responses to those prompts for the situation described in the text. The numbering indicates the order in which the annotations should be read.

Table 4.2 A sample of prompts relating to the different nodes of the decision ladder for examining workers' errors, and their error-detection and error-recovery processes

Nodes of the decision ladder	What was the error? Why didn't the workers detect the error? Why didn't the workers recover from the error?	How did the workers detect the error?	How did the workers recover from the error?
Observation	Did the workers fail to observe critical information?	What information or cues did the workers observe?	
Diagnosis	Did the workers find it difficult to diagnose the situation?	What was the workers' diagnosis of the situation?	
Option evaluation	Did the workers find it difficult to evaluate options?		What options did the workers consider? Why did the workers select/reject options?
Prioritization of goals	Did the workers give precedence to alternative goals?		What were the workers' goals?
Planning of tasks and resources	Did the workers find it difficult to plan the tasks and resources required for dealing with the situation?		What tasks and resources did the workers plan to use to recover from the error?
Planning or selection of procedures	Did the workers find it difficult to plan or select procedures for dealing with the situation?		What procedures did the workers formulate to recover from the error? What standard procedures did the workers select to recover from the error?
Execution	Did the workers fail to execute the procedure as intended?		What procedures did the workers execute to recover from the error?

what errors by workers placed the system in an undesired state (second column of table), why workers did not detect or recover from their errors (also in second column of table), or how workers detected (third column of table) and recovered from (fourth column of table) their errors.

Figure 4.3 shows the decision ladder representation that resulted from this step for the first critical event in the situation described earlier. From this representation, it becomes clear that the pilot did not execute a procedure correctly (i.e., the pilot performed a maneuver manually without first disengaging the autopilot) and then had difficulty in detecting this error. The difficulty in detecting the error did not occur because the pilot did not observe critical information (i.e., the pilot was aware that it was hard to perform the maneuver), but because he or she was unable to diagnose why it was hard to execute the maneuver manually. If the pilot had made the correct diagnosis, he or she would probably have detected the error (i.e., realized that he or she had forgotten to disengage the autopilot) and subsequently recovered from the error (i.e., disengaged the autopilot while applying appropriate force to the controls of the aircraft).

The final step is to define training requirements for error management on the basis of the preceding analysis. The training requirement that may be derived from Figure 4.3 is to provide pilots with the opportunity to experience the undesired state that would result from performing a maneuver manually with the autopilot still engaged, specifically, the difficulty of executing the maneuver manually under those conditions. Pilots should also have the opportunity to practice recovering the aircraft to a desired state by disengaging the autopilot while applying appropriate force to the controls of the aircraft, so that it is not thrown into a hazardous state when the autopilot relinquishes control. Then, if pilots forget to disengage the autopilot before performing a maneuver manually on a real flight, they are more likely to recognize the cues that the autopilot is still engaged and to disengage the autopilot to regain control of the aircraft without throwing it into a hazardous state.

Application

In the fourth case study, control task analysis was used to identify training requirements for error management for F-111 aircraft in the Australian Defence Organisation. The F-111 is a long-range strike aircraft that can take off and land at relatively low speeds and fly near to the ground at supersonic speeds in order to avoid detection. Its suite of physical devices includes digital flight controls, terrain-following radar, and inertial navigation and integrated weapons systems. The F-111 aircraft, which is crewed by a pilot and a navigator, is used for delivering weapons to both land and maritime targets and also for tactical reconnaissance.

On this project, 3 F-111 accidents and 21 F-111 incidents were analyzed to identify training requirements for error management. The F-111 accidents resulted in six training requirements and the F-111 incidents resulted in eight training requirements (several of the incidents involved the same types of errors). To evaluate these training requirements, interviews were conducted with a number of F-111 aircrew (both pilots and navigators) and training instructors. Interviewees were asked (1) whether or not the training requirements were already being met, (2) whether or not the training requirements would be useful for helping them to manage errors on real missions if they were met, and (3) whether or not they had been in unsafe situations that were similar to the accidents or incidents that had led to the training requirements. Table 4.3 displays the number of interviewees who answered "yes" to these questions against the total number of responses for each training requirement.

The table shows that in every case, the majority of interviewees responded that the training requirements were not currently being met. A few interviewees indicated that some training requirements had been met because they had previously received such training when the opportunity arose naturally through aircrew error. However, none of the training requirements were addressed systematically or documented in the F-111 training syllabus. It should be emphasized that these results indicate the novelty of the training approach to error management that is proposed here, and not deficiencies in the F-111 training system. Table 4.3 also shows that

Table 4.3 The number of interviewees who responded "yes" to the interview questions against the total number of responses for each training requirement

Training requirements	(1) Already being met? Number that responded "yes"(total responses)	(2) Useful? Number that responded "yes" (total responses)	(3) Similar unsafe situation? Number that responded "yes" (total responses)
1	0 (10)	10 (10)	8 (10)
2	1 (14)	13 (14)	7 (14)
3	0 (13)	14 (14)	8 (13)
4	1 (13)	12 (13)	4 (13)
5	4 (14)	14 (14)	13 (13)
6	5 (12)	11 (11)	9 (12)
7	1 (12)	12 (12)	6 (12)
8	1 (12)	11 (12)	8 (12)
9	0 (12)	12 (12)	11 (12)
10	0 (12)	12 (12)	5 (12)
11	2 (12)	11 (12)	10 (12)
12	1 (12)	12 (12)	4 (12)
13	0 (12)	12 (12)	9 (12)
14	0 (12)	12 (12)	5 (12)

almost all of the interviewees thought that the training requirements, if they were met, would be useful for helping them to manage errors on real missions. A few of the interviewees indicated that some of the training requirements would not be useful because they were unlikely to make the types of errors that had motivated those training requirements. Nevertheless, for every training requirement, several interviewees reported that they had been in unsafe situations that were similar to the accidents or incidents that had led to the training requirements, suggesting that the errors are not impossible.

The interviewees also raised some general concerns about the implementation of the training requirements. One of the biggest challenges that they highlighted was that current-generation simulators may not be able to simulate the undesired states that are necessary to provide training in error management. For instance, a simulator may not be able to replicate how an aircraft would respond or perform if it is flown manually with the autopilot engaged. This issue too indicates the novelty of the training approach that has been proposed. Before investing in the development of simulators that can support this approach or implementing this approach in real systems, empirical evaluation of whether or not such training leads to better performance in detecting and recovering from errors is desirable.

Outcomes

Impact. On the basis of the interviews that were conducted with F-111 aircrew and training instructors, control task analysis was judged to be useful for defining training requirements for error management. Some obstacles to the implementation of the training requirements were identified, in particular, that many current-generation simulators may not be able to simulate the requisite undesired states. Further research into the implementation of the training approach for error management is necessary.

Uniqueness. Standard approaches to error management include the design of error-tolerant interfaces (Billings, 1997; Kontogiannis, 1999; Noyes, 1998; Rasmussen & Vicente, 1989) and crew resource management (Gunther, 2002; Helmreich, 2001; Helmreich, Klinect, & Wilhelm, 1999; Helmreich, Merritt, & Wilhelm, 1999). Error-tolerant interfaces aim to make errors and their effects visible and reversible. Crew resource management, which is popular in the aviation domain, entails training flight crew in non-technical skills for error management such as maintaining vigilance, cross-checking team members, and leadership and communication skills. In contrast, the approach described in this study focuses on training workers in technical skills for detecting and recovering from errors (Naikar & Saunders, 2003).

It should be acknowledged that aspects of the training regime in the Australian Defence Organisation resemble the approach presented here. First, equipment failures are regularly introduced into simulator-based training sessions so that air-crew can practice processes for managing those failures. Second, aircrew also receive training in managing human error, although usually not to the extent that is advocated here.

Feasibility. The application of control task analysis to define training requirements for error management was achieved within the schedule, staff limitations, and financial budget of the F-111 project.

Conclusion

Although cognitive work analysis is well established as a framework for ecological interface design, the relevance of this approach to other applications is not widely recognized. By presenting four case studies of projects for the Australian Defence Organisation, this chapter has demonstrated that the first two phases of cognitive work analysis can be applied to a variety of problems in industrial settings. The first case study showed that work domain analysis can be used to evaluate competing design proposals for complex sociotechnical systems. The second showed that work domain analysis and control task analysis can be used to design teams for future, first-of-a-kind systems at the early stages of development. The third showed that work domain analysis can be used to examine training needs and training-system requirements. The final case study showed that control task analysis can be used to define training requirements for error management.

As well as highlighting the range of problems to which the first two phases of cognitive work analysis are relevant, this chapter has established the usefulness and feasibility of applying work domain analysis and control task analysis in industrial settings. Of the four projects reviewed in this chapter, the applications of work domain analysis and control task analysis (1) influenced practice on three projects (in the fourth project, the results of control task analysis were judged to be useful, but could not be implemented without conducting further research in other related areas); (2) made a unique contribution to all projects relative to standard techniques that are used in industry; and (3) were completed within the schedule, staff, and financial restrictions of all projects.

It must be recognized, however, that it has not been demonstrated empirically that work domain analysis and control task analysis lead to sound or effective solutions in the areas of application proposed in this chapter. To conduct such studies in industrial settings would require substantial resources. For instance, to establish whether or not work domain analysis identified the best design proposal for AEW&C, one approach would be to build three systems, one based on each of the three design proposals, and to test the performance of the three systems in a set of relevant scenarios. This option is highly impractical. Another approach would be to develop computer-based models of the three design proposals and to test their performance in a range of scenarios using Monte Carlo simulation. Normally, this option would also be impractical. On the AEW&C project, this option was implemented, to a degree, in the form of the operational evaluation. The aim of the operational evaluation was not to provide a test of work domain analysis, but to offer an additional approach for evaluation. Although the operational evaluation identified the same proposal as work domain analysis as offering the best design for AEW&C, only certain aspects of the three design proposals were represented in the computer-based models. Hence, a strong claim cannot be made that the commonality in the results of the two approaches indicates that work domain analysis alone identified the best design proposal for AEW&C.

Feasible options for determining with reasonable certainty whether or not work domain analysis and control task analysis led to sound or effective solutions in the other projects discussed in this chapter were also not available. For example, in the second project, it was not feasible to test whether or not these analyses resulted in an effective team design for AEW&C. One reason for this is that suitable mock-ups, prototypes, computer-based simulations, or human-in-the-loop simulations of the AEW&C technical system did not exist at that stage of the project. During system development, when mock-ups, prototypes or simulations of the AEW&C technical system become available, or when the AEW&C system is eventually brought into service, it may be possible to evaluate whether or not the team design for AEW&C is effective. However, it will be difficult, if not impossible, to isolate the contribution of the first two phases of cognitive work analysis to the performance of the team design from the contribution of other aspects of the system's design, such as its interface design.

In a similar vein, it should be acknowledged that it has not been established empirically that work domain analysis and control task analysis lead to better solutions than standard techniques in the areas of application proposed in this chapter. Conducting such studies in industrial settings also requires substantial resources. For instance, to determine whether or not work domain analysis and control task analysis led to a better team design for AEW&C than standard techniques, one approach would be to develop several team designs for AEW&C using different techniques and to test the performance of the team designs in a range of relevant scenarios. Aside from the usual schedule, staff, and financial constraints faced by many industrial projects, which made such a study impossible on the AEW&C project, there were additional specific restrictions. For example, to apply standard techniques for team design that rely on descriptive methods of work analysis on the AEW&C project, it would have been necessary to have access to current-generation systems, such as the AWACS or E2C system, which are not available in Australia.

Such challenges with the empirical validation of techniques in industrial settings are not limited to work domain analysis and control task analysis. They are also relevant to other techniques. Hence, techniques that are applied in industrial settings are rarely evaluated in a formal way (Czaja, 1997). Instead, the principle criterion that is employed for evaluation is usefulness (Vicente, 1999; Whitefield, Wilson, & Dowell, 1991). In this chapter, usefulness was defined in terms of impact and uniqueness. Thus, together with feasibility, three criteria were used for evaluating the applications of work domain analysis and control task analysis. Out of four projects, work domain analysis and control task analysis fulfilled all three criteria on three of the projects. On one project, the ability of control task analysis to influence practice could not be ascertained.

Another shortcoming of this chapter is that all of the case studies are limited to military systems. The extent to which the proposed applications of work domain analysis and control task analysis are relevant to other types of systems is therefore unknown. Future research should evaluate the suitability of these applications of work domain analysis and control task analysis to other types of systems.

To implement the applications of work domain analysis and control task analysis proposed in this chapter, analysts may require more guidance than is supplied here. Burns and Hajdukiewicz (2004) and Naikar (forthcoming) offer guidelines for performing work domain analysis, and Naikar, Moylan, and Pearce (2006) offer guidelines for performing control task analysis. Naikar (forthcoming) also provides detailed explanations of how work domain analysis and control task analysis can be used to evaluate design proposals, design teams for future, first-of-a-kind systems, and examine training needs and training-system requirements. Finally, Naikar and Saunders (2003) provide a detailed explanation of how control task analysis can be used to define training requirements for error management.

Acknowledgments

I am grateful to the subject matter experts who contributed to the four projects, particularly Group Captain Chris Westwood, Wing Commander Antony Martin, Squadron Leader Carl Zell, Squadron Leader Paul Nowland, and Ms Tracey Bryan; Mr. Dominic Drumm, formerly of the Defence Science and Technology Organisation, and Mr. Ben Hall from the Aerospace Simulators Systems Support Office for clarifying the details of some of the case studies; Dr. Greg Jamieson from the University of Toronto and one anonymous reviewer for their useful comments on this chapter; Dr. Kate Branford from Dédale Asia Pacific and Dr. James Meehan from the Defence Science and Technology Organisation for their help with editing this chapter; and Mr. Adam Woollett from Boeing Australia Limited for producing the figures for this chapter.

References

Billings, C. E. 1997. *Aviation automation: The search for a human-centered approach.* Mahwah, NJ: Lawrence Erlbaum Associates.

Burns, C. M., & J. R. Hajdukiewicz. 2004. *Ecological interface design.* Boca Raton, FL: CRC Press.

Charlton, S. G., & T. G. O'Brien. 1996. The role of human factors testing and evaluation in systems development. In *Handbook of human factors testing and evaluation*, Ed., T. G. O'Brien, and S. G. Charlton, 13–26. Mahwah, NJ: Lawrence Erlbaum Associates.

Childs, J. M., & H. H. Bell. 2002. Training systems evaluation. In *Handbook of human factors testing and evaluation*. 2nd ed. Ed., S. G. Charlton, and T. G. O'Brien, 473–509. Mahwah, NJ: Lawrence Erlbaum Associates.

Czaja, S. J. 1997. Systems design and evaluation. In *Handbook of human factors and ergonomics*. 2nd ed. Ed., G. Salvendy, 17–40. New York: John Wiley.

Davis, L. E., & G. L. Wacker. 1982. Job design. In *Handbook of industrial engineering*, Ed., G. Salvendy, 2.5.1–2.5.31. New York: John Wiley.

Davis, L. E., & G. L. Wacker. 1987. Job design. In *Handbook of human factors*, Ed., G. Salvendy, 431–452. New York: John Wiley.

Department of Defence. 1992. *The capital equipment procurement manual*. Canberra, Australia: Defence Publishing Service.

Department of Defence. 1999. *Defence procurement policy manual*. Canberra, Australia: Defence Publishing Service.

Department of the Air Force. 1993. *Instructional system development*. Washington DC: Headquarters U. S. Air Force. (AF Manual 36-2234)

Dick, W., & L. Carey. 1990. *The systematic design of instruction*. 3rd ed. Tallahassee, Glenview, IL: Scott, Foresman/Little, Brown Higher Education.

Dick, W., L. Carey, & J. O. Carey. 2005. *The systematic design of instruction*. 6th ed. Boston: Pearson.

Dieterly, D. L. 1988. Team performance requirements. In *The job analysis handbook for business, industry, and government*. Vol. 1. Ed., S. Gael, 766–777. New York: John Wiley.

Dormann, T., & M. Frese. 1994. Error training: Replication and the function of exploratory behavior. *International Journal of Human–Computer Interaction* 6(4): 365–372.

Dubrovsky, V., & A. Piscoppel. 1991. Toward a framework for structured job-collaboration design. In *Proceedings of the Human Factors Society 35th Annual Meeting*, 959–963. Santa Monica, CA: Human Factors Society.

Flach, J. M., & J. Rasmussen. 2000. Cognitive engineering: Designing for situation awareness. In *Cognitive engineering in the aviation domain*, Ed., N. B. Sarter, and R. Amalberti, 153–179. Mahwah, NJ: Lawrence Erlbaum Associates.

Frese, M., & A. Altmann. 1989. The treatment of errors in learning and training. In *Developing skills with information technology*, Ed., L. Bainbridge, and S. A. R. Quintanilla, 65–86. Chichester, West Sussex: John Wiley.

Frese, M., F. Brodbeck, T. Heinbokel, C. Mooser, E. Schleiffenbaum, & P. Thiemann. 1991. Errors in training computer skills: On the positive function of errors. *Human–Computer Interaction* 6: 77–93.

Gabb, A. P., & D. E. Henderson. 1995. *A review of Navy's technical and operational tender evaluation practices*. Salisbury, South Australia: Electronics and Surveillance Research Laboratory. (Defence Science and Technology Organisation Tech. Rep. No. DSTO-TR-0194)

Gabb, A. P., & D. E. Henderson. 1996. *Technical and operational tender evaluations for complex military systems*. Salisbury, South Australia: Electronics and Surveillance Research Laboratory. (Defence Science and Technology Organisation Tech. Rep. No. DSTO-TR-0303)

Gagne, R. M., W. W. Wager, K. C. Golas, & J. M. Keller. 2005. *Principles of instructional design.* 5th ed. Belmont, CA: Thomson Wadsworth.

Gunther, D. 2002. Threat and error management training counters complacency, fosters operational excellence. *ICAO Journal* 57(4): 12–13.

Hackman, J. R., & G. R. Oldham. 1980. *Work redesign.* Reading, MA: Addison-Wesley.

Hays, R. T., & M. J. Singer. 1989. *Simulation fidelity in training system design: Bridging the gap between reality and training.* New York: Springer-Verlag.

Helmreich, B. 2001. A closer inspection: What really happens in the cockpit. *Flight Safety Australia* 5 (January–February): 32–35.

Helmreich, R. L., J. R. Klinect, & J. A. Wilhelm. 1999. Models of threat, error, and CRM in flight operations. In *Proceedings of the Tenth International Symposium on Aviation Psychology,* 677–682. Columbus, OH: The Ohio State University.

Helmreich, R. L., A. C. Merritt, & J. A. Wilhelm. 1999. The evolution of Crew Resource Management training in commercial aviation. *The International Journal of Aviation Psychology* 9(1): 19–32.

Kirwan, B., & L. K. Ainsworth, Eds. 1992. *A guide to task analysis.* London: Taylor & Francis.

Klein, G. 1997. Developing expertise in decision making. *Thinking and Reasoning* 3(4): 337–352.

Klein, G. A., R. Calderwood, & D. MacGregor. 1989. Critical decision method for eliciting knowledge. *IEEE Transactions on Systems, Man, and Cybernetics* 19(3): 462–472.

Kontogiannis, T. 1999. User strategies in recovering from errors in man–machine systems. *Safety Science* 32:49–68.

Lehner, P. E. 1991. Towards a prescriptive theory of team design. In *Proceedings of the IEEE International Conference on Systems, Man, and Cybernetics: Decision aiding for complex systems,* 2029–2034. New York: Institute of Electrical and Electronics Engineers.

Malone, T. B. 1996. Human factors test support documentation. In *Handbook of human factors testing and evaluation,* Ed., T. G. O'Brien, and S. G. Charlton, 101–116. Mahwah, NJ: Lawrence Erlbaum Associates.

Medsker, G. J., & M. A. Campion. 1997. Job and team design. In *Handbook of human factors and ergonomics.* 2nd ed. Ed., G. Salvendy, 450–489. New York: John Wiley and Sons, Inc.

Meister, D. 1996. Human factors test and evaluation in the twenty-first century. In *Handbook of human factors testing and evaluation,* Ed., T. G. O'Brien, and S. G. Charlton, 313–322. Mahwah, NJ: Lawrence Erlbaum Associates.

Mundel, M. E. 1985. *Motion and time study: Improving productivity.* Englewood Cliffs, NJ: Prentice-Hall.

Naikar, N. 2006. Beyond interface design: Further applications of cognitive work analysis. *International Journal of Industrial Ergonomics* 36:423–438.

Naikar, N. Forthcoming. *Work domain analysis: Concepts, guidelines, and cases.* Boca Raton, FL: Taylor and Francis Group, LLC.

Naikar, N., A. Moylan, & B. Pearce. 2006. Analysing activity in complex systems with cognitive work analysis: Concepts, guidelines, and case study for control task analysis. *Theoretical Issues in Ergonomics Science* 7(4): 371–394.

Naikar, N., B. Pearce, D. Drumm, & P. M. Sanderson. 2003. Designing teams for first-of-a-kind, complex systems using the initial phases of cognitive work analysis: Case study. *Human Factors* 45(2): 202–217.

Naikar, N., & P. M. Sanderson. 1999. Work domain analysis for training-system definition and acquisition. *The International Journal of Aviation Psychology* 9(3): 271–290.

Naikar, N., & P. M. Sanderson. 2001. Evaluating design proposals for complex systems with work domain analysis. *Human Factors* 43(4): 529–542.

Naikar, N., & A. Saunders. 2003. Crossing the boundaries of safe operation: An approach for training technical skills in error management. *Cognition, Technology & Work* 5:171–180.

Niebel, B. W. 1988. *Motion and time study.* Homewood, IL: Irwin.

Noyes, J. M. 1998. Managing errors. In *Proceedings of the UKACC International Conference on CONTROL '98*, 578–583. London: Institution of Electrical Engineers.

O'Brien, T. G. 1996. Preparing human factors test plans and reports. In *Handbook of human factors testing and evaluation*, Ed., T. G. O'Brien, and S. G. Charlton, 117–134. Mahwah, NJ: Lawrence Erlbaum Associates.

Perrow, C. 1984. *Normal accidents: Living with high-risk technologies.* New York: Basic Books.

Rasmussen, J. 1974. *The human data processor as a system component: Bits and pieces of a model.* Roskilde, Denmark: Danish Atomic Energy Commission Research Establishment Risö. (Rep. No. Risö-M-1722)

Rasmussen, J., A. M. Pejtersen, & L. P. Goodstein. 1994. *Cognitive systems engineering.* New York: John Wiley.

Rasmussen, J., & K. J. Vicente. 1989. Coping with human errors through system design: Implications for ecological interface design. *International Journal of Man-Machine Studies* 31: 517–534.

Reason, J. 1990. *Human error.* Cambridge: Cambridge University Press.

Sundstrom, E., K. P. De Meuse, & D. Futrell. 1990. Work teams: Applications and effectiveness. *American Psychologist* 45(2): 120–133.

Vicente, K. J. 1999. *Cognitive work analysis: Toward safe, productive, and healthy computer-based work.* Mahwah, NJ: Lawrence Erlbaum Associates.

Vicente, K. J. 2002. Ecological interface design: Progress and challenges. *Human Factors* 44(1): 62–78.

Wallace, P., L. Walkden, & G. Lintern. 1999. An analysis of F/A-18A pilot training tasks. In *Proceedings of the Fourth International SimTect Conference*, 15–19. Melbourne, Australia: The SimTect 99 Organising and Technical Committee.

Whitefield, A., F. Wilson, & J. Dowell. 1991. A framework for human factors evaluation. *Behaviour & Information Technology* 10 (1): 65–79.

Chapter 5

Control Task Analysis: Methodologies for Eliciting and Applying Decision Ladder Models for Command and Control

Tab M. Lamoureux and Bruce A. Chalmers

Contents

Overview

Defence R&D Canada–Atlantic is investigating technologies such as data fusion and advanced operator–machine interfaces to support naval operators of Canadian warships in their command and control (C2) work, focusing specifically on their work areas of maritime tactical picture compilation (MTPC, or "the Picture Compilation project"; Chalmers, Webb, & Keeble, 2002; Sartori, Keeble, Bandali, Boothby, Bos, Rehak, & Lamoureux, 2006; Zobarich, Lamoureux, Keeble, Sartori, Bandali, & Boothby, 2006) and tactical planning and response management (TPRM, or "the Response Management project"; Chalmers, 2003; Bos, Rehak, Keeble, & Lamoureux, 2005). Canadian warships are generally deployed as part of a Canadian task group (TG) or as an integral member of a multinational task force. The tactical picture, compiled from Above Water Warfare, Underwater Warfare, Tactical Data Link, and Wide Area Picture data inputs, is the situation picture shared among all engaged units that underlies all aspects of C2 decision making over an area of interest of the maritime commander for the units. TPRM is concerned with the development, integration, and management of tactical plans within and across the air, surface, and subsurface warfare areas.

A key aspect of this exploratory investigation is determining the requirements for effective computer-based tools for picture compilation and response management. The research literature on existing complex sociotechnical work environments that share many of the characteristics of C2 provides strong evidence for the merit of designing cognitive and collaborative support systems for such environments on the basis of an in-depth understanding of the work system and the specific work demands that operators have to deal with (Patterson, Woods, Tinapple, & Roth, 2001). In such a design approach, some form of work analysis to model the operators' demands with regard to the work environment inevitably emerges as a critical consideration. We have been investigating such an approach in our defense setting based on emerging concepts in the field of Cognitive Systems Engineering (Vicente, 1999).

In the following sections, we describe three projects that used the Control Task Analysis (ConTA) work-modeling approach of Cognitive Work Analysis (CWA; Vicente, 1999) within a design framework to identify the work demands of C2 operators aboard Canadian naval vessels and to uncover potential design solutions to support these demands. A significant and novel contribution of the ConTA approach is that it permits developing a traceable design thread that directly links knowledge elicitation and work analysis outputs to specific design hypotheses for supporting operator work demands. The resulting design concepts are briefly described, and the approaches adopted during the projects are compared.

Exploratory Design Framework

The primary purpose of our exploratory design framework is to allow developing and testing, from a work-centered perspective, of design hypotheses to support operators with advanced computer-based capabilities in their individual or collaborative cognitive work. The framework encompasses various activities aimed at developing an increasing understanding of the work's demands, and using this knowledge to develop and test specific hypotheses about ways to support operators with these demands. Figure 5.1 illustrates the framework, showing, for concreteness, one specific activity trajectory within it (there can be many), and the activity nodes on that trajectory and their potential linkages in terms of inputs and outputs.

An essential part of the framework is a work analysis that explicitly models work demands or constraints in the work environment as a basis for design. This analysis is conducted along the lines of Rasmussen and Vicente's CWA framework (Vicente, 1999). CWA is a systems-oriented approach to analyzing a work environment aimed at capturing the behavior-shaping constraints for that environment. Such constraints delimit an envelope within which all productive work occurs. This underlies the formative focus that CWA brings to design, by providing a modeling capability that deals with demands across a broad spectrum of situations, from familiar ones that operators encounter routinely, to unfamiliar, but anticipated ones

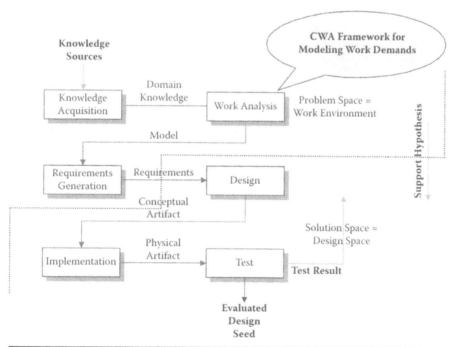

Figure 5.1   Methodological framework for eliciting design seeds.

(i.e., anticipated by system designers, policy, doctrine, tactics, and procedures), to ones that are unfamiliar and unanticipated.

Hypotheses for supporting work are rendered as design seeds, each seed effectively instantiating some specific support hypothesis that needs to be tested. Figuratively, work analysis serves to "seed" promising design concepts or support concepts that may turn into design "nuggets," which when integrated could be used in defining complete capability for supporting work demands. Referring to these instantiations as design seeds is meant to capture their essential role as one of seeding or jumpstarting an exploratory design process. The terminology is borrowed from Patterson et al. (2001). A seed represents some specific and relatively independent design concept to support some specific aspect of the work. Testing the validity of a support hypothesis for a seed might range from obtaining initial subjective subject matter expert (SME) feedback to the seed to conducting objective performance tests using it. In this manner, therefore, complex aspects of the work's demands are decomposed into manageable portions for design purposes. In addition, increasingly realistic prototypes of design seeds can be iteratively developed, refined, integrated, and tested, leading eventually to a coherent support capability.

Analysis Framework

ConTA is the second phase of analysis of a CWA (Vicente, 1999), and is used to show what decisions are made and the states and processes that may be encountered in control tasks (see Vicente, 1999 for a definition of "control tasks"). In our work, ConTA was done using a decision ladder approach (Rasmussen & Jensen, 1974), which is a modeling tool that presents a generic sequence of data-processing activities and the states of knowledge involved in a decision-making function "folded" to reinforce the notion that processing activities can move between nonsequential points in the framework. The template is shown in Figure 5.2 (adapted from Vicente, 1999).

In the figure, the steps in rectangles represent data-processing activities (e.g., Observe); the steps in circles represent states of knowledge gained from previous data-processing activity (e.g., Set of Observations). The entire process illustrates a single decision and, therefore, may take only seconds to complete. Indeed, experts at a task may skip steps or take even less time to complete the process. Moreover, several ladders may occur simultaneously, and it has been argued that depending on the grain of analysis adopted, each step may in fact contain its own ladder.

There are two different kinds of shortcuts in a ladder outlined by Vicente (1999). The first type is called a *shunt*. A shunt occurs when a process (rectangle) leads to an advanced state of knowledge (circle), for example, an Observe to Procedure shunt occurs if a decision maker perceives his/her observations of the system in terms of actions that need to be performed. The second type of shortcut outlined by Vicente

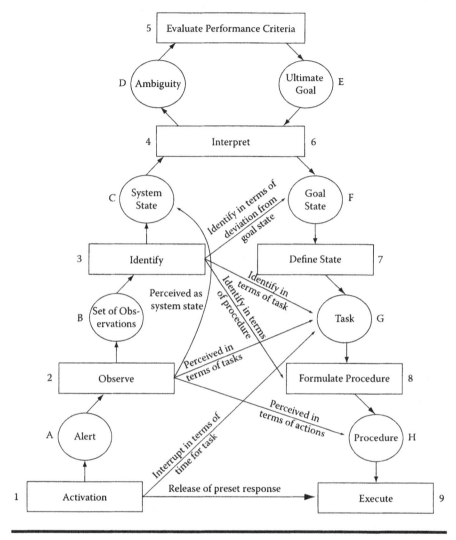

Figure 5.2 Decision ladder template. (Adapted from Vicente, 1999.)

is called a *leap*. A leap is where two states of knowledge are automatically associated (circle to circle) with no processing in between. Figure 5.2 graphically depicts some of the more common shunts.

There are two other types of shortcuts not explicitly outlined by Vicente: process to process and state of knowledge to process. Though process to process was not originally discussed by Vicente, there is one example shown on Vicente's diagram (e.g., Release of preset response) that represents a process associated with another process (reflexes are another good example). Similarly, though state-to-process shortcuts were never discussed by Vicente, such transitions may also be necessary to adequately describe a control task. The possibilities of each of these various types of shortcuts have been exploited in the identification of design seeds.

The Projects

Although four different data collection approaches are described as part of knowledge acquisition in our work, only three specific R&D projects will be introduced here. Furthermore, these three projects adopted two significantly different approaches to ConTA. The first two projects, maritime tactical picture (MTP) and TPRM, applied ConTA by mapping SME reports of their decision-making processes onto the decision ladder framework described earlier. Because this analysis was based on SME reports, the process maps show opportunistic movement through the ladder. Whereas the MTP project dealt specifically with picture compilation at the level of a single Canadian Navy HALIFAX Class frigate, the third project, TG MTP (task group MTP), looked at tactical picture compilation for a naval TG. With TG MTP, ConTA was applied by creating an inventory of all possible activities or processes that might correspond to a generic data processing activity and all possible information or constructs that might represent an ensuing state of knowledge. A more detailed discussion of the decision ladder inventory is presented in the following section on the TG MTP data analysis approach.

Practically, the MTP and TG MTP projects addressed similar issues: the aggregation of information from a variety of sensors into a coherent representation of the combat team's operational environment. However, the combat team could comprise several people colocated in the same operations room, sharing a hardwired data network (the case for the MTP project), or could, in addition, comprise several subteams of several people distributed across several types of TG platforms (the case for TG MTP), each subteam located in one of the TG platforms, coordinating data across a heterogenous low bandwidth data link network. The TPRM project focused on examples in which a platform's team planned its tactical responses for an event that took place between 10 minutes and 7 hours in the future. The event needed to involve trade-offs between different factors, such that there was no single obvious solution, but rather a variety of plans could be reached or adopted depending the people involved.

Data Collection Approach

We have used a variety of approaches in our work to date to elicit domain knowledge from SMEs for the CWA modeling shown in Figure 5.1. In one approach, described in Chalmers et al. (2002), a detailed paper-based description of a tactical scenario (including representations of the tactical situation on the C2 system) was prepared, and teams of operators described their own and their team's activities in terms of their goals, information needs, information sources, information transfers, processing activities, strategies, and the collaboration that might be involved. In a subsequent work, we adopted a detailed scenario-based approach (Chalmers & Lamoureux, 2004) that used a combination of operators dealing with a tactical scenario in a training simulator, combined with scheduled "freezes" in the scenario and group debriefing, to capture relevant data. This approach was adopted because of limitations to the paper-based data collection approach.

A further effort adapted Klein's Critical Decision Method (CDM; Bos et al., 2005; Klein, Calderwood, & MacGregor, 1989). This change stemmed from a concern that, presenting SMEs with a scenario restricted the breadth of information that could be collected. This would mean that any support concept eventually developed would address a layperson's understanding of what the naval operator does, rather than that of the SME's. Because the layperson cannot have the SME's breadth of domain experience and understanding, issues identified and their associated support concepts might not address the more critical operator work demands. CDM has permitted SMEs to describe what they considered to be the most difficult situations they encountered. Once the problem space was bounded in this way, analysts could systematically investigate each facet of it. This approach provided enough data to construct a detailed account of the actions undertaken by the operators in the scenario they described, including the triggers, information requirements, and outputs. Using this approach, it was felt that the resulting support concepts would be more relevant and acceptable to operators. Interestingly, Naikar and her colleagues (Naikar, Moylan, & Pearce, 2006) have recently used a similar data collection approach independently to conduct a ConTA for a new military system.

Using CDM with several teams of SMEs is time and resource intensive (similar to using a simulator), and a more pragmatic approach to data collection was needed. The most recent work (Sartori et al., 2006; Zobarich et al., 2006) followed a "workshop" approach that included one SME working actively with a team of analysts for an intensive 3-day session involving both data collection and preliminary analysis. During this session, the work domain was described in physical terms, organizational terms, and functional terms (to a fine level of detail). This information was then structured into critical task sequences, and then, selected critical task sequences were subjected to a work domain analysis (WDA) and a ConTA. The SME was involved in the conduct of both the WDA and the ConTA. Following a longer period of analysis, the structured data and analysis results were presented to a wider group of SMEs for validation. This approach proved to be extremely efficient in producing data to the level required to perform CWA (Sartori et al., 2006) and subsequently engage in design activities (Zobarich et al., 2006).

MTP and TPRM Data Analysis Approach

Having collected data on the work performed by operators, the data was first structured in three ways:

- As scenario-based chronological task sequences
- As a part–whole decomposition of the work domain according to the framework of an Abstraction–Decomposition Space (ADS; Vicente, 1999)
- As exemplars for each intersection of a specific level of the part–whole decomposition with a specific level in the abstraction hierarchy in the ADS

The scenario-based chronological task sequences were then entered in a spreadsheet, with each row representing a discrete task (with significant scenario events inserted to preserve the context of the tasks). Because tasks were described at varying levels of detail (indeed, some cannot be considered tasks, and are more properly described as conclusions or states of knowledge), it was decided to use the term *statement* to describe the task outline entered at the beginning of each row. Each statement was mapped to a specific cell in the ADS (to identify the level at which the subject was operating), and traced as a route an expert might take through the ladder. The data (i.e., analysis output) was entered on the same row. The details of the ConTA portion of these mappings are given in the following sections.

Decision Ladder Notation

To simplify the analysis of scenarios during this project, a coding system for the ladder was adopted. Data-processing activities were coded as numbers, and states of knowledge were coded as letters in the ladder (Figure 5.2). The coding notation was 1–9 for data-processing activities, counted from the bottom left, up the ladder, counting Interpret as numbers 4 and 6, and then down the right-hand side of the ladder. Notation was A–H for states of knowledge.

The Interpret rectangle was intentionally coded twice (4 and 6). This described whether the operator was interpreting the consequences of the System State (4) or the consequences of the Ultimate Goal (6). The coding that was developed for each task (its mapping on the ladder) described where the task enters the ladder, how it moved through the ladder, and where it left the ladder. This coding structure allowed for the easy recognition of leaps and shunts. Shunts are shown as a number combined with a nonsequential letter (e.g., 1D), and leaps are shown as a number and two consecutive letters (e.g., 1AG).

Analysts' Interpretation Guidance

After the coding structure was outlined, further definition of each step in the ladder was agreed upon by all the analysts (a total of three). This was felt to be necessary because Vicente (1999) provides very little guidance on the content of the ladder's steps. The formulation of additional interpretation guidance helped to ensure that all the analysts had the same interpretation of the different steps, leading to greater interanalyst agreement in their analyses when mapping tasks onto the steps of the ladder. Table 5.1 outlines the interpretations used (Column 3).

With an agreed-upon interpretation of the steps of the ladder, each analyst independently analyzed every task outlined by the SMEs. Control tasks associated with statements were mapped onto the ladder, one statement at a time. Letters and numbers were assigned in order, coding the steps involved in the task on the decision ladder.

Analysis Process

Analysts considered each statement in the scenario chronology from the perspective of how the activity enters the ladder (i.e., where), how it leaves the ladder, and then what happened in between (what data-processing activities and what states of knowledge are passed through). This "in between" portion of the mapping included noting leaps and shunts. Analysts also attempted to track through the analysis when a control task might be interrupted and picked up again. Related statements (i.e., representing that a task was interrupted and then resumed) were linked through notes in the margin.

As is often the case, there were limited resources and time available to perform this analysis. However, analysts benefited from the ADS analyses performed earlier in the project. Part of this analysis was the identification of the tasks that were critical to the successful performance of the scenario. If certain tasks were analyzed as a priority during the ADS analysis, the same priority was afforded to the task in the decision ladder analysis and these tasks were analyzed first.

Analysts worked independently of each other to encode all the SME statements onto the ladder, coming together at the end to discuss their findings for each scenario and settle upon a single, agreed-upon, decision ladder interpretation. Independent working was adopted so analysis was more likely to capture all the decision ladder elements of a task. While conducting the analysis, impromptu discussions between analysts about difficulties encountered, uncertainties in analysis, etc., were encouraged in order to improve the consistency of analysis.

Table 5.1 Interpretation of decision ladder steps

Decision ladder step	Vicente (1999) guidance	Bos et al. (2005) guidance	Sartori et al. (2006) guidance
Activation	Detection of need for action	Perception	All possible ways that an operator can be alerted to the need for an activity
Alert	What is going on?	Realization	All possible examples of what it "looks like" when someone has noticed the need to act
Observe	Information and data	Display of contacts and other information	All possible processes through which observations are collected
Set of Observations	What lies behind?	A corpus of information	Collectively, all the information known about a single contact
Identify	Present state of the system	Consider the information	All possible inferences based on known information
System State	What is the effect?	What does this mean?	Into which standard ID category does this contact fit, how detailed is my recognition assessment?
Interpret	Consequences for current task, safety, efficiency, etc.	How does this fit in my "idealized" progress toward my goal?	All possible implications of the contact on our mission
Ambiguity	Which goal to choose?	I do not know how this fits	All possible reasons why we might still be unsure what this means for the mission. Where is the uncertainty?
Evaluate Performance Criteria	—	How does it need to fit?	All possible to interpret this data, so we can get a probabilistic idea of what it means

—continued

Table 5.1   Interpretation of decision ladder steps (*continued*)

Decision ladder step	Vicente (1999) guidance	Bos et al. (2005) guidance	Sartori et al. (2006) guidance
Ultimate Goal	Which is then the goal state?	This has this effect on my ultimate goal	All possible ways this ambiguous information can change my goal? How?
Interpret	Consequences for current task, safety, efficiency, etc.	What steps do I need to add to get my progress back on track?	All possible outcomes in progress toward accomplishing goals
Goal State	Which is the appropriate change in operating conditions?	Know what you want to achieve	All possible ways that subgoals may be modified/added/removed to accomplish the Ultimate Goal
Define Task	Select appropriate change of system conditions	Determine what you need to do	What possible ways can I achieve my new goals from my current starting point?
Task	How to do it?	Know what you need to do	Collective list of the tasks identified in the Define Task step
Formulate Procedure	Plan sequences of actions	Plan how to do it	Determine the specific actions involved in accomplishing tasks
Procedure	—	Know how to do it	Knowing what to do
Execute	Coordinate manipulations	Do it	Do it

MTP/TPRM Results

Once a scenario was analyzed, the analysts met to discuss their findings and agree upon a "final" analysis. Although a great deal of debate resulted from these meetings, it was felt that, in general, analysts agreed on the sequence of ladder steps for each statement. Disagreement tended to stem from differences in the analysts' understanding of the domain. Some analysts proceeded from a more "micro" level of understanding of what the subject of a statement was about, whereas others relied strictly on the language and information used in the statement, which by itself was already the subject of one person's interpretation. This potential source of variance was addressed in the subsequent TG MTP project.

The identification of design seeds, however, did not seem to suffer from any disagreements between analysts. Many design seeds were possible from the ConTA, which were ultimately combined to produce an integrated, coherent design concept.

Example Analyses

To concretely illustrate the ConTA process that the analysts went through, five example statements from the TPRM project are presented here. These examples were selected because of the range of activities they represent:

■ *Information Management Director (IMD) monitors text messaging because CHAT and other communications make the Operations Room officer (ORO) too busy.* This first statement explains the creation of a new position called the IMD position. This operator is in charge of certain communication systems, and was created to relieve the ORO of some of the time-consuming tasks. The perspective used to conduct the analysis of this statement was a focus on the IMD monitoring text messaging. This statement was represented by the ladder coding "1A2BG." First, the activation stage (1) occurs with the arrival of a text message that alerts (A: Alert) the IMD. The IMD reads that message (2: Observe), and considers it in the context of the other messages he is monitoring (B: Set of Observations). The IMD immediately knows who to tell about the message (G: Task), usually the ORO.

■ *Determine trade-off to defend others and put self at undue risk.* This statement was represented by the coding "4D5E6." Considering the task, the ORO must already have a set of observations and know the system state; he is now using this information to "determine trade-off." The ORO is interpreting the information he has (4: Interpret). However, because he is determining trade-offs, we can assume that there will be some ambiguity, an assumption supported by the rest of the statement "to defend others and put self at undue risk" (D: Ambiguity). The ORO must then evaluate the performance criteria, not only for the frigate (e.g., Can I realistically defend myself if we are actually attacked? What is the likelihood that we will be attacked?) but also for the mission (e.g., Do I really have to defend this other vessel? What are my orders?). This may lead to an altered ultimate goal, meaning the ORO has passed through 5 (Evaluate Performance Criteria) on his way to E (Ultimate Goal). The ORO is left considering what this new, altered ultimate goal means for the current task (6: Interpret). Because of the nature of the statement (i.e., the ORO is only determining trade-offs, not deciding what to do about them) and those following it in the scenario (according to the story SMEs told), the ORO has not actually formed an opinion about what intermediate goals (F: Goal State) arise from this activity and can therefore not progress on to defining the task (7: Define Task).

■ *Determine appropriate weapons load.* The third statement was represented by "2BC4F7G." Considering the task, the operator must already be alert to the fact that there is some threat in the local area, so it enters the ladder at 2 (Observe). The operator must consider what is going on external to the frigate (B: Set of Observations), and the operator leaps to knowledge about the system state (C: System State) based on practice and experience. The operator must then interpret the consequences of the contacts surrounding the frigate in terms of their possible impact on the mission (4: Interpret). Once the possible impact of the contacts has been established, the operator knows what needs to be changed to succeed in the mission (F: Goal State; there is unlikely to be ambiguity or requirement for higher-level evaluations). The operator can then determine what is needed in the weapons load to maximize mission success (7: Define Task). Finally, the operator knows what changes need to be made to ensure that the appropriate weapons load has been deployed to meet the threat (G: Task).

■ *A line is drawn on chart.* The fourth statement was represented by "9." The decision to draw the line, considerations of where to put the line, and the formulation of a procedure to draw the line (a skill-based task) are statements that precede this one. This is a very simple and straightforward statement that specifies only the final actions executed (9: Execute).

■ *Monitor small boats overtaking.* In the scenario, there were small taxi-like boats with children on board, and the potential for the ship's wake to cause them to sink was an area of concern. This statement was represented by "2BF7H9." This statement starts at 2 (Observe), as the alert that monitoring is needed has arrived prior to the statement. A set of observations about the small boats is then formed (B). The ORO knows what the contacts are and also knows the system state (of the frigate, the small boats, and the frigate and small boats together). Because the monitoring activity is affected by the location of the frigate relative to the small boats, but the ORO already knows this information, the next step in the ladder is F (Goal State). In other words, the observations lead directly to knowing how the goal state of the frigate needs to change because of the small boats. The ORO must then decide what to do to meet the new goal state (7). Again, this is a well-trained and practiced procedure, so the ORO does not have to engage in any complex problem solving and decision making to formulate a procedure. Essentially, the monitoring must continue if a small boat is in a problematic location, as opposed to the cessation of monitoring if there are no boats nearby (H: Procedure, 9: Execute).

Identification of Design Seeds

The decision ladder analysis resulted in a number of design seeds being identified. However, the nature of this analysis, that is, by focusing on information-processing flow through the steps of the ladder framework, means that it is difficult to correlate design seeds with specific data-processing activities and states of knowledge. Although some design seeds related to specific steps on the ladder, there were many that cut across steps. In fact, it was found more expedient to aggregate steps on the ladder into broader groupings, allowing design seeds that are broader in scope to be identified. Wickens' model of information processing (Wickens, 1984) offered a structure for this approach.

Wickens' model posits three broad stages in human information processing: perceptual encoding, central processing, and responding. By and large, the ladder framework fits into this model, with the possible addition of working memory to bridge the gap between perceptual encoding and central processing (working memory is part of central processing in the Wickens' model). The mapping of the ladder with the model of human information processing is shown in Table 5.2.

By conducting a decision ladder analysis, it became apparent to the analyst whether the emphasis of the processing around a statement lay at an early (perceptual encoding), middle (working memory and central processing), or

Table 5.2 Mapping of decision ladder steps to human information-processing stages

Decision ladder step	Human information-processing stage
Activation	Perceptual encoding
Alert	
Observe	
Set of Observations	Working memory
Identify	Central processing
System State	
Interpret	
Ambiguity	
Evaluate Performance Criteria	
Ultimate Goal	
Goal State	
Define Task	Responding
Task	
Formulate Procedure	
Procedure	
Execute	

Note: Data-processing activities are shaded.

late (responding) stage of information processing. Predictably, those statements mapped largely to an early stage of information processing resulted in design seeds that focus on bringing a stimulus to the operator's attention, or on retaining that stimulus in a location for the operator to access and use quickly and easily in decision-making and problem-solving processes. Practically, these design seeds are auditory or visual cues using display objects to convey the new information so that the operator remembers that it exists without running the risk of it being forced out of working memory.

The next stage in the mapping just described is working memory. A design seed that recurred frequently in this case took the form of a representation aid that externalizes the burden of remembering the relevant information. Traditionally, this would be a list or set of discrete display objects. However, this application domain also suggested composite, synergistic display objects that are a result of data fusion, the consideration of the information being conveyed, the cognitive operation to be performed on that information, and the mental model to be triggered. These features can help to overcome working memory limitations, and move information seamlessly from perceptual encoding, through working memory, to central processing.

The central processing stage of information processing encompasses the realization of the current system state, the knowledge of what state the system needs to be in, and the resolution of any ambiguity. The resolution of this ambiguity and the determination of what the system state should be represent significant challenges to automated systems. The decision ladder analysis (in contrast to the abstraction hierarchy in a WDA) helps focus a designer's attention on the precise aspects of the information-processing activity that might need to be supported with a decision-aiding type of design seed. For instance, in the central processing stage, how can the operator determine the system state or resolve ambiguity? What information is required by the operator to arrive at the next state of knowledge? What process is adopted by the operator to arrive at a new state of knowledge?

The final stage of information processing is broadly classed as responding. The processing focus at this point in the ladder is concerned with determining how to change the system state to match the goal state and then executing actions to effect the change. Often, navy operators will engage in highly routine but time-consuming processes to effect the change. This was a frequent source of design seeds: the automation of work associated with making a change. It is understood that some of this work has evolved with the navy and is bound up with the formal command hierarchy, but it may be necessary to change the process to "inform" another officer rather than "seek permission of." However, in many cases, operators immediately know how to change the system state and begin executing that course of action at once. In such cases, the design seed attempts to overcome the processes and procedures that may delay the action, especially when it is highly unusual for the higher authority to refuse permission. To cater for this rare case, the provision of a "command override" function would suffice.

Perhaps the most helpful factor for identifying design seeds, decision ladder analysis identifies leaps and shunts (formally defined as *shortcuts* in the ladder) that almost certainly become the basis of design seeds. As a simple example, when tasked with escorting a tanker, operators already know that not being with the tanker means they must first plot a course to it. A design seed to support such a shortcut in the ladder is to automatically link mission objectives with work activities (i.e., without need for operator involvement). Carrying this further, the system could monitor other systems' states and invoke the appropriate subroutine or decision support tool (see Table 5.3).

Many of the design seeds that resulted from the decision ladder analysis focused on the automation of work activities; for instance, automation of route plotting, automation of determining what ammunition to use, automation of searching for things that might affect a plan and so on. In comparison to developing design seeds based on a work domain analysis, it was found that the decision ladder analysis tended to result in design seeds that were much more focused on placing the operator in a passive role, by taking away the active problem-solving and decision-making tasks. This may not always be the most appropriate use of the analysis, as this may

Table 5.3   Decision ladder design seeds

Task description	Design seed	Interface element
Generate plan for a rendezvous with an escort tanker	Automated planning assistant focusing on some or all of: risk, efficiency, navigation, etc.	Course line, pick lists, thermal display, fast forward
The ORO will discuss the approach with the Commanding Officer (CO) and the Tanker Captain	Graphical method of communicating (supplement verbal communication of spatial information)	Course line
Consider risks in escort	Risk trade-off tool	Thermal displays
A line is drawn on chart	Automatic drawing and/or translation of the line in terms of coordinates and speeds	Linking of course line to ship's navigation system
Operations room team considers territorial waters as threats, and alters course for tactical considerations	Graphical display of course considerations	Course line
De-risk plan	Risk trade-off tool or graphical depictions of risks	Thermal displays

—continued

Table 5.3 Decision ladder design seeds (*continued*)

Task description	Design seed	Interface element
The CO does "what-if" planning to pick apart the plan	Graphical display of course considerations	Thermal displays, notable features on display (e.g., ships, missile sites), fast forward, course line
Debating process conducted between the ORO and warfare directors		
Weigh trade-offs regarding time and proximity to hazards (e.g., missile batteries)		
Plan contingencies	Combination of graphical display of course considerations and risk trade-off tool	Thermal displays
Command team decides upon the tactical considerations, what the threats are, and how to mitigate those threats (Intel, doctrine, and capabilities)		
Optimal plans are established	Automatic calculation algorithm replacing human thought	Thermal displays
Manage underwater environment: the underwater environment is a significant complicating factor because very little is "known"	Graphical display of anti-submarine warfare "no go" areas	Notable features on display
Carry out route planning to avoid mine or subthreat (or torpedo avoidance distractive means)		
Conduct ASW		
Trade off requirement for silence with ASW Threat	Graphical depiction of trade-offs in terms of reaction time, risk, etc.	Thermal displays
The ORO makes a risk assessment and decides on basis of the available options	Risk bar charts/advisor	Thermal displays

deskill operators and render them less able to conduct novel problem solving. However, if the focus is on supporting the operator in achieving the various activities in the ladder, rather than on actually automating them, a decision ladder analysis can be seen to support a design approach that is complementary to the human operator.

Resulting Interface Concept

Figure 5.3 illustrates the tactical display portion of a conceptual interface developed on the basis of the design seeds identified during the TPRM project, including those from the decision ladder analysis. This was part of a more extensive concept for supporting cognitive, metacognitive, collaborative, and communication demands of the ship's operations room team. As indicated earlier, the TPRM project focused on the decision-making activity of the tactical coordinator on a frigate, its ORO, in making trade-offs among different factors as a basis for responding to tactical events over a near-term planning horizon.

The interface permits the ORO to access all the information on-screen near to the primary tactical point of interest (i.e., near to the own-ship). The ORO can select factors he/she is most concerned with (the checklist) and have them represented on a thermal display (shown in the form of a colored polygon with each vertex corresponding to a single factor, denoted by corresponding words or symbology on "mouseover"), which presents the level of risk for each factor, calculated according to predetermined algorithms. The cumulative level of risk for the ship is represented on the line denoting ship's course (again as thermal shading, i.e., moving through a gradient from white "hot" to "cold" blue), on

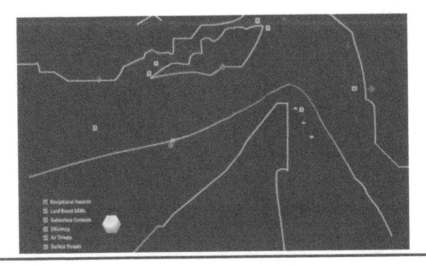

Figure 5.3   Interface concept for tactical planning and response management (TPRM) project.

which the operator can click to highlight the significant contributors to that risk. The ORO can also click and drag the course line to change course, in order to avoid or address factors. The resultant course would be fed directly into the ship's navigation system based on the ORO's decision. The ORO can also activate a "fast forward" simulation mode in the display, allowing him/her to visualize the course of the ship and the most likely courses of other contacts (e.g., aircrafts, ships). The interface concept thus:

- Reduces workload by rendering relevant information in one place and creating fused data representations
- Enables the operator to maintain situation awareness at Level 3 projection (Endsley, 1988)
- Provides feedback
- Is configurable according to personal and situational needs
- Alleviates memory load by presenting cues and system-generated start points for response selection
- Enhances decision making by presenting relevant information in an easily digestible form and allowing the operator to see the likely impact of different decisions

Table 5.3 links tasks (each being the subject of a separate decision ladder analysis), design seeds, and the interface element shown in Figure 5.3.

TG MTP Data Analysis Approach

The TG MTP ConTA work began with a 3-day workshop, during which analysts worked with an SME to collect data about the work domain and tasks associated with TG maritime tactical picture compilation, as well as to construct an ADS for analysis purposes. Day 3 of the workshop comprised deciding on which work activities involved in picture compilation would be analyzed in the ConTA. This decision followed a procedure outlined in Naikar et al. (2006). The procedure involves selecting representative work activities on the basis of a combination of work functions and situations. Using this method, it is possible to deconstruct a work activity (i.e., maritime TG tactical picture compilation) across meaningful dimensions. According to Naikar et al. (2006), "the various combinations of work situations and functions will impose qualitatively different sets of cognitive demands on workers," allowing both breadth and depth to the analysis. As a result of this procedure, the decision was made to study two activities in addition to the overall tactical picture compilation activities: Asset Allocation, concerned with deploying the most suitable asset to carry out a specified task; and Single Contact activities, which focuses on aggregating relevant information about a single contact. The analysis process was reviewed with all analysts, then the analysts worked jointly with the SME to gain a common understanding of the processes involved within each activity and of the application of the ConTA. The analysts and the SME agreed that the three activities represented all relevant TG MTP compilation activities, and that they overlapped and diversified to a degree that corresponded with the recommendations of Naikar et al. (2006).

Differences between the MTP/TPRM and the TG MTP ConTA Approach

The decision to use a different analysis methodology was based on our belief that there were opportunities for improvement in the analysis approach used in our previous studies. In the TPRM work (see the section "Analysis Process"), the decision ladder output was aimed at representing actual (i.e., the path an expert would take through the decision ladder) information-processing activities and the states of knowledge passed through to effect the processing. In that study, an activity selected for analysis was described according to its processing flow using the template of the ladder. The desire in the TG MTP study was to obtain a comprehensive picture of what information processing activities could and should occur during TG MTP compilation, and understand the ways in which they could be accomplished. As such, the approach taken in this study was for the analysts (working together) to develop an exhaustive list or inventory of all possible primary and secondary activities, states, information, information sources, procedures, processes, interpretations, etc., that could be relevant at each step in the ladder. Column 4 of Table 5.1 outlines the interpretations of the ladder's steps used to build these inventories. This was done for three specific work functions: Interpreting the Overall Tactical Picture, Interpreting a Single Contact, and Asset Allocation, chosen using the method (Naikar et al., 2006) described earlier. As well as ultimately being used to identify design seeds, the items in these inventories represent "shopping lists" that designers can use to build design seeds for broader support concepts. The approach also meant that the ladder coding scheme used in our earlier ConTA work was not used in the TG MTP project.

For illustration purposes, Table 5.4 shows some examples of inventory data from the TG MTP. In particular, it shows some possible actions involved in the Observe step of the Asset Allocation ladder and the set of observations obtained from that step.

Validation of TG MTP ConTA

A 1-day validation session was held with four groups of serving naval operators (six per session). The validation of the decision ladder model consisted of the operators examining the inventories that had been generated by the analysts for the various

Table 5.4 Example of "Observe" and "Set of Observations" activities for asset allocation

Observe	The process through which observations are collected	Radio calls, sensors (e.g., radars, sonar, EW), various information technology systems (both local and wide area), communication between TG members, external communications, information gained through hailing contacts visually, intelligence communication with shore authority, Web sites, messages.
Set of Observations	Collectively, all the information known concerning asset allocation	List of factors to be taken into account: member of TG running low on resources (e.g., food, fuel); location and characteristics of suspicious or hostile contacts; location of the area that TG requires more information about; current areas of interest; mission and its subgoals; equipment available. This state also includes background information on the capabilities of each asset, and past experience with asset allocation.

work functions for possible modifications. No prior training of the operators in ConTA was required. Because these inventories were meant to be relatively exhaustive, operators were instructed not to delete any item unless the possible activities at a particular step were clearly misrepresented. Overall, the inventories were quickly validated. Very few items were deleted totally (N=4), a handful of items were reworded slightly to better reflect the point being made (N=6), and many new items were added (N=37; see Table 5.5 for the total numbers of inventory items listed). These results illustrate that the analysts had developed an excellent understanding of the activities involved at each of these steps even prior to validation, a finding which supports the effectiveness of the method adopted for this work, including ConTA.

Table 5.5   Inventory frequencies for all three decision ladders

Decision ladder step	Aggregate information on a single contact	Asset allocation	Interpret overall tactical picture	Total N[a]
Activation	13	16	6	35
Alert	8	7	9	24
Observe	21	16	11	48
Set of Observations	17	11	3	31
Identify	10	3	4	17
System State	4	2	7	13
Interpret	6	7	5	18
Ambiguity	14	9	5	28
Evaluate Performance Criteria	7	3	4	14
Ultimate Goal	8	3	6	17
Interpret	8	3	4	15
Goal State	8	5	2	15
Define Task	7	3	3	13
Task	8	3	3	14
Formulate Procedure	8	6	3	17
Procedure	9	5	1	15
Execute	11	5	1	17
Total	167	107	77	351

[a]   Number of Inventory items listed.

TG MTP Results

As indicated earlier, three work functions (as distinct as possible and, therefore, not exhibiting much, if any, overlap) for ConTA were identified using the method given by Naikar et al. (2006). This produced three sets of decision ladder inventories, one for each function: Interpret Overall Tactical Picture, Asset Allocation, and Aggregate Information on a Single Contact. The frequency count of the number of items in the inventory of each step of the ladder for each work function is tabulated in Table 5.5. This allowed examining where the majority of inputs, options (i.e., methods using which the operator can perform a task), and potential patterns of clusters might lie within each design ladder and across all three ladders. Among other things, the tabulation helped analysts to determine potential issues regarding design seeds. A total of 351 items were identified across all three ladders.

The relations between the ladder steps and the inventory frequencies generated across all three scenarios are depicted in graphical form in Figure 5.4. Also shown are the "lines of best fit" for the patterns of each graph.

These ConTA results provided two significant qualitative inputs for the determination of design interventions: the consideration of the frequency of options and the flow or patterns in the frequencies of options. The results suggest that many inventory options are available for activities involving information gathering and diagnosis, and we hypothesize that most of the cognitive effort is likely to be expended by operators in these early processing stages. However, it must be acknowledged that, on the basis of this data, experiments are needed to confirm this effect. Conversely, the fewest number of options lie at bottom right side of the ladder, specifically at Procedure and Execute work activities. This suggests that there are rather limited planning and execution options for operators in these specific activities. The pattern in the middle of the ladder suggests that more options are available to an operator at the beginning of a task or decision, but that options decrease as movement through the ladder proceeds. This progressive decrease in the number of available options may be an artifact of ConTA, given that the purpose of a ladder is to show arrival at a decision. Regardless, it appears that as an operator becomes more certain about how to proceed, his/her list of options likely becomes more specific and declines in number. In terms of directions for design seeds, it may be useful to focus on how the cognitive workload is distributed across an entire ladder. Taking a step further, this suggests that the single contact ladder is most in need of design intervention as its ConTA results indicate relatively consistent and high numbers of options throughout all steps of the ladder.

The patterns resulting from each ladder are very similar as seen, for example, in the slopes of the lines of best fit for the numbers of available options (represented as "Linear" in Figure 5.4) in the three sets of decision ladder inventories. This general pattern indicates that the potential for cognitive overload most likely exists at the beginning, that is, during activities involving information gathering and diagnosis (although it must be acknowledged that it is unlikely all these options would exist every time). If operators experience significant demand for cognitive resources at the beginning of an activity, they may not have a great deal of it at their disposal for the remainder. This illustration suggests the potential for overload, and thus error is highest at the front end, and then declines throughout. Consequently, in terms of developing design seeds to support operators' work demands, designers should concentrate first on the beginning steps of these ladders and then on ways to proceed quickly to Execution stage.

In summary, findings such as these suggest that a good design objective would be to develop aiding capabilities that either decrease the number of options available to operators in the beginning steps of the ladder or help them identify the few best choices quickly, paying less attention to all others.

Figure 5.5 illustrates a flow of operator activities. Two of the activities (single contact and asset allocation) exhibit similar flows. Further work might conclude

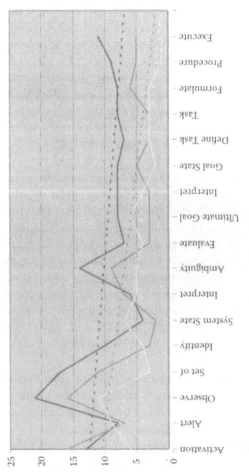

Figure 5.4   Graph of frequencies across decision ladders.

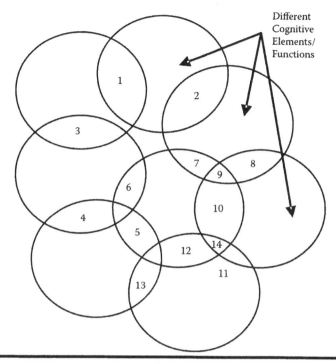

Figure 5.5   Grouping of design seeds on the basis of cognitive elements and functional similarity.

that developing a consistent pattern in the flow for different activities may make tasks more intuitive, reducing the cognitive workload. However, making the patterns too similar may result in increasing cognitive demand should two ladders occur simultaneously. An alternative interpretation of the pattern of peaks and dips may be that natural breaks are built into the processes, which may somewhat mediate the risk of overload. Each ladder showed an ebb and flow of greater and fewer options. Because extreme peaks in options tended to be followed by extreme dips, these dips may represent a sort of natural (i.e., unplanned) cognitive break. That is, it does not seem desirable to consistently consider high numbers of options.

Identification of Design Seeds and Development of Support Concept

The use of a decision ladder inventory in the TG MTP project, as opposed to tracing the information processing through the steps of the ladder as was done in the MTP/TPRM projects, necessitated a different approach to identifying design seeds. For instance, there was no need to aggregate ladder steps into superordinate categories. Using these inventories, the analyst was able to find design seeds in an opportunistic as well as a systematic fashion. For example, some items inventoried in a step might represent information already available in the operations room but needs to be displayed to operators using some interface, whereas other elements might not even be available and the designer could then find ways to get the information and make it available to the operations room personnel. The decision ladder inventories served to identify design seeds as well as advance their development for the purpose of better realizing (i.e., elaborating) the appropriate design intervention and supporting the cognitive constraints in TG MTP. However, the analysis on its own was not sufficient for translation into solid design. We now discuss in more detail why we considered other perspectives, and how this translation was done.

Consideration of Cognitive Elements

The designer considered the decision ladder inventories for each data-processing activity and state of knowledge with respect to a list of specific cognitive elements (e.g., perception, working memory, long-term memory, decision making, response selection/execution, feedback, attentional resources, etc.). Such a list has been developed over several years by the authors, but it is expected that analysts following this approach could readily generate their own lists to serve particular design purposes. If the inventory documented for each step, considered in the light of the purpose of that step, along with the various cognitive elements, holds some sort of implication for task performance, a design seed would result. For instance, suppose the step Activation could be triggered by up to 15 different sources of information, it may hold implications for perception, working memory, decision making, response selection, situation awareness, workload, and error. A design seed addressing all these concerns would thus be appropriate. Such a design seed could be "a single fused stimulus, combining the most relevant of the 15 information sources, should be presented to the operator. This stimulus should make its meaning apparent to the operator, and convey a suggested response" (Zobarich et al., 2006). Elaborating this seed could involve the use of data fusion and decision support algorithms combined with display coding.

Supporting Expert Performance

The decision ladder inventories could also be used to help identify opportunities to support expert performance. The identification of a leap (state of knowledge to state of knowledge) and shunt (information processing activity to state of knowledge) is indicative of expert performance. The designer would consider each step of the ladder individually, considering what that step attempts to achieve or represent, in addition to the cognitive elements that would support that function. The designer would then consider the information, processes, etc., that form the inventory for that step, and systematically consider how operators might leap or shunt to any and all other steps in the ladder. At this point, the designer is specifically interested in determining whether the leap or shunt is supported by the cognitive elements engendered by the ladder step of origin, the information, processes, etc., inventoried for that step.

The net result of this particular search for design seeds would necessarily be limited to two types of output: design seeds that prohibit leaps and shunts, and those that encourage leaps or shunts. This way, the designer can help ensure that the expert engages in the necessary information processing to progress toward an accurate outcome, while also providing appropriate opportunities to economize their activities. The designer can also guard against "use without design" whereby, if there is a particularly dangerous or inappropriate shortcut, the designer can intentionally force the operator to follow the ladder sequence, at least until a permissible shortcut is reached. The specific composition of the design seed (i.e., beyond whether it prohibits or encourages a leap or a shunt) would vary according to the specific needs of the ladder step and its associated inventory. If the desirability of a leap or shunt depends on the situation at the time, design seeds could reflect this by specifying adaptive automation.

Elaborating Design Seeds

To elaborate a design seed, each identified design seed would be considered against the inventory to determine what specific information, processes, etc., might support its evolution and/or implementation. Because the cognitive elements are key to developing the design seeds, elaborating a seed in this manner would focus on the ladder step that most closely corresponds to the cognitive elements that were the basis of the seed. The mapping of ladder steps to cognition used for this purpose corresponded to that shown in Table 5.2. If it were determined that the inventory did not contain information, processes, etc., to support the eventual implementation of the design seed, this probably means that new information or processes would have to be invented. The nature of this invention would then be described as part of the evolved design seed.

Developing an Integrated Support Concept

An integrated support concept (i.e., one that aggregates a number of different concepts for support functions into a single consistent presentation) for TG MTP evolves from a systematic consideration of multiple design seeds and a grouping of support concepts. At this stage, the designer would seek to identify sets of design seeds that contribute to large support concepts, whereby elemental support concepts implied by the various design seeds are perceived to interact, overlap, and give rise to an integrated support concept. As more "flesh" gets added to the design seed "bones," the integrated support concept framework would grow to a robust and comprehensive one for the key activities involved in TG MTP.

We considered three primary methods of aggregating design seeds into an integrated support concept. These methods are: grouping design seeds on the basis of the cognitive elements that they support (e.g., do they all support working memory?); grouping design seeds on the basis of functional similarity (i.e., do they support different cognitive elements of the same task?); and grouping design seeds on the basis of their contribution to a sequence of activities (e.g., one design seed enables another design seed, etc., in a related sequence of activities). The grouping of design seeds on the basis of cognitive elements and functional similarity is illustrated in Figure 5.5; the grouping principle described by a sequence of activities is illustrated in Figure 5.6. Numbers in the figures refer to distinct design seeds. Integrated support concepts could arise from those design seeds that fall into the same circle (i.e., the grouping principle).

The last grouping method implies that the designer would need to consider factors other than strictly the design seeds that have been identified. This requires the operators' goals to be considered in a more holistic manner, and supports the belief that a higher-level organizing principle is required to effectively develop an integrated support concept. If only the grouping techniques just described were followed, the development of a truly integrated support concept would be a matter of chance. At some point, the development of an integrated support concept must become directed. The direction used may correspond to higher-level operator goals (e.g., manage sensors, detect all contacts, consider trends in data, etc.) or higher-level cognitive elements (e.g., optimize workload, maximize situation awareness, minimize error, create scalable mappings, etc.). The higher-level organizing principle can then be broken down into constituent parts that the analyst can use to structure a systematic search for design seeds that could support this principle in the manner dictated by the constituent parts. For instance, managing sensors might include knowing the weather and sea conditions, knowing the properties of the contact, knowing blind spots, and effectively trading off between all factors.

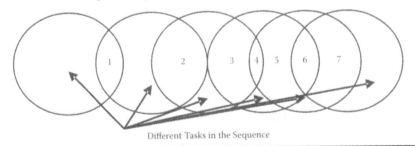

Different Tasks in the Sequence

Figure 5.6 Grouping of design seeds on the basis of task sequence.

The analyst would search for design seeds that would be relevant to both sensors and the knowledge of these other factors. Similarly, for example, workload could be broken down according to modality (visual, auditory, cognitive, and psychomotor). The analyst would then look for design seeds that are relevant to both workload and the modality.

Resulting Interface Concept

Figure 5.7 provides a glimpse of a conceptual tactical display that was presented to naval operators. It was developed on the basis of the design seeds identified in the TG MTP project, including those from the decision ladder analysis.

The figure illustrates three support concepts: overlays, corridors, and track summary. The overlays concept refers to the information that aids situation awareness such as geographical information, conceptual information, etc. There are three subclasses of this concept related to the air, surface, and submarine environments. The corridors concept refers to the provision of specific information pertaining to shipping or air corridors (in this example, they are denoted by the filled areas), such that the operator can quickly consider and respond appropriately to contacts that seem to be deviating from expected patterns of behavior (patterns that are actually provided by the overlays concept). The track summary concept shows all information accumulated by the TG about a particular contact. In Figure 5.7, the contact's

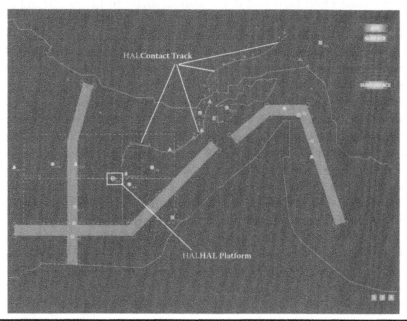

Figure 5.7   Interface concept for task group maritime tactical picture project (TG MTP).

track (the line with arrows approaching the platform labeled HAL) represents the summary support concept, and the arrows shown at various points of the contact's track indicate tactically significant events in the life cycle of the track based on information accumulated from across the TG; the operator can click on the arrows for more information to help understand patterns of behavior of the contact, and then extrapolate to its likely actions in the future.

Space precludes an explicit listing of the design seeds represented in Figure 5.7 developed on the basis of the ConTA (as in Table 5.3). However, it is worth noting that over 45 design seeds are represented in the support concept in Figure 5.7. These design seeds address all steps in the ladder from Activation (e.g., fusing secondary cues with the appearance of a contact to create a higher-intensity stimulus) to Execute (e.g., dragging a radio frequency to a radio icon to call an aircraft on the appropriate air traffic frequency). The design seeds were aggregated into integrated support concepts (i.e., overlays, corridors, and track summary) using all the strategies suggested in the previous section: support to common cognitive elements, functional similarity, and contribution to an activity sequence.

Comparison of the Two ConTA Approaches

Consistent with the methodological framework of Figure 5.1, both WDA and ConTA methods have been used in our work as means to an end; the end being the development of support concepts for shipboard C2. This has meant that less emphasis has been placed on reliability and validity, and more emphasis has been placed on utility, as measured by the number of design insights developed. Our ConTAs, which are the specific focus of this chapter, have followed two general approaches: the first one based on mapping information-processing activity of experts onto the steps in a ladder; and the second based on developing an inventory of all possible primary and secondary activities, states, information, information sources, procedures, processes, interpretations, etc., that could be relevant at each step in the ladder. Both approaches have been extremely successful. In particular, both types of ConTA analysis resulted in many design seeds, some of which were incorporated into a broader, integrated support concept. Both types of analysis also proved to be very time efficient. For example, the TPRM project took eight working days from the beginning of data collection to the end of identification of design seeds (two analysts; no validation or report), and the TG MTP project took eight weeks from the beginning of data collection to the end of identification of design seeds (four analysts; including time to conduct an SME validation and produce a report on the CWA phase of the work).

Comparing the two ConTA approaches, the expert path approach began with analysts working independently and then coming together to agree on the final output. The inventory approach permitted the analysts to work together from the outset and negated the need for an exhausting session to settle upon a single analysis outcome. From this perspective, the inventory approach is more efficient. The inventory approach also caters for every conceivable task under a broader class of activity, but for every discrete task, only a subset of the inventory would then be applicable. The expert path approach is both specific to a discrete task, and describes the way it is actually performed. For large or ill-defined development programs, a hybrid approach may be preferable. This hybrid approach would begin with the identification of the broad activity groups and development of the associated inventory. Then, the discrete tasks could be defined in terms of the path the expert would typically take through the ladder for that discrete task and the subset of inventory items that the expert would exploit (this may also suggest adopting a hybrid data collection approach). This is not what Naikar et al. (2006) proposes, but seems closer to the approach described by Vicente (1999). Because the MTP, TPRM, and TG MTP projects were exploratory, the inventory approach seems better. A more mature stage of development, however, may benefit more from the expert path approach to analysis.

The inventory approach also did not suffer from variability in the data and in the analysts' understanding of the domain, unlike the expert path approach. Variability in the data (due to the data collection process and analyst's own interpretation) can lead to one ladder mapping being performed at a different level of detail than another, leading to an incomplete design solution. Likewise, differences in the analysts' understanding can also lead to different ConTAs exhibiting different levels of detail, and also to significant loss of time for resolving differences in the analyses of different analysts. To overcome these difficulties with the expert path approach, a significant amount of extra time and resources would ideally be allocated to the task, a luxury that is generally not available to human factors analysts. Another alternative suggested by the authors to overcome this problem is the use of a formal grammar to render the data consistent, but this too would add significant costs to the work. Ultimately, if both approaches result in equally specific proposals for design seeds but one approach has greater methodological consistency, that approach must be considered preferable from the perspective of the final design solution. Thus, the inventory approach seems the most adequate for future work.

Both ConTA approaches were implemented in our projects by novice analysts, that is, novice to both the analysis approach and to the work domain of naval C2. The CDM and workshop approaches to data collection meant that SMEs themselves develop and structure the work activities, without undue influence from the analysts. The analyst must merely facilitate the input of the SMEs by structuring the information provided by them. This means that the interested analyst, company, or research organization can direct other analysts who are relatively new to a domain in investigating the domain for the purposes of pursuing design outcomes having reasonable confidence of project success. This outcome is one of the greatest ConTA successes of the three projects (and two approaches) described. In particular, the inventory approach to analyze complex, dynamic cognitive work permitted novice analysts, with little or no domain experience, to collect data and analyze it in such a way that meaningful outputs resulted. This perspective was supported by the validation session undertaken with SMEs based on the degree of their concurrence when presented with details about the analysis method, the data collected, and the analysis outputs.

The identification of design seeds was arguably the most successful part of using the methodological framework for exploratory design because so many were identified. Although data collection, structuring and analysis of the data, and work modeling were successful, the measure of this work's success was whether or not any value is derived in terms of developing support concepts, and so the identification of design seeds is one of the best indicators of success in this endeavor. Traditionally, the identification of design seeds has been an implicit process, subjugated to the development of a support concept. However, by identifying the elements that comprise a support concept, the developer can consider them individually for the manner in which they will complement the overall concept. This eliminates the addition of functionality to a support concept during the development process, which often results in subtly disjointed elements of a supposedly integrated design.

Through a combination of structuring the data, analysis and modeling, and a consideration of cognitive, metacognitive, perceptual, and collaborative demands at a variety of levels of aggregation, design seeds could be identified for every statement, task or activity, and every step in a ladder (or subset of steps in the case of the expert path approach). This should lead to a wealth of possible ideas from which to develop a single support concept. Further, several design seeds might be possible for each ladder if the perspectives of different analysts, or the perspectives of SMEs, were brought to this process. The identification of design seeds from the inventories required a greater level of insight on the part of the analyst because the inventories were necessarily broad. This was nonetheless a reasonable expectation to place on the analyst who was a novice in both the analysis approach and the application domain, assuming he/she is trained in human factors, or has some acquaintance with psychology and ergonomics. Beyond these precursors, there is no need for the analyst to look more broadly than the analysis outputs.

It should be clear that the applications of ConTA in this chapter were extremely successful. These results plainly show several worthwhile paths toward generating useful design seeds. As such, ConTA, and CWA more generally, come well recommended for use in future research concerning complex, dynamic cognitive work domains. In addition, the fact that neophytes to this analysis, SMEs and analysts alike, could grasp CWA and its outputs as quickly and easily as they did lends further support to the validity of this relatively new cognitive research approach. Future research might seek to generate a more concrete ConTA recipe, however, so that the approach could become more standardized and thus, could be more consistently taught in human factors and engineering courses.

References

Bos, J. C., Rehak, L. A., Keeble, A. R., and Lamoureux, T. M. (2005). Unpublished manuscript.

Chalmers, B.A. (2003). *Supporting threat response management in a tactical naval environment.* Proceedings of the 8th International Command and Control Research and Technology Symposium, National Defense University, Washington, DC.

Chalmers, B. A. and Lamoureux, T. M. (2004). Unpublished manuscript.

Chalmers, B. A., Webb, R. D. G., and Keeble, A. R. (2002). *Modeling shipboard tactical picture compilation.* Proceedings of the Fifth International Conference on Information Fusion, Loews Annapolis Hotel, Annapolis, MD.

Endsley, M. (1988). *Design and evaluation for situation awareness enhancement.* Proceedings of the Human Factors and Ergonomics Society 32nd Annual Meeting, pp. 96–101.

Klein, G., Calderwood, R., and MacGregor, D. (1989). *Critical decision method for eliciting knowledge.* IEEE Transactions on Systems, Man, and Cybernetics, 19, 462–472.

Naikar. N., Moylan, A., and Pearce, B. (2006). *Analysing activity in complex systems with cognitive work analysis: concepts, guidelines, and case study for control task analysis.* Theoretical Issues in Ergonomics Science, in press.

Patterson, E. S., Woods, D. D., Tinapple, D., and Roth, E. M. (2001). *Using cognitive task analysis (ConTA) to seed design concepts for intelligence analysts under data overload.* Proceedings of the Human Factors and Ergonomics Society 45th Annual Meeting, pp. 439–443.

Rasmussen, J., and Jensen, A. (1974). *Mental procedures in real-life tasks: A case study of electronic troubleshooting.* Ergonomics, 17, pp. 293–307.

Sartori, J. A., Keeble, A. R., Bandali, F., Boothby, R., Bos, J. C., Rehak, L. A., and Lamoureux, T.M. (2006). Unpublished manuscript.

Vicente, K. J. (1999). *Cognitive Work Analysis: Toward Safe, Productive, and Healthy Computer-Based Work.* Erlbaum and Associates, Mahwah, NJ.

Wickens, C. D. (1984). *Engineering Psychology and Human Performance.* Merrill, Columbus, OH.

Zobarich, R. M., Lamoureux, T. M., Keeble, A. R., Sartori, J. A., Bandali, F., and Boothby, R. (2006). Unpublished manuscript.

Chapter 6

Understanding Cognitive Strategies for Shared Situation Awareness across a Distributed System: An Example of Strategies Analysis

---

Emilie M. Roth

Contents

Overview

Work domain analysis characterizes the constraints and affordances in a domain. Subsequent analyses are required to uncover the work situations and task functions that arise in the domain, the range of strategies available to accomplish task functions, and the strategies actually used by domain practitioners that lead to effective and ineffective performance. Within the Cognitive Work Analysis (CWA) framework, *control task analysis* is conducted to uncover *what* needs to be accomplished, and *strategies analysis* is conducted to obtain a "process" description of *how* it can be accomplished (Vicente, 1999). This chapter provides an introduction to strategies analysis.

The chapter begins by describing the goals of strategies analysis as defined by Vicente (1999) and Rasmussen (Rasmussen, Pejtersen, & Goodstein, 1994). An overview of empirical knowledge acquisition methods for performing strategies analysis is then presented. This is followed by a description of a specific case study that looks at the strategies employed by railroad workers to maintain shared situation awareness across the distributed system. The study provides a concrete illustration of how a strategies analysis can be performed and how the results can be used to inform system design.

Strategies Analysis within CWA

One of the foundational elements of a CWA is work domain analysis. It characterizes the constraints and affordances in a domain and helps to identify the goals and possibilities for action for domain practitioners. However, it does not in itself reveal what needs to be accomplished, how it can be accomplished, and what knowledge and strategies are used by actual domain practitioners. As a concrete example, a work domain analysis of a power plant can reveal the goals of the engineered system (e.g., produce energy and maintain safety) and the physical subsystems available to achieve those goals (e.g., systems for controlling mass and energy balances). However, work domain analysis by itself does not reveal the major tasks that need to be performed to operate the plant. These include the need to start up the plant, change energy production to match demand, and shut down the plant. Similarly, work domain analysis does not specify the processes by which a plant can be started up, power levels changed, or plant shut down. It provides no direct insight into the knowledge, and cognitive and collaborative strategies that domain practitioners possess that allow them to function effectively in the domain. Nor does it provide direct insight into the limitations in knowledge or cognitive processing that might result in suboptimal performance and error.

Within CWA, it is control task analysis coupled with strategies analysis that provides the means for uncovering what needs to be accomplished in a domain, how it can be accomplished, and what knowledge and strategies used by actual domain practitioners lead to effective and ineffective performance. Vicente (1999) defines *control task analysis* as the means for identifying what needs to be accomplished independent of the strategy (how) or actor (who), and *strategies analysis* as a process description of the various ways these control tasks can be accomplished.

Rasmussen et al. (1994) provide a slightly different characterization. They distinguish work domain analysis from *activity analysis* and indicate that activity analysis involves three stages of decomposition and shift of conceptual level. The first level defines prototypical work situations and task functions. The purpose of this conceptual level of analysis is to define the objectives that are active and resources that are available to actors for achieving those objectives. The second conceptual level characterizes activity in terms of the control functions that the actors need to carry out to achieve the objectives. The third conceptual level provides a description in cognitive terms of the mental strategies that can be used to achieve the control functions.

Although Vicente and Rasmussen and his colleagues, provide slightly different characterizations of the transition from a description of the domain to a description of the activity in the domain, they agree on the end state as being a cognitive-level description of the strategies required to operate in the domain. This is the working definition of strategies analysis adopted here.

Formative and Empirical Approaches to Strategies Analysis

There are two approaches that can be taken in strategies analysis. One approach is to derive the space of possible strategies in a domain analytically, based on work domain analysis. This approach produces a formative characterization of the strategies afforded by the work domain for accomplishing control tasks (Vicente, 1999). An alternative approach is to employ empirical methods to examine the knowledge and strategies that domain practitioners actually use to cope with domain demands. This produces a descriptive characterization of practitioner strategies.

Formative characterizations of strategies can be derived without having to specify what agent or agents (human and/or machine) will accomplish the tasks (Vicente, 1999). Roth and Woods (1988) provide an early example of this approach. They combined formative and descriptive methods to understand why controlling feedwater during start-up of a nuclear power plant, a seemingly highly procedural task, was cognitively challenging. They began with a work domain analysis of the power plant from which they derived a formative characterization of the strategies required to control feedwater during start-up, which they referred to as a *competence model*. The objective of a start-up is to increase energy production by increasing the amount of steam generated. The control process involves increasing the reactor power level, while simultaneously increasing feedwater to the steam generators to match the outflow (in the form of steam) so as to maintain steam generator mass. This formative analysis revealed that controlling feedwater to match steam flow during start-up is a challenging task for any agent— human or machine— because it requires consideration of multiple interacting processes controlled by multiple independent agents, complex process dynamics with long lags, and poor state information. For example, accurate measures of steam flow and feed flow are not available at low power. As a result, neither human nor machine controllers have any direct way of knowing whether steam flow and feed flow are in balance, or how much of a change in feed flow is required to bring them into balance.

Formative strategies analysis is particularly valuable early in a design endeavor as a basis for specifying requirements for displays, training, or automation. Understanding the space of possible strategies makes it possible to develop displays to enable people to apply those strategies, training to teach individuals the strategies, and/or automated agents that can exploit the strategies. For example, the results of feedwater control analysis became a driving force in developing several new kinds of support systems, including new forms of information to enhance a controller's ability to anticipate process behavior, new integral representations to better reveal the dynamics of the process, and new automated controllers (Woods & Roth, 1988).

More recently, Nehme, Scott, Cummings, and Furusho (2006) conducted a formative strategies analysis to develop information and display requirements for futuristic unmanned systems, for which no current implementations exist. They used a decision ladder formalism to map out the monitoring, planning, and decision-making activities that would be required of operators of these systems. Callouts were then used to specify information and display requirements in order to support the corresponding cognitive tasks.

Empirical approaches to strategies analysis offer an attractive alternative and/or complement to analytic approaches. First, in many cases, it can be more efficient to uncover effective strategies by examining the performance of highly experienced domain practitioners than by attempting to derive them analytically through work domain analysis (Klein, 1998). Second, although an analytic approach can generate a range of hypothetical strategies, there is no guarantee that actual domain practitioners utilize, or would be able to utilize, those strategies even in principle. One reason is that different agents (human or machine) are subject to different processing constraints (e.g., perceptual limits, working memory capacity limits, attention limits, and computational limits) that will impact the extent to which any given (hypothetical) strategy can be employed. A formative analysis that attempts to derive strategies independent of "who" is to perform the strategies can result in specifications of strategies that are not compatible with the knowledge and mental processing capabilities of actual domain practitioners. As a consequence, it is often beneficial to perform empirical strategies analysis in conjunction with formative analysis. This approach was taken in the case of feedwater control task strategies analysis (Roth & Woods, 1988).

Empirical approaches to strategies analysis can reveal the knowledge and skills required for expert performance in a domain, as well as the suboptimal strategies and vulnerability to error of less experienced domain practitioners. For example, Roth and Woods (1988) performed empirical strategies analysis to examine the knowledge and strategies used by actual operators in controlling feedwater during power plant start-up. They brought together nine "expert" operators, each from a different power plant, to examine the knowledge and strategies that differentiated experts from less experienced operators. Using a variety of knowledge acquisition techniques, including retrospective analysis of actual past cases of successful and unsuccessful power plant start-ups, and observation of performance in a high-fidelity simulator, they were able to identify distinct differences between expert control room operators and less experienced operators in their mental models of process dynamics and the process control strategies that they used for raising power levels. Expert operators had developed sophisticated strategies for extracting process-state information from indirect sources and team communication, and coordination strategies for manipulating process dynamics to smoothly bring the plant up in power. In contrast, less experienced operators had poor mental models of process dynamics, and ineffective communication and coordination strategies. As a consequence, they were unable to accurately assess the process state or predict the impact of actions on process behavior. Inexperienced operators were more likely to take actions that would create process disturbances, causing automatic systems to prematurely terminate plant start-up. The results of the analysis were used to define training requirements that would enable less experienced operators to develop the mental models and cognitive and collaborative strategies required for effective process control during plant start-up. The results were also used to provide recommendations for new visualizations that would reduce cognitive and collaborative demands on operators by providing more direct information on process dynamics (Woods & Roth, 1988).

The preceding example illustrates the multiple uses of strategies analysis. These include the ability to:

■ Understand the sources of performance difficulty and contributors to human error
■ Define knowledge and skill requirements for expert-level performance
■ Provide the basis for training program development
■ Define requirements for information displays and decision-support systems to reduce cognitive and collaborative burden on domain practitioners, and provide flexible support for multiple effective strategies

Empirically based strategies analysis can be conducted to uncover the strategies employed by domain practitioners in the current environment. They can also be used to examine the strategies that would likely be required to operate future systems that are still under development. For example, a strategies analysis for a future system can be accomplished by examining the strategies employed by people under conditions that simulate the "envisioned world" (Woods & Dekker, 2000).

An overview of empirical methods for performing strategies analysis is provided in the next section.

Empirical Methods for Conducting Strategies Analysis

Empirical methods for conducting strategies analysis are similar in objectives and scope to what has traditionally been referred to as Cognitive Task Analysis (CTA). They both aim to uncover the knowledge, skills, and strategies that enable domain practitioners to operate at an expert level, as well as the cognitive factors that limit the performance of less experienced individuals (e.g., incomplete or inaccurate mental models). Bisantz and Roth (2008) provide a broad survey of different CTA approaches.

A variety of specific techniques for uncovering practitioner knowledge and strategies have been developed that draw on basic principles and methods of Cognitive Psychology (Cooke, 1994; Ericsson & Simon, 1993; Hoffman, 1987; Potter, Roth, Woods, & Elm, 2000; Roth & Patterson, 2005). These include structured interview techniques such as the Applied Cognitive Task Analysis method (Militello & Hutton, 1998) and the Goal-Directed Task Analysis method (Endsley, Bolte, & Jones, 2003); critical incident analysis methods, which investigate actual incidents that have occurred in the past (Dekker, 2002; Flanagan, 1954; Klein, Calderwood, & MacGregor, 1989); cognitive field observation studies, which examine performance in actual environments or in high-fidelity simulators (Roth & Patterson, 2005; Woods, 1993; Woods & Hollnagel, 2006); "think-aloud" protocol analysis methods, in which domain practitioners are asked to *think aloud* as they solve actual or simulated problems (Gray & Kirschenbaum, 2000); and simulated task methods, in which domain practitioners are observed as they solve analog problems under controlled conditions (Patterson, Roth, & Woods, 2001).

Whereas most CTA methods focus on understanding the knowledge and skill employed by domain practitioners in current environments, it has also been employed to explore how changes in technology and training are likely to impact practitioner skills, strategies, and performance vulnerabilities. The introduction of new technologies can often have unanticipated effects (Woods & Dekker, 2000). New, unanticipated complexities can arise to create new sources of workload, problem-solving challenges, and coordination requirements. In turn, individuals in the system will adapt, exploiting the new power provided by the technology in unanticipated ways and creating clever workarounds to cope with technology limitations so as to meet the needs of the work and human purposes. A number of techniques have been developed to explore how people are likely to adapt to future worlds. For example, in one study, a high-fidelity training simulator was used to explore how new computerized procedures and advanced alarms were likely to affect the strategies used by nuclear power plant crews to coordinate activities and maintain shared situation awareness (Roth & Patterson, 2005). In another study, a *future incident technique* was used to explore the potential impact of proposed future air traffic management architectures on the cognitive demands placed on domain practitioners (Dekker & Woods, 1999). Controllers, pilots, and dispatchers were presented with a series of future incidents to be jointly resolved. By examining their problem-solving and decision-making strategies, it was possible to uncover the dilemmas, tradeoffs, and points of vulnerability in the contemplated architectures, enabling practitioners and developers to think critically about the requirements of effective performance for these envisioned systems.

The next section uses a specific case study to illustrate how empirical strategies analysis can be used to uncover undocumented strategies that contribute substantively to safe and efficient performance in a complex high-risk domain.

Illustrative Case Study: Using Strategies Analysis to Inform Technology Deployment

The case study involved a comprehensive effort to examine the cognitive and collaborative activities of railroad workers and the potential impact of new technology on their ability to coordinate work and maintain safe operations. This section summarizes selected results of the study relating to information extraction and communication strategies that have been developed to support roadway worker safety. The objective is to illustrate how empirical strategies analysis can be used to reveal domain practitioner strategies that are important to safe and efficient performance, yet are undocumented and not readily articulated by the practitioners themselves.

Overview of the Railroad Operations Domain

Railroad operations are an example of a highly distributed organization that must coordinate work among individuals widely distributed in space. These include train crews that operate trains across multiple territories that can be owned and operated by different railroads; roadway workers that maintain tracks, signals, and related infrastructure; and dispatchers that manage track usage, allocating time on the track to different trains and roadway worker activity as required. These individuals rely heavily on analog radio communication to maintain awareness of each other's location, coordinate work, and maintain safe operations (Roth, Malsch, & Multer,

2001; Roth, Malsch, Multer, & Coplen, 1999; Roth, Multer, & Raslear, 2006; Roth & Patterson, 2005).

New technologies are emerging that have the potential to drastically change railroad operations. These include digital communication technologies, which offer the possibility of more reliable communication than analog radios, as well as an opportunity to move away from the broadcast "party-line" aspect of analog radio to private communication lines; global positioning systems (GPS), which have the potential of providing much more accurate information on the location of trains and roadway workers; and positive train control (PTC) systems, which offer the possibility of providing more precise train location information as well as automated control. Positive train control technology allows the location and speed of a train to be monitored and compared with train movement permissions and directives provided by the train dispatcher (referred to as *movement authority*). The positive train control system can take automated action to stop the train in cases where movement authorities have been violated (e.g., exceeding speed limits or entering territory for which authority has not been granted).

The Federal Railroad Administration (FRA) embarked on a series of studies to understand the knowledge and strategies of dispatchers, train crews, and roadway workers to maintain safe and efficient operation in today's environment with the explicit goal of using the results to inform the design and evaluation of emerging technologies (Roth et al., 1999). By understanding the cognitive and collaborative strategies currently employed, it becomes possible to identify ways they could be more effectively supported. Equally important, it ensures that features of the existing environment that support important cognitive and collaborative activities are not inadvertently eliminated without providing alternative means to support these functions.

Roadway workers inspect, maintain, and repair railroad facilities and equipment, including track, signals, communications, and electric traction systems. They may work alone or as part of a multi-person group that must coordinate their work in order to accomplish a common task. Some jobs require working at a particular location on the track (e.g., changing a rail or troubleshooting a malfunctioning signal). Other jobs require moving across tracks, for example, to perform track inspection. Because the activities of roadway workers are performed on or near railroad tracks, they are at risk of being struck by a train or other on-track equipment.

A primary goal of the research was to understand the factors that affect roadway worker safety in today's environment so as to provide guidance for design and introduction of new technologies to enhance roadway worker safety.

Knowledge Acquisition Approach

The study combined interviews with field observations. Site visits and interviews of roadway workers were conducted at five locations in the United States, and included passenger and freight rail operations. A total of 26 individuals were observed and/or interviewed, including 13 trackmen who are responsible for inspection and maintenance of track; 8 signalmen who are responsible for inspection and maintenance of signal systems; and 5 dispatchers who control track usage. At two of the sites, interviews occurred while accompanying a trackman on high-rail track car rides used to traverse track for the purpose of inspecting the track.

The interviews covered factors that impact roadway worker safety; the needs for the communication and coordination with dispatchers, train crews, and other roadway workers; the challenges that arise; and how portable digital-based communication devices might impact their work. These included discussion of both formally prescribed communication protocols and informal communications.

The interviews were conducted with individuals or groups of up to five people. They lasted approximately 2 hours and were tape-recorded with the permission of the individuals being interviewed. There were two or more interviewers representing complementary areas of expertise (e.g., a cognitive engineer and a systems engineer). The primary interviewer led the interview sessions using a set of predefined interview questions that served as a "checklist" of topics to be covered. Actual questions asked and their order varied, depending on participant responses. The set of predefined questions are presented in Roth and Multer (in preparation).

The tape-recorded interviews were transcribed and analyzed with the goal of identifying recurrent themes across interviews as well as specific actual incidents described by interviewees to illustrate the themes. The analysis focused on identifying cognitive and collaborative demands in the current environment that contribute to performance difficulties and errors; strategies that expert practitioners have developed to build and maintain shared situation awareness, avoid or catch errors, and/or improve efficiency and enhance safety; opportunities to enhance performance and/or improve safety through the introduction of new technologies; as well as concerns relating to potential new problems that could emerge with the introduction of new technologies.

Strategies Analysis

The study identified active strategies that dispatchers, roadway workers, and train crews employ to build and maintain shared situation awareness. Strategies include (1) taking advantage of the "party-line" aspect of radio communication to share relevant information across multiple parties simultaneously; (2) monitoring party-line radio communication directed at others to extract relevant information about the activities and intentions of other parties; and (3) proactive communication strategies, where individuals actively contribute to each other's situation awareness by sharing safety-relevant information about the location, activities, and intentions of others in the vicinity.

Roadway workers and dispatchers emphasized the importance of maintaining shared awareness of the physical location where the work is taking place to insure that the roadway workers are properly protected. Railroad workers maintain shared situation awareness of the location and activities of roadway workers who may be widely distributed geographically, by taking advantage of the party-line aspect of radio communication. For example, one roadway worker supervisor (referred to as the employee in charge or EIC) mentioned that he has individuals in his charge (e.g., people in track cars, flagmen, etc.) listen in over the radio when he obtains a track authority from the dispatcher. This increases efficiency, fosters shared situation awareness of the location and activities of the roadway workers, and reduces the potential for communication error.

Dispatchers also actively work to maintain awareness of the location and activities of roadway workers in their territory. One way dispatchers keep track of the location of roadway workers is to monitor radio communication among the roadway workers in their territory. One dispatcher we interviewed mentioned that he routinely listens in when the EIC gives permission for someone (e.g., a track car) to come into the work zone. This allows him to keep track of who is in the work zone and what activities they are engaged in. We were told of several instances when dispatchers caught errors, such as roadway workers unintentionally working outside the limits of authority for which protection was granted, by overhearing radio communications meant for others.

Roadway workers particularly need to keep track of trains operating in their vicinity. Interviews with roadway workers and dispatchers indicate that the former actively engage in building and maintaining awareness of trains in their vicinity to help them predict when trains are likely to approach and in what direction. Roadway workers are able to anticipate regularly scheduled trains based on review of train bulletins, time tables, and their own experiences on the territory. Anticipating unscheduled trains can be more challenging.

Roadway workers have developed information extraction strategies to help them anticipate unscheduled trains. For example, the EIC will routinely monitor the road channel for train communication, such as locomotive engineers calling out signals as they reach them or conversations between locomotive engineers that can provide indirect indication of their location and intentions. This provides another example of exploiting the party-line aspect of radio communication to extract information important to building and maintaining a broad situation awareness.

In addition, informal, cooperative practices have grown whereby others routinely alert roadway workers of trains that may be approaching them. Both roadway workers and dispatchers mentioned that the latter routinely call the former to let them know if a train will be coming by on an adjacent track, particularly if the train is unscheduled (e.g., a freight train or a work engine), coming on a different track than usual, at a different time, or in an unanticipated direction. As one dispatcher stated, "I let them know what my plan is so that they are not startled." This call is not strictly mandated by operating rules. We were repeatedly informed that such calls were considered "courtesies."

Similar informal communications that provide an important safety function have been observed among train crews. For example, if a train crew passes a roadway worker group working by the side of the track, he or she may call over the radio to alert other trains passing through the territory of the presence of the roadway workers. We also observed cases where roadway workers traveling on track cars called other groups they had passed earlier to alert them to a train heading their way. These informal communications were also treated as courtesies.

The results of the study highlight the active cognitive and collaborative processes that workers engage in to develop and maintain shared situation awareness of each other's location, activities, and intentions across a distributed system. These processes include active strategies for extracting relevant information by listening in on radio communications directed at others. These active listening processes enabled individuals in the distributed organization to identify information that had a bearing on achieving their own goals or on maintaining their safety. It also enabled them to recognize situations in which information in their possession was relevant to the performance or safety of others and needed to be communicated. We heard about several instances where third parties intervened to prevent accidents after overhearing conversations.

We also uncovered informal cooperative communication practices that went beyond the requirements of formal operating rules and served to facilitate work and enhance on-track safety. Interestingly, these communication practices were frequently referred to as courtesies, highlighting their optional nature and positive contribution to overall safety.

Implications for Technology Deployment

The results of the analysis clearly indicated the importance of shared situation awareness among roadway workers, train crews, and dispatchers with respect to the location, activities, and intentions of roadway workers and trains operating within each other's vicinity. Further, it revealed that current technology does not provide effective support for this important cognitive function.

Roadway workers, train crews, and dispatchers had to rely on indirect means, monitoring the communication directed at others that occurred over party-line radio, to extract relevant information. Although listening-in strategies, which depend on the party-line aspect of radio communication, contribute to shared situation awareness, they are inefficient and impose costs in terms of attention demands. Most communication over radio channels is not relevant, but there are multiple radio channels that are potentially relevant, forcing the EIC to scan across several channels for potentially relevant information. Digital technology can be used to support shared situation awareness, reducing the need to attend to irrelevant information.

Enhancing Shared Awareness of the Location of Roadway Workers and Trains

The study highlighted the importance of roadway workers maintaining a broad situation awareness of their own location in relation to those of other roadway workers and trains in the vicinity. Dispatchers also need to maintain a broad situation awareness of the location and activities of roadway workers and trains in the territory they are controlling. This is important so as to facilitate their own decision making with respect to track allocation, as well as to provide a redundant layer of safety for roadway workers. Similarly, train crews work to maintain awareness of the location of roadway workers in the territory they are crossing so that they can blow their whistle to let the latter know that a train is approaching.

GPS technology, coupled with graphics display technology, can be used to create displays that enable roadway workers, dispatchers, and train crews to share common awareness of the location and activities of roadway workers and trains in a given vicinity. For example, a graphic display could be provided that shows the location of a roadway worker relative to work authority limits. The same graphic display of roadway worker location relative to work authority limits could be provided to both the roadway worker (on a portable display unit) and the dispatcher (on a display in the dispatch center) to facilitate shared understanding of the location information being communicated and to reduce the potential for communication error.

PTC technology, which provides information on the location of trains, could also be used to enhance the roadway worker situation awareness of trains in the vicinity. For example, roadway workers could carry portable graphic devices that depict train location and movement relative to their own locations.

Similarly, a graphic display could be developed for dispatchers to enable them to keep track of the location and dispersion of roadway workers, trains, and equipment in the territory they control. It would similarly be useful to enable EICs to electronically track roadway workers, other work groups, or lone workers protected by their authority.

In addition to the graphic means of fostering broad situation awareness, digital technology can be used to provide active audio alerts to direct attention to potential safety problems. For example, roadway workers could be alerted to approaching trains on tracks they are working on or on adjacent tracks. Audio alerts could also be provided in cases where roadway workers (e.g., on a track car) are approaching the limits of the physical track segment for which they have movement authority so as to prevent an unintended excursion beyond the authority limits. Alerts can also be provided when the time limit of authority to occupy a segment of track is about to expire.

Enhancing Shared Situation Awareness
of Activities and Intentions

The results of the study revealed that railroad workers rely on the party-line aspect of analog radio communications not only to extract location information of others in the vicinity but also information about their current activities, plans, and intentions. Depending on how digital communication systems are implemented, this potential indirect benefit of party-line radio communication could be lost. For example, if digital communications are deployed in a manner that only allowed communication between two parties, then there would be no opportunity to listen in as plans for work were discussed, permissions were granted, or problems and delays encountered were communicated. This would severely limit the ability of individuals to anticipate and head off potential problems based on understanding of others' actions and intentions. In contrast, digital technology can be deployed with explicit consideration of the need for broadcast capability. The initiator of the communication could explicitly specify multiple receivers (equivalent to conference calls, or cc's on e-mail messages), enabling dispatchers, train crews, and roadway workers to maintain shared awareness not only of the location of others in their vicinity but also of their activities, plans, and intentions. Examples include allowing the dispatcher to broadcast messages to multiple parties simultaneously to foster shared situation awareness of their activities and intentions in a given vicinity. As one dispatcher put it, "I would want people to know what is going on in the locations they are working in." This broadcast capability would have the effect of reproducing the "common ground" that is fostered by the party-line feature of radio communication. Similarly, it is important to enable roadway workers to broadcast information to multiple individuals simultaneously. This could be other roadway workers in the work group or trains in the vicinity.

Systems that provide some of these support functions are beginning to emerge. For example, a handheld digital communication device with integrated GPS technology designed for roadway workers has been developed and tested by Multer and his colleagues (Malsch, Sheridan, & Multer, 2004; Masquelier, Sheridan, & Multer, 2004; Oriol, Sheridan, & Multer, 2004). The prototype device operates on a cell phone with integrated personal digital assistant capabilities coupled with a GPS receiver. It enables roadway workers to obtain real-time train and territory status information as well as request and receive work authorization from dispatchers. The integrated GPS technology provides an accurate means to identify and communicate roadway worker location information. The prototype handheld communication device allows information to be broadcast to multiple designated receivers, and provides a promising model for leveraging the efficiency and reliability of digital communication while maintaining the ability to foster shared situation awareness that characterizes analog radio communication.

General Points Illustrated by the Case Study

The roadway worker case study illustrates several important points about strategies analysis. First, it provides a concrete illustration of how structured interviews, coupled with field observations, can be used to uncover important strategies that are otherwise undocumented and difficult for domain practitioners to spontaneously describe. In the roadway worker example, we observed information extraction and proactive communication strategies that enabled railroad personnel to maintain awareness of each others location and intentions, enhancing overall safety and efficiency of operations. Railroad workers did not spontaneously volunteer these strategies during interviews. They were initially identified via field observations. When the observed strategies were mentioned to other railroad workers during subsequent interviews, they readily confirmed that these were routine practices.

The example highlights the importance of drawing on multiple sources of information (e.g., observations and interviews) and the role of analysis in uncovering domain practitioner strategies. Strategies analysis is not merely a matter of documenting what domain practitioners say are their strategies. Domain practitioners are not necessarily capable of reflecting on their own strategies. Strategies analysis requires active analysis of a corpus of concrete cases derived from multiple sources, including direct observation and interview descriptions of actual cases, to generate more abstract-level descriptions (Roth & Patterson, 2005; Woods & Hollnagel, 2006). The objective is to identify common patterns across a corpus of illustrative cases and to draw generalizations that have applicability beyond the particular cases examined.

Roth and Patterson (2005) describe the general abstraction process by which generalizations can be drawn from a corpus of specific cases. This abstraction process is illustrated in Figure 6.1. They discuss several techniques that can be used to broaden the sample of observations to hone in on the commonalities across cases and identify boundary conditions for generalization. The techniques include:

- Broadly sampling the domain practice (e.g., multiple shifts, multiple practitioners with different types and levels of experience, multiple sites)
- Use of multiple converging techniques (e.g., field observations, structured interviews, questionnaires, and review of past incidents and critical cases)
- Use of multiple observers/interviewers who are likely to bring different conceptual frameworks

Several strategies can be used to strengthen the validity of the analysis and guard against potential analyst biases. First, analysts need to take the stance that their conceptualizations are inherently tentative and subject to revision. Each new observation or interview provides an opportunity to generate new conjectures as well as test conjectures generated in prior observations. This approach allows for the discovery of critical factors not predicted in advance as well as opportunities to discard earlier

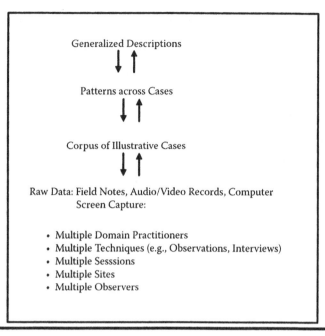

Figure 6.1   Logic used to abstract general descriptions of domain practitioner strategies through analysis of a corpus of illustrative cases. (Adapted from Roth & Patterson [2005], In H. Montgomery, R. Lipshitz, & B. Brehmer (Eds.), *How professionals make decisions* **(pp. 379–393). Mahwah, NJ: Lawrence Erlbaum Associates. With permission.)**

conjectures that are not supported by later observations. The ability to sample the domain of practice broadly (e.g., multiple practitioners, multiple levels of domain expertise, multiple sessions, and multiple sites) provides an opportunity to look for commonalities across cases as well as divergences (contrasting cases) that reveal interesting insights.

As Woods and Hollnagel (2006) have argued, extracting generalizable patterns from observations fundamentally requires repeated shifts in levels of abstraction to develop and test tentative conceptualizations. When reviewing particular cases, the objective is to try to understand how the specific example is an instantiation of a more abstract pattern. Similarly, each observation should be viewed as an opportunity to test the validity and generality of conceptualizations derived from earlier cases. As Woods and Hollnagel point out, moving up and down across levels of abstraction help to balance the risk of being trapped in the details of specific situations, people, and events with the risk of being captured by a set of concepts that provide an incomplete or inaccurate characterization of the domain of practice.

The railroad worker case study also illustrates the importance of capitalizing on contrasts between "prescriptive" descriptions of how work is "supposed to be done," that is, how the work is described in formal procedures, manuals, and rule books, and how it is actually accomplished. Identifying homegrown "artifacts," work-arounds, and undocumented processes provide important cues to domain goals and practitioner strategies that are not currently well supported (Mumaw, Roth, Vicente, & Burns, 2000; Roth, Scott et al., 2006). In the case of railroad workers, there are "rule books" that prescribe formal communication protocols. The roadway workers followed these prescribed communication protocols. However, they also engaged in extra informal communications, which proved to contribute substantively to roadway worker safety. These informal proactive communications were referred to as "courtesies," providing a linguistic marker highlighting the discretionary nature of these communications.

Note that the railroad worker proactive communication strategies could not easily have been derived from a formative strategies analysis of the railroad domain. Formative analysis would indicate the need to maintain awareness of each other's location, actions, and intention but not reveal that railroad workers would spontaneously develop proactive cooperative techniques for sharing that information with each other.

Finally, the railroad worker case study highlights the role of strategies analysis in guiding the design and evaluation of emerging technologies. In many cases, domain strategies are identified that have evolved to compensate for lack of appropriate support in the existing environment. For example, railroad workers have developed information extraction strategies that exploit the party-line aspect of radio communication to identify the location, actions, and intentions of others that have relevance to their own work and the safety and efficiency of the entire railroad system. In moving to new technologies, the goal should not be to faithfully mimic existing ones so as to preserve current strategies. The results of strategies analysis should be used to understand the functional objectives of the strategies. New technologies can then be deployed to more directly support these functional requirements. In the case of the railroad worker listening-in strategies, the functional requirement is to be able to maintain awareness of the location, actions, and intentions of others in the vicinity. As was discussed in the section on technology deployment, the need for shared situation awareness of each others location and intentions can be supported more directly through alternative technologies (e.g., graphic displays with location information, messages directed to multiple parties, etc.).

Summary and Conclusions

In conclusion, the objective of strategies analysis is to find out what needs to be accomplished in a domain, how it can be accomplished, and what knowledge and strategies are used by actual domain practitioners that lead to successful as well as erroneous performance.

There are two approaches to strategies analysis. Formative strategies analysis uses analytic methods to uncover the strategies afforded by the work domain. It is useful early in design, particularly for first-of-a-kind systems, as a way to establish requirements for displays, training, and/or automation. Empirical approaches to strategies analysis examine the actual strategies used by domain practitioners. They provide an efficient way to uncover knowledge and skills that underlie expert performance, strategies that have evolved to compensate for limitations in the current environment (e.g., workarounds), and "buggy knowledge" and suboptimal strategies that characterize less-skilled performance that can lead to inefficiencies and error. The results can point to opportunities for improvements through new training and support systems.

There is a wide range of analytic and empirical techniques that can be used to support strategies analysis. Ideally, multiple complementary techniques that combine analytic and empirical methods should be used. The results of strategies analysis provide the basis for:

- Understanding the sources of performance difficulty and contributors to human error
- Defining knowledge and skill requirements for expert-level performance (e.g., to guide development of training)
- Identifying strategies that domain practitioners have developed to exploit affordances in the environment and compensate for limitations (e.g., development of homegrown artifacts and workarounds)
- Defining support requirements to reduce cognitive and collaborative burden on domain practitioners and provide flexible support for multiple effective strategies
- Guiding deployment of new technologies to more effectively support cognitive and collaborative work

References

Bisantz, A. and Roth, E. M. (2008). Analysis of cognitive work. In Deborah A. Boehm-Davis (Ed.) *Reviews of Human Factors and Ergonomics Volume 3*. Santa Monica, CA: Human Factors and Ergonomics Society, 1–43.

Cooke, N.J. (1994). Varieties of knowledge elicitation techniques. *Int. J. Human–Computer Studies, 41*, 801–849.

Dekker, S.W.A. (2002). *The Field Guide to Human Error Investigations*. London: Ashgate.

Dekker, S.W.A., & Woods, D.D. (1999). To intervene or not to intervene: The dilemma of management by exception. *Cognition Technology and Work, 1*(2), 86–96.

Endsley, M.R., Bolte, B., & Jones, D.G. (2003). *Designing for Situation Awareness*. Boca Raton, FL: Taylor & Francis.

Ericsson, K.A., & Simon, H.A. (1993). *Protocol Analysis: Verbal Reports as Data* (Second Edition). Cambridge, MA: The MIT Press.

Flanagan, J.C. (1954). The critical incident technique. *Psychological Bulletin, 51,* 327–358.

Gray, W.D., & Kirschenbaum, S.S. (2000). Analyzing a novel expertise: An unmarked road. In J.M.C. Schraagen, S.F. Chipman, & V.L. Shalin (Eds.), *Cognitive Task Analysis* (pp. 275–290). Mahwah, NJ: Lawrence Erlbaum Associates.

Hoffman, R. (1987). The problem of extracting the knowledge of experts from the perspective of experimental psychology. *AI Magazine, 8* (Summer), 53–67.

Klein, G.A. (1998). *Sources of Power: How People Make Decisions.* Cambridge, MA: The MIT Press.

Klein, G.A., Calderwood, R., & MacGregor, D. (1989). Critical decision method for eliciting knowledge. *IEEE Transactions on Systems, Man, and Cybernetics, 19*(3), 462–472.

Malsch, N., Sheridan, T., & Multer, J. (2004). *Impact of Data Link Technology on Railroad Dispatching Operations.* DOT/FRA/ORD-04/11. Washington, DC: U.S. Department of Transportation, Federal Railroad Administration.

Masquelier, T., Sheridan, T., & Multer, J. (2004). *Supporting Railroad Roadway Worker Communications with a Wireless Handheld Computer: Volume 2: Impact on Dispatcher Performance.* DOT/FRA/ORD-04/13II. Washington, DC: U.S. Department of Transportation, Federal Railroad Administration.

Militello, L.G., & Hutton, R.J.B. (1998). Applied Cognitive Task Analysis (ACTA): A practitioners toolkit for understanding task demands. *Ergonomics, 41*(11), 1618–1641.

Mumaw, R.J., Roth, E.M., Vicente, K.J., and Burns, C.M. (2000). There is more to monitoring a nuclear power plant than meets the eye. *Human Factors, 42*(1), 36–55.

Nehme, C.E., Scott, S.D., Cummings, M.L., & Furusho, C.Y. (2006). Generating requirements for futuristic heterogeneous unmanned systems. In *Proceedings of the Human Factors and Ergonomics Society 50th Annual Meeting* (pp. 235–239). Santa Monica, CA: Human Factors and Ergonomics Society.

Oriol, N., Sheridan, T., & Multer, J. (2004). *Supporting Railroad Roadway Worker Communications with a Wireless Handheld Computer: Volume 1: Usability for the Roadway Worker.* DOT/FRA/ORD-04/13.I. Washington, DC: U.S. Department of Transportation, Federal Railroad Administration.

Patterson, E.S., Roth, E.M., & Woods, D.D. (2001). Predicting vulnerability in computer-supported inferential analysis under data overload. *Cognition Technology and Work, 3,* 224–237.

Potter, S.S., Roth, E.M., Woods, D., & Elm, W.C. (2000). Bootstrapping multiple converging cognitive task analysis techniques for system design. In J.M. Schraagen, S.F. Chipman, & V.L. Shalin (Eds.), *Cognitive task analysis.* Mahwah, NJ: Erlbaum.

Rasmussen, J., Pejtersen, A.M., & Goodstein, L.P. (1994). *Cognitive systems engineering.* New York: John Wiley.

Roth, E.M., Malsch, N., & Multer, J. (2001). *Understanding How Train Dispatchers Manage and Control Trains: Results of a Cognitive Task Analysis.* Washington, DC: U.S. Department of Transportation/Federal Railroad Administration.

Roth, E.M., Malsch, N., Multer, J., & Coplen, M. (1999). Understanding how train dispatchers manage and control trains: A cognitive analysis of a distributed planning task. In *Proceedings of the Human Factors and Ergonomics Society 43rd Annual Meeting.* Santa Monica, CA: Human Factors and Ergonomics Society.

Roth, E. and Multer J. (in preparation). Communication and coordination demands of railroad roadway worker activities and implications for new technology. Washington, DC: U.S. Department of Transportation, Federal Railroad Administration (DOT/FRA/ORD-07/XX)

Roth, E. M., Multer, J., & Raslear, T. (2006). Shared situation awareness as a contributor to high reliability performance in railroad operations. *Organization Studies, 27*(7), 967–987.

Roth, E.M., & Patterson, E.S. (2005). Using observational study as a tool for discovery: Uncovering cognitive and collaborative demands and adaptive strategies. In H. Montgomery, R. Lipshitz, & B. Brehmer (Eds.), *How Professionals Make Decisions* (pp. 379–393). Mahwah, NJ: Lawrence Erlbaum Associates.

Roth, E.M., Scott, R., Deutsch, S., Kuper, S., Schmidt, V., Stilson, M. et al. (2006). Evolvable work-centered support systems for command and control: Creating systems users can adapt to meet changing demands. *Ergonomics, 49*(7), 688–705.

Roth, E.M., & Woods, D.D. (1988). Aiding Human Performance I: Cognitive Analysis. *Le Travail Humain, 51*, 39–64.

Vicente, K.J. (1999). *Cognitive Work Analysis: Toward Safe, Productive, and Healthy Computer-based Work*. Mahwah, NJ: Lawrence Erlbaum Associates.

Woods, D.D. (1993). Process-tracing methods for the study of cognition outside of the experimental psychology laboratory. In G.A. Klein, J. Orasanu, R. Calderwood, & C.E. Zsambok (Eds.), *Decision Making in Action: Models and Methods*. Norwood, NJ: Ablex.

Woods, D.D., & Dekker, S.W.A. (2000). Anticipating the effects of technological change: A new era of dynamics for Human Factors. *Theoretical Issues in Ergonomic Science, 1*, 272–282.

Woods, D.D., & Hollnagel, E. (2006). *Joint Cognitive Systems: Patterns in Cognitive Systems Engineering*. Boca Raton, FL: Taylor & Francis.

Woods, D.D., & Roth, E.M. (1988). Aiding human performance: II. From cognitive analysis to support systems. *Le Travail Humain, 51*(2), 139–171.

Chapter 7

A Cognitive Work Analysis of Cardiac Care Nurses Performing Teletriage

Catherine M. Burns, Yukari Enomoto,
and Kathryn Momtahan

Contents

Overview

Cardiac care nursing is complex, and requires managing patient health in an environment of limited resources. In this chapter, we discuss a Cognitive Work Analysis (CWA) of this domain. The CWA consisted of three analysis phases: work domain analysis (WDA), control task analysis, and strategies analysis. Nursing has historically functioned in a rule-based environment, but modern nursing, particularly as practiced by advanced practice nurses, leans much more toward knowledge-based reasoning. The present research adds to our understanding of how to make CWA a useful analytical tool for understanding cognition in complex contexts. Furthermore, our project was somewhat unique in that it was not limited to a single CWA procedure, as in the case of WDA. Thus, it is a case study in a fuller program of CWA. We discuss these three phases of the CWA, how we collected the information that was used in the CWA, and how these phases contributed to our understanding of the cognitive work of cardiac nursing coordinators (NCs) performing telephone triage.

Introduction

Cardiac care nursing has, to this point, not been the beneficiary of decision support technologies (Lewis & Sommers, 2003; Tooey & Mayo, 2003). Yet, nurses are often the first line in most health care systems, and are largely responsible for decision making in the first triage of a patient. In the Canadian healthcare system, although nurses do not make medical diagnoses or prescribe medications, they suggest courses of actions based on their assessment of the severity of a patient's condition, and they can make adjustments to treatment regimes. This requires accurate assessment, rich knowledge, clinical skill, and strong problem-solving abilities.

In Canada, technology has made possible indirect health care consultation through telephone or e-mail. These "tele-health systems" are appealing because they allow patients to receive advice without leaving their homes, often at any time of the day or night. For a nurse however, these systems present unique challenges in that the patient must be advised on the basis of their self-report, often without clinical observation or test data. Although the safest route is for the nurse to refer the patient to an emergency room, for these systems to work effectively, nurses must be able to refer patients to less critical care facilities or provide treatment information with confidence.

Currently, most telephone triage nurses give advice based on their training and experience, so these positions are reserved for senior nurses. This can overwork the senior nursing staff. It also means that triage strategies and questions can vary based on each nurse's training, experience, and past incident experiences. This can result in uneven quality of care as well as an unsubstantiated process, should any questions or challenges arise in the care of the patient. For many situations, it is possible to determine best practices in teleconsultation to ensure that high-priority questions are asked first, and symptoms are documented thoroughly.

We worked with NCs at the University of Ottawa Heart Institute (UOHI), a prestigious cardiac care facility in Canada, to determine best practice strategies for their particular domain of teleconsultation, which is cardiology and cardiac surgery patients. We used the CWA for several reasons. First, we had colleagues in an independent team using Cognitive Task Analysis methods on the project, and secondly, CWA was our area of expertise. We were optimistic that CWA could help us understand the complex work environment.

This chapter reports the results of the CWA. Following Vicente's (1999) description of the full program of CWA research and analysis, the present project conducted three phases of CWA: WDA, Control Task Analysis, and strategies analysis. It can be contrasted to many other projects that have used only WDA. Knowledge-based behavior for nurses operating in a hospital environment is constrained not only by hospital policies and procedures, but also by the scope of nursing practice set out by licensing bodies whose policies do not always keep pace with modern practice. This type of behavior, which is constrained by the environment, influences and challenges the CWA.

When we say that nursing is a predominantly rule-based environment, we certainly do not mean to suggest that nursing does not involve knowledge-based behavior. As nurses interact with patients in the triage situation, they are certainly solving problems. In challenging or novel triage situations, they do reason analytically about the situation. However, as experienced professionals, they bring many more cognitive skills to the table. They make use of their work experience and critical incidents experience (consistent with naturalistic decision making; Klein, Orasanu, Calderwood, & Zsambok, 1993; Bogner, 1994), as well as tools and techniques that they are taught during training. We include this kind of problem-solving behavior with the term rule-based behavior.

In the coming sections, we discuss the information gathering phase of our CWA. In the section "Analysis," we discuss the three analytical phases of WDA, decision ladder analysis, and strategies analysis. We pay particular attention to how the context of work is represented in each of these three levels. Finally, we discuss how the three levels influenced our design recommendations and ideas.

Method

The full program of CWA as described by Vicente (1999) has five levels of analysis: WDA, control task analysis, strategies analysis, social-organizational analysis, and worker competency analysis. These five levels describe different aspects of cognitive work. In any project where one does not perform all the five phases, there needs to be an awareness that there are elements of the cognitive work picture that may not be captured. For example, many ecological interface design projects have primarily or even exclusively relied upon WDA (Sharp & Helmicki, 1998; Burns, 2000; Vicente, 2002; Burns & Hajdukiewicz, 2004 are or contain examples). Although these displays show the complex relationships revealed from the WDA, they often miss other kinds of cognitive support, such as support for procedures or rule-based information.

In this project, we strategically chose certain CWA analyses to meet our project objectives. We wanted our NCs to be able to solve unanticipated problems, so we chose to perform a WDA. We learned early that the NCs had rules and a work flow among themselves, with the patient, and with other workers (such as physicians); therefore, a control task analysis made sense so that we could map the actions of these different decision makers. In interviews with the NCs, the richness of their strategies became quickly apparent, and this indicated that mapping their strategies would be worthwhile and, indeed, necessary in order to provide a decision support solution. Our solution would not cover organizational interactions (such as understanding management goals and hospital economic objectives), so we did not opt to do a social-organizational analysis. It was also a domain in which our workers (NCs) had a high level of experience, and the competency level in terms of patient care and cardiac knowledge was consistent across the group. An NC by definition is an advanced practice nurse who must have completed and excelled at cardiac care nursing for several years before being promoted to an NC, so we did not do a formal worker competency analysis. We did consider their range of competency with personal digital assistants (PDAs) and computers.

The Information-Gathering Phase

Information for the analyses were collected from textbooks, by analyzing call sequences, through semi-structured interviews, and by analyzing records from past calls.

Knowledge from Textbooks

We studied texts to learn about cardiac care and various treatment algorithms. The purpose of this effort was to understand the basic physiology and symptoms of cardiac conditions and their basic treatment. This was very useful for our abstraction hierarchy description and for building the domain expertise to communicate effectively with the NCs. On our extended team we also had a cardiologist and a cardiac surgeon. The physicians were able to review our abstraction hierarchy descriptions and the treatment algorithms that were generated.

Call Sequences

We asked the NCs to generate typical call sequences for common cardiac conditions. This was not quite as specific a technique as clinical case recall (Rikers, Loyens, & Schmidt, 2004). The intention was to recall generic triage processes and questioning strategies, not specific case details. In Figure 7.1 is one of the hypothetical call scenarios generated by an NCs.

In all, the NCs generated 25 different call scenarios. The calls identified the range of problems patients were calling in with, and were consistent with the types

NC: Tell me about your pain
Patient: It's sharp. In my neck back and shoulder.

NC: Did you have an chest pain/angina before your OR?
Patient: Yes.

NC: Is this pain similar to that?
Patient: No

NC: Is this pain like the pain you had after surgery?
Patient: A little, but not really.

NC: When did the pain start?
Patient: The day before yesterday.

NC: So you've had this pain for about 2 days?
Patient: Yes.

NC: You described the pain as sharp, in the neck and shoulder. Is there anything else you want to tell me about the pain?
Patient: No.

NC: What makes the pain worse?
Patient: Taking a deep breath.

NC: What makes it better?
Patient: Nothing.

NC: What have you done to treat your pain?
Patient: I took some pills.

NC: How much medicine did you take?
Patient: Three or four Tylenol Extra Strength but it didn't help.

NC: Tell me again, you have more pain when you take a deep breath?
Patient: Yes, it feels like a knife in my chest.

NC: How is your breathing?
Patient: A little tight, but not bad.

NC: I would like you to try something – lean forward from the waist and take a deep breath. Tell me how you feel.
Patient: OK. I can take a deep breath and it doesn't hurt.

NC: Tell me the names of the medicines you are taking.
Patient: *The patient lists the medications he is on.*

NC: Have you had any problems with your stomach?
Patient: No.

NC: Any allergies?
Patient: No.
The NC then suggests a recommended course of action.

Figure 7.1   A call scenario generated by an nurse coordinator (NC).

of calls analyzed over a 1-year period. The scenarios showed that, although the types of questions asked by the NCs were similar, they followed no consistent questioning procedure. There were clearly rules and strategies behind the questions, but these were deeper and driven by knowledge and understanding and not from procedures or rote-learning.

Interviews

We interviewed each NC to get information on workflow, challenging problems, and novice–expert differences. These interviews were structured interviews, and contained a set of questions on triage tasks and domain relationships. Although the questions were set and presented in a consistent order, they were open ended, and interviewers were allowed to probe opportunistically during the response to a particular question. The interviews were conducted by a team that included two human factors researchers, one with a nursing background and the other, a graduate student.

To give a sample of some of the questions that were asked, we have included example questions, the motivation behind each question, and the role this question would play in a CWA in Table 7.1.

Call Data

We examined records from calls handled by NCs from January 2, 2002 to December 12, 2003. This covered 1523 cardiac surgery and 552 cardiology telephone consultations. The records had been taken by NCs during their normal course of work using a paper documentation sheet. These records could include the date and time of the call, patient name and phone number, the nurses assessment of the problem, and how the problem was resolved. However, as NCs rarely fill out the forms completely, this generated an incomplete data set. We attempted to extract patterns using statistical clustering techniques, and to derive decision trees using computational decision tree induction methods (Enomoto, 2005). The relative noise in the data, sparse sampling patterns, and the relatively small size of the sample made these methods unreliable.

Analysis

We performed three particular CWA stages: a WDA, a control task analysis, and a strategies analysis. Each of these stages and their resulting models are described in the following sections.

Table 7.1 Interview questions and their role in CWA

Question	Motivation	Role	Results
We see some NCs comparing "pain before the operation" and "pain after the operation." Is it simply asking if it is improving or worsening, or is there any significance (especially related to the type of operation)?	Understand the meaning of event-related pain as a cue.	Work domain relationships of pain in cardiac treatment. Strategies of how NCs use pain to understand the situation.	6/7 NCs used this comparison, and one did not. 3 NCs indicated that this question comes after the patient describes his/her pain. *Comments:* 1 NC explained that for a cardiology patient, dangerous pain is constant, severe, and cannot be relieved by rest or nitroglycerin. For a cardiac surgery patient, you expect a certain pattern of pain and changes in pain as the patient recovers.
Have there been cases when you received unexpected answers to open-ended questions that led to a better assessment?	Probe for high-value cues. Collect critical incidents.	Extraction of less-obvious work domain relationships.	5/7 NCs agreed that unexpected answers happen and lead to new assessments. 2/7 NCs felt this does not happen and they mostly stick to direct questions. *Comments:* The NCs that did find this occurs tend to use "tell me more" questions to probe the patient deeper. An example of an unexpected answer was when a patient mentioned dizziness with chest pain. Until that point, the most likely condition had been myocardial infarction (heart attack) but the mention of dizziness suggested arrhythmia to the NC. NCs mentioned that patients do not always call in about their most important symptoms.

(continued)

Table 7.1   Interview questions and their role in CWA (continued)

Question	Motivation	Role	Results
If you are calling back a patient who called in on another NC's shift, what is the most important information that you need them to pass on to you? What makes for an easy-to-follow callback record? What makes for a callback record that is difficult to follow?	Understand information flow between NCs. Extract key cues.	Decision ladder flow.	NCs transfer basic information, patient name, physician, issue, etc. *Comments:* It is most useful to have the previous NC's record, and just add information to it. Poor handwriting makes this difficult, as well as the use of nonstandard acronyms. If a record has gaps, the NC will access the patients chart and records if these are still available.
Is there a major difference between cardiology and cardiac surgery patients when you go through pain (or any other) scenario (i.e., any special question sequence for surgery patients)?	Reveal work flow processes and large patterns consistent with common case types.	Explore major control tasks. Strategy analysis. Probe for basic strategy differences between cardiology and cardiac surgery patients.	6/7 NCs agreed that they used different strategies in these situations, whereas 1 NC felt that all cases could be handled with the same approach to questioning. *Comments:* The NCs who did use different strategies expanded and gave examples. With a cardiology patient, they are looking for changes in pain and its frequency intensity or duration. They are trying to differentiate between angina, a myocardial infarction (heart attack), pericardial pain, or pain that is not cardiac related. With a cardiac surgery patient, they are looking more directly for signs of, for example, postpericardiotomy syndrome, pleural infusion, or an infection of their incision. The NC who felt that questioning strategies were the same for cardiology and cardiac surgery patients argued that patients often cross boundaries between cardiology and cardiac surgery. For example, a cardiac surgery patient who is several years postoperative could have a condition related to their surgery or a new cardiology condition. Rather than consider this a dissenting view, we felt it revealed more of the cognitive challenges of the work, that patient state is dynamic and shifts with time and events.

Work Domain Analysis (WDA)

The first phase of a WDA is to determine the system boundary. This specifies the scope of the analysis, and can influence the models. A tight work domain boundary will leave out connections and interactions outside the boundary and a broad boundary can dilute the design effort, as time is spent in areas that are less critical.

In this case, the operator was the NC, and we chose to restrict the domain to those things she could control, even if that control was only partial or indirect. We included the patient as well as the various resources to which the NC can refer the patient. These resources needed to be part of the work domain because the NC can use these resources in the solution of the triage problem. Even if she herself does not interact with the emergency department, when she sends the patient there, she has indirectly activated that option. We also chose to include medications, an issue that has been debated in the CWA literature (Miller, 2004).

From this boundary analysis, we built descriptions that focused on two objectives: improving patient health and using resources appropriately. In each section, we describe the general contents of these descriptions and how they were placed within an abstraction hierarchy.

Objective 1: Improving Patient Health

A patient phones in with an irregular heart beat. The NC questions the patient and learns that she is a cardiology patient of Dr. Smith's. The irregularity has been ongoing for about a month but seems to be happening more frequently. The patient is on a pharmaceutical regimen of an ACE inhibitor and a cholesterol-lowering statin. In questioning the patient further, the NC learns that the patient has been self-medicating with a herbal remedy and due to a recent cold, been taking a decongestant containing pseudoephedrine from the drugstore. She recommends that the patient stop using the herbal remedy and the decongestant, maintain her other medications, and book an appointment with her doctor. She explains to the patient that certain herbal remedies can interfere with her medications, that pseudoephedrine can cause heart rhythm irregularities, and that there may be better choices for a decongestant. She encourages the patient to call back at anytime if the condition worsens or still concerns her.

The case above demonstrates some of the complexities of managing patient health. The patient's physiological functions are operating, but often in a compromised state. Patients are likely being treated with medications and may be self-medicating as well. They may call in at the first event or the NC may be asked for advice on a problem that has been recurring. The objective of this description is to look at the complex relationships between anatomy, physiology, medications, activity, and health.

To model patient health, we looked at patient physiology and anatomy and the relationships these play in patient health and physiological functions. The analysis performed by Hajdukiewicz provided us with a basic model of patient health that we used as a starting point (Hajdukiewicz, Vicente, Doyle, Milgram, & Burns, 2001). Hajdukiewicz's model structures patient anatomy, physiological processes, and the principles behind maintaining homeostasis in a five-level abstraction hierarchy. We supplemented this basic model with knowledge specific to cardiology and information from our interviews with the nurses. For example, the nurses regularly ask if a patient has been able to keep to their postsurgical exercise plan. Being unable to keep to this plan may indicate slow recovery or additional problems. In terms of work domain structure, this indicates two processes, one related to activity and one related to healing. (It should be noted that as Hajdukiewicz's model was specific to anesthesiology, these particular processes were not there in his model.) Although the basic idea that patient health is related to activity level is not surprising, when we map this into an abstraction hierarchy description we see other relationships—for example, that activity level may be influenced not only by cardiac difficulties, but also by certain medications. It can make sense for an NC to probe further and determine the patient's pharmaceutical regime as well, when faced with a complaint of low energy.

Another significant difference from Hajdukiewicz's model was that we included medications in our abstraction hierarchy description. We included medications because they act on the patient and the cardiac system, in particular, in ways that have certain effects and alter the functionality and behavior of organs, systems, and the body. This is consistent with the views expressed by Miller (2004).

Our abstraction hierarchy had five levels consistent with that of Rasmussen's (1985), and is shown in Figure 7.2. We described Purposes at the highest level, then Principles and Balances, Processes, Physiological Functions, Anatomical Details, and Attributes of Objects. For patient health, we used three levels of decomposition: the body, systems, and organs. These levels are discussed in the following sections.

Purposes

The purpose of the healthcare system when considering the whole body is to "maintain and improve patient's health." In terms of body systems, this means improving circulatory and respiratory systems functions.

Principles and Balances

The main principle of cardiac care management is to maintain the patients' cardiac workload within a manageable range and ensure proper recovery from the heart problem. This principle of managing cardiac recovery is expressed in textbooks, and

	Body	Systems	Organs
Purposes	Maintain and improve patients' health	Regulate circulatory and respiration systems. Improve circulatory and respiratory function	Regulate and improve organ function
Principles, Priorities and Balances	Balance recovery and restore cardiac workload to normal levels Balance oxygen and nutrition supply and demands	Balances of oxygen and nutrition in the systems. Maintain balances of blood flow and water.	Balances of oxygen and nutrition to the heart and lungs. Maintain balances of blood flow and water in the heart, lungs, and blood vessels.
Processes	Healing Activity Homeostatic processes	At the System Level: Healing Workload Oxygen exchange Carbon dioxide exchange Nutrient and waste exchange Flow of blood and water	At the Organ Level: Healing Metabolism Oxygen exchange Carbon dioxide exchange Nutrient and waste exchange Flow of blood and water
Physiology and Objects	Body as a whole medications	Cardiac and respiratory systems Active ingredients of medications	Heart Lungs Blood vessels Active ingredients of medications
Anatomical Details and Attributes of Objects	Age, weight, gender of patient Type, dose, concentration of prescribed medications	Condition of systems Composition of medications and concentration in the body	Condition of organs Degree of disease Key physical dimensions (e.g. Minimum internal diameter of blood vessels) Chemical influence of medications

Figure 7.2   Abstraction hierarchy for patient health.

confirmed through our interviews with the nurses. Nurses regularly look for clues such as activity level, expected recovery pattern, and time frame when assessing the severity of the patient's condition. At the level of body systems and organs, these principles coalesce into maintaining the primary balances of life and homeostasis, oxygen, energy, and water balances. The details of these balances can be found in textbooks but their practical relevance was understood from the interviews with the nurses. Examples of how these balances influence cardiac care can be seen when determining postsurgical exercise levels, which depend on energy and oxygen balances, and in the use of diuretics, which commonly influences water balances in cardiac patients.

Processes

This level includes processes such as healing, activity, or workload, and the maintenance of other physiological processes such as oxygen exchanges, and nutrient and waste exchanges.

Physiology and Objects

At this level, we describe the body as a whole (at the level of the body); the cardiac and respiratory systems (at the level of systems); and the heart, lungs, and blood vessels (at the level of organs). We also included medications and their influence on cardiac function. The reason for describing medications here is to show that they act as controllers for various organs and units. To a certain degree, the NC, within the scope of the physician's medication orders, can confirm and advise on the proper administration of these medications. NCs do not prescribe or change the physician's instructions unless there is prior written approval to, for instance, titrate medications such as furosemide based on the patient's symptoms. They can also ask the patient to withhold a medication while a physician's advice is sought.

Anatomy and Attributes

At this level, condition of the organs and systems, and degree of disease are considered. Key physical details such as the size of an incision or the diameter of a narrowed artery could also be considered. The age, weight, and gender of the patient are important. Similarly the type, concentration, and dosage of medications are also important.

As an example of using the information at this level, the condition of an incision may be described by its color, smell, the amount of drainage, or whether it is swollen. These attributes of an incision help nurses to determine whether the patient's discomfort is caused by an incision or the heart itself. The change in discomfort in relation to the body position also helps identify the internal condition. Nurses often ask if the discomfort is better when lying flat or sitting forward. Changing the body position increases or decreases pressure in a different part of the body and that might be the cause of discomfort. For example, patients with cardiac tamponade often experience shortness of breath especially when lying flat because of the pressure caused by the fluid accumulation in the pericardium. These are examples of information obtained from the interview process that influenced the abstraction hierarchy.

Objective 2: Using Resources Appropriately

A patient phones in with moderate chest pain that has been ongoing for several hours. The NC decides the patient should be seen immediately and quickly. The patient lives on the outskirts of the city and his specialist's office is on the opposite side of the city. She determines that the patient has a car and his wife may be able to drive him. After talking briefly to the patient's wife, she determines that the wife is quite agitated over her husband's condition and unlikely to be able to drive safely. The NC decides to order an ambulance to take the patient to the emergency room of the closest hospital.

The case above highlights another dimension of the telephone triage problem. The NC must, along with determining and providing recommendations for the patient's health, decide if the patient can remain at home or should seek additional treatment. This particular decision can depend on the state of patient, whether the patient has family or lives alone, the distance to a facility, and the location and expertise of facilities nearby.

We built another abstraction hierarchy description looking specifically at the complexities of matching the patient to various healthcare resources. This description is shown in Figure 7.3. This work was loosely based on the work of Chow

	Institution-Level Resources	Patient-Level Resources
Purposes	Maintain and improve patients' health within available institution resources	Maintain and improve patients' health within available patient resources
Principles, Priorities and Balances	Balance priorities of: Quality of care Patient outcomes Time to treat # of patients treated Maintain patient inflow and outflow	Priorities of timely treatment appropriate treatment
Processes	Admitting Treating Release ordering Planning Assessing	Phoning a healthcare professional Driving to a facility Modifying home routines
Objects	Availability and state of: Facilities Personnel	State and Capability of Patient Their Family Their Family Physician Local ER Accessible Medications
Details and Attributes of Objects	Location and hours of facilities, ER, physicians	Location of patient Availabilities of car, ambulance, support, medication

Figure 7.3 Abstraction hierarchy for using resources appropriately.

(2004) who studied emergency ambulance display management. In both cases, the primary task is to match a person who needs help to a resource that can give help: in one case, someone who has called for an ambulance to the appropriate ambulance; and in the other, someone with a health concern to the appropriate caregiver (doctor, hospital, or phone advice).

Purposes

On this side of the problem, triage nurses are concerned with matching the patient to the healthcare professional or location that can give them the best help. More often than not, though, they may also have to consider which resources suit the patient's location and mobility, and which resources (hospitals, physicians, or particular specialists) are available. Often, one of the main reasons for having a telephone triage or similar service is to be able to utilize medical resources properly and to encourage timely and appropriate access to healthcare. Triage services are designed for the purpose of helping the patient to decide if they can remain at home or need to access specialized resources.

Principles, Priorities, and Balances

The triage nurse must try to match the patient to the most appropriate resources. Achieving this match can mean balancing multiple factors such as the availability of medical resources, the needs of the patient, and the ability of the patient to reach the various resources. The patient's priorities are to obtain the appropriate treatment, within the needed time. Institutions must manage priorities of meeting or exceeding goals in terms of quality of care, time to treat, number of patients treated, and the quality of patient outcomes. These priorities drive the need to match patients properly with healthcare resources.

Processes

To access resources, patients can phone for advice, drive to help, or take an ambulance. Within the facility, patients move through several processes such as admission, treatment, assessment, and release. Although the NC would not perform these processes, she can advise the patient on them and recommend actions that modify these processes. For example, a patient with a severe cardiac condition may be better managed by traveling to a hospital in an ambulance so that some preliminary assessment and treatment can be performed by paramedics, and CPR and defibrillation are available if necessary.

Objects

In the description of objects, the capabilities of the patient, specialists, and the healthcare resources are considered. For example, this level would consider factors such as a patient's ability to drive to the hospital, have a family member drive him/her, or to call an ambulance. Certain hospitals may have specialty units such as cardiac or stroke units and have a greater ability to take care of certain patients.

Details and Attributes of Objects

The nurse also considers the location of the patient, the location of the hospital, and the condition of the patient before making a recommendation. For example, a patient having a potential for heart attack may not be in a condition to drive to a hospital, and sending an ambulance may be a more appropriate recommendation.

Control Task Analysis

Following the abstraction hierarchy description, we performed a control task analysis. Although the abstraction hierarchy description maps the scope of the decision-making space, it does not give information on how tasks are to be performed. A control task analysis can look at the modes of operation, the sequencing of steps, and shortcuts between tasks, information that is not represented in the abstraction hierarchy.

In terms of modes of operation, a patient at the UOHI can be in preoperation, hospitalized (postoperation), or in postoperation postdischarge recovery at home. Of these, preoperation and postoperation postdischarge are distinctly different modes that are important to how the patients are managed and triaged. (Hospitalized care is not seen by the triage nurse.) Physicians are divided along these modes as well: cardiologists treat presurgical patients and extensively postsurgical patients, and cardiac surgeons treat patients scheduled for an operation or in recovery from one. The NCs at the hospital also specialize in either cardiology or cardiac surgery with the exception of night or weekend staff who are expected to handle both classes of calls.

The distinctions between the two modes lie in several areas. Cardiology patients will have to follow instructions while waiting for surgery. There may be medications to suppress or manage symptoms and control their activities. A cardiology patient who phones in to the hospital may be experiencing some worsening of their cardiac function. Postsurgical patients, in contrast, are in recovery programs that control cardiac workload, gradually increasing activity as recovery proceeds. They may still be on medications to improve cardiac function, but may also have specific postsurgical medications that may be temporary. When these patients phone in, many of the questions may relate to postsurgical management or reduced progress on their

Table 7.2 Modes of operation

Modes	Stages	Primary goals and control tasks	Primary decision makers	Secondary decision maker
Cardiology	Preoperation	Suppress and control symptoms	Cardiologist Patient	Family physician NCs
Cardiac surgery	Operation	Surgical goal (e.g., replace a valve)	Surgeon and surgical team	—
	Hospitalized postoperation	Improve cardiac function to a manageable level Stabilize surgical intervention	Surgeon Nurses	—
	Postdischarge recovery	Increase cardiac workload Complete surgical healing	Surgeon Patient	NCs
	Stable	Maintain and monitor internal balances	Surgeon Patient	NCs Family Physician
Cardiology	Stable over a long period of time	Maintain and monitor internal balances	Family physician Patient	Cardiologist

recovery programs. The two modes of cardiology and cardiac surgery are related and, yet, quite distinct in terms of patient care and expected triaging concerns.

In the interviews with the NCs, seven out of eight NCs felt that patients were triaged differently based on these modes. The mode boundary becomes less clear when postsurgery patients phone in a long time after their surgery. After a certain time, these patients may be more appropriately treated as cardiology patients as new conditions arise that are not related to their surgery. In fact, it was this possibility (i.e., a mode-based view could be incorrect in the case where a post-surgical patient develops a new cardiac condition unrelated to their previous surgery) that caused one NC to disagree with the two-mode classification of their work.

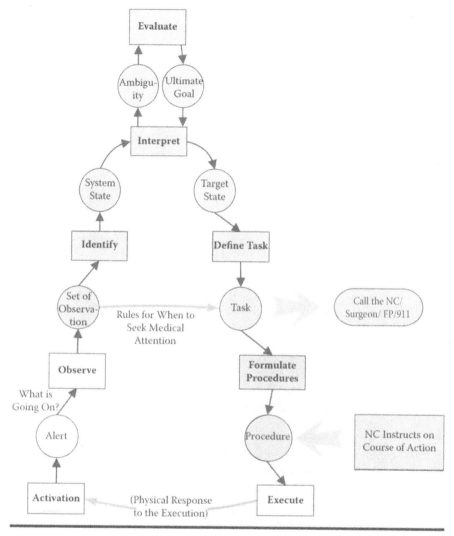

Figure 7.4 Decision ladder for the patient.

Probing further, we asked how often the NCs assumed that complaints arose from a recent surgery. Four of the eight NCs confirmed that this was usually their first hypothesis, three disagreed that they use this hypothesis, and one NC mentioned that this depends on the length of time elapsed after the surgery. If the surgery was in the last two weeks, the complaint is likely due to the surgery. If more time has elapsed, there is a greater probability that it is a different condition.

We also built decision ladders (shown in Figures 7.4 to 7.6) for each of the three main decision makers in the triage process: the patient (Figure 7.4), the NC (Figure 7.5), and the surgeon or cardiologist (Figure 7.6). The filled stages indicate

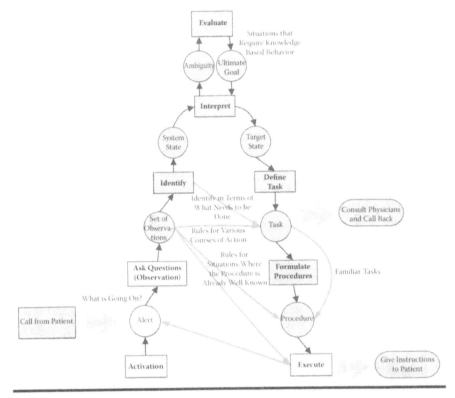

Figure 7.5 Decision ladder for the nurse coordinator (NC).

the most commonly used parts of the decision ladder. Arrows are shown for common shortcuts. Finally, because all three decision makers are involved in making a correct decision, information that transfers from the patient to NC to physician and back is shown.

Strategy Analysis

Within this task of consulting with the patient, there are several strategies that the NCs might take. We identified these strategies from our semistructured interview process discussed earlier and the NC's responses to typical questioning scenarios. We observed that NCs may ask the patient open-ended questions, work from a standardized question list, question for symptoms topographically, build a hypothesis of the situation, and test for confirming or refuting the information, or use a strategy of deliberately ruling out possibilities. We discuss these strategies in the following sections.

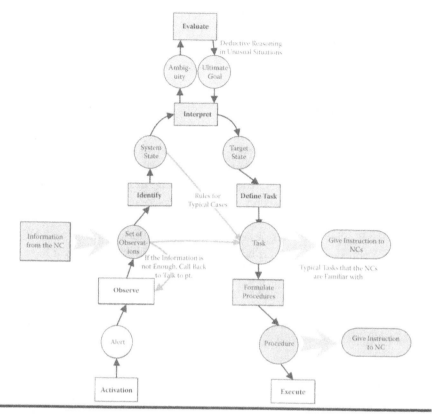

Figure 7.6   Decision ladder for the physician.

Open-Ended Questions

Open-ended questions such as "Describe your discomfort" are often used to gather key information from the patient to build a mental model of the patient's state. The purpose of these questions is to allow the patient to use their own words to describe the condition, and to extract symptoms without the NC leading the patient. This is typically done at the beginning of the consultation process so that the decision maker does not form a biased hypothesis; all NCs confirmed that they use open-ended questions. The advantage to using such questions is that the patient may offer information that the NC might not have probed. Six of eight NCs mentioned that open-ended questions had revealed information that they would not have expected through standardized probing. As an example, when one patient called in about diarrhea, an open-ended discussion revealed that his surgical incision was infected as well, eventually suggesting a case of *C. difficile*, a challenging bacterial infection that can sometimes be contracted from hospital stays.

Open-ended questions also allow the NC to perceive less tangible aspects of the situation such as the patient's anxiety level, ability to articulate, or breathing pattern. The disadvantage to open-ended questions is that they may be less efficient, and some patients, without prompting, may be unable to provide a useful description of the problem.

Standardized Question List

Standardized question lists are often taught in training, and provide efficient capture of information. OLDCART is one such decision aid taught in the training of nurses, and reminds the NC to ask about onset, location, duration, characteristics, associated symptoms or aggravating factors, relieving factors, and treatments. This strategy is in essence a mental checklist procedure. It does not replace other questioning strategies but rather provides a backup or check that appropriate questions have been answered. As an example, when NCs were asked about triaging difficult patients, one NC said that if open-ended questions were not working she would "OLDCART it."

As another example of standardized questions, NCs would regularly have the patients compare their current pain level to "before the surgery" and "after the surgery." Six of eight NCs regularly used pain-scale comparisons, one NC said she never did, and one NC said she was more focused on how the pain affected the patient's activity level.

Topographical Search

In topographical search (*topography* in ancient Greek refers to the detailed description of a place), the NC probes for the details of where and what places on the body the patient is experiencing symptoms. When the patient reports one problem (e.g., chest pain), the NC may ask about other related problems (e.g., neck pain, shoulder pain, arm pain, etc.). This reflects an understanding that systems are connected, and symptoms are related. Secondary symptoms and their locations can provide important information in deciding the severity of the problem. Two of the eight NCs referred to trying to visualize patients and where the pain is on their body when making an assessment.

Hypothesis, Test, and Rule Out

All NCs mentioned that they form hypotheses regarding the patient's condition through the call. Many (six out of eight) mentioned that the hypothesis is formed at early stages of the consultation process, often within one or two questions, and they proceed to question the patient to confirm or disconfirm it. Although it was not unexpected to learn that NCs form hypotheses, we were surprised about how rapidly the first hypothesis is developed during the triage sequence. As an example, a patient with chest pain may be asked if nitroglycerin relieves it. This question

Table 7.3 Various questions that arise from different questioning strategies

"I have chest pain ..."	
Strategy	Next question
Open-ended	Describe your pain to me.
Standardized	When did it start? (onset)
Topographical	Do you have any pain in your shoulders?
Hypothesis and test	Is it relieved by nitroglycerin?
Ruling out possibilities	Do you have any problems breathing?

reflects the hypothesis that the patient may have angina. The strategy of hypothesis and test reflects extensive previous experience and deep knowledge of the associated symptoms of various conditions.

As part of this strategy, the NC may deliberately probe to rule out very serious conditions that would require that the patient end the call and proceed to emergency. For example, if a patient calls with chest pain, they may immediately ask whether they also have shortness of breath as these two symptoms in combination mean the patient should be seen immediately. They may also rule out very common conditions (colds and the flu) that can be dealt with at a local pharmacy or a less-specialized institution. In Table 7.3, we demonstrate these strategies in response to a hypothetical patient who has called in and said, "I have chest pain."

The difference between these strategies is often in the motivation underlying them. In several of the strategies, the same questions would likely be asked, but they would appear at different times. The strategies influence how the NCs prioritize the various questions, and this prioritization reflects their training and past experiences. An NC who has had a past experience with an unusual but critical case may be more likely to use a ruling out strategy for that kind of condition. In the interviews, several NCs who had experienced highly unusual cases, particularly with negative results, confirmed a tendency to check (and hopefully rule out) these conditions in the questioning process.

NCs also switch strategies frequently in adaptive response to the context of the patient and the condition. Figure 7.7 shows a hypothetical but fairly common example of a strategy switch. NCs may start a call with an open-ended question to allow the patient to describe their situation and listen to less tangible clues, and then change to more standardized probes to complete the information gathering and collect details. They may switch to hypothesis-driven questions as cues come in (in the case given in the next section, chest pain shortly after surgery is most often related to recovery from the surgery).

We used the call sequences to examine in detail how the NCs switch their questioning strategies. In Figure 7.8, a question sequence chart for a surgery patient experiencing chest pain submitted by one NC is shown. She included a comment: "Once I have ruled out angina and PPS and infection … the next place I would

Strategy

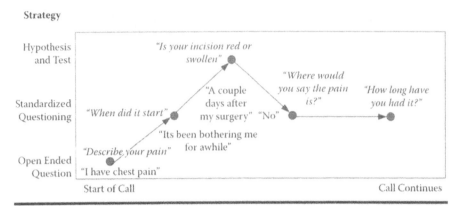

Figure 7.7 **Strategy switching during a typical call.**

		1	2	3	4	5	6	7	8	9	10	11	12	13	14	15
	Activity Level															
	Medication										▓		▓	▓		
	Relieving Factors										▓					
O L D C A R T	Aggravating Factors											▓				
	Associated Symptoms						▓									
	Characteristics			▓	▓			▓								
	Location															
	Onset and Duration					▓										
	History		▓													
	Open ended	▓														
	Question number	1	2	3	4	5	6	7	8	9	10	11	12	13	14	15

Figure 7.8 **A questioning sequence generated by an NC.**

go to is pain management strategies." In the chart, the NC starts from an open-ended question, asks about the location (taken from the OLDCART strategy), collects history information, asks more routine questions from the OLDCART sequence with another open-ended question in between, and then concentrates on medication-related probing for the remainder of the session. Note that even within the OLDCART questions, the NC selectively adapts the question order to suit her probing. It is not a rote application of the rule.

A key clue to differentiating strategies from other aspects of cognitive work is that strategies differ between workers and, in some cases, with work context. Different workers may apply different strategies to the same problem. In addition, a single worker may show strategic shifts in response to the work situation (such as changes in workload level, changes in worker fatigue, or feedback on the success of the strategy). In contrast, work domain constraints hold true across all work, and control tasks identify modes of operation that are also common to all workers. Some strategies are used by some workers and not others, and even with the same worker, there may be opportunistic shifts in strategies as required. Strategies can be thought of as an energy state; the worker, given his or her background, experiences, and context, will work in that state until some perturbation arises that changes the state of work. At this time, the worker may shift into a different strategic pattern.

It is difficult with strategy analysis to confirm that you have identified all possible strategies. Certainly, through an observational approach as demonstrated here, you cannot claim that all strategies have been identified, merely that the most prevalent strategies have been seen. New strategies could emerge, or less-frequently used strategies may not have occurred. In some physically constrained systems, such as DURESS, strategies can be analytically determined by mapping out the possible ways that the system can be operated. In these cases, it may even be possible to identify possible strategies that have not appeared in observations of work with the system. However, this analytical approach is most likely to be successful in systems with a finite and manageable number of solutions.

A second challenge with identifying strategies is distinguishing between them, particularly in situations where strategies shift dynamically. Vicente (1999) refers to strategy analysis as a "binning" exercise, in which various action sets are binned in strategy categories. This requires a determination of the bin boundaries. Once again, a physically constrained system may have physical constraints that determine bin boundaries (i.e., a valve is open or it is closed). An intentional or less-constrained system requires a different approach to determine strategic boundaries. In this example, we used the cognitive categorizations provided by the workers themselves, or built the categorizations from worker-described actions and approaches. Although this needs more exploration, it does have cognitive validity, and may be appropriate in intentional systems.

We provided the two maps in Figures 7.7 and 7.8 to show the richness of studying strategies. An exploration of strategies opens a whole world of adaptive cognitive behavior. Strategies shift and weave, as workers adapt to the problem at hand. The choice and application of strategies, even rule-based strategies, is often a knowledge-based decision, showing the richness and adaptivity of cognitive work.

Conclusion

A CWA is a set of analytical methods for producing representations of cognitive work. In its full representation, CWA should have five phases of description, which would include worker competency and social-organizational analyses that were not performed here. Although a five-phase CWA is ideal, it is reasonable to pick and choose between the analyses and create the right combination for a particular project. When adopting a pick and choose approach to CWA, the analyst must always be aware that they have not developed a full description, and that their description may as such be open to critique. But a practical approach to CWA requires analytical flexibility in choosing and using the various phases.

We have also learned from various projects using CWA that the benefits derived from each phase can vary with the work being analyzed and the objectives of the project. In a project that seeks to develop visualizations of complex relationships in an engineered industrial process, WDA and the use of the abstraction hierarchy is a useful approach for developing a structure and asking the questions that reveal the complex relationships and why they are important. In this current project, we feel we derived the most benefit from the strategies analysis, which gave us several ideas for providing decision support that might be useful to the triage nurses.

The methods for learning about cognitive work are various. In industrial systems, a high emphasis may be placed on referring to plant design and the underlying physics or chemistry of the processes that occur. In healthcare systems, which are less engineered and where knowledge is still developing, these methods may be best supplemented with the CTA methods of interviewing and observing.

There are still areas for improvement in CWA. In this project, we found the categories of rule- and knowledge-based behavior to be limiting. In reality, the triage nurses move flexibly between rules and strategies, experience and past critical incidents (as in naturalistic decision making), and knowledge-based reasoning. In categorizing the behavior, the richness seems to be lost.

Acknowledgments

We are thankful for a grant from the Primary Healthcare Transition Fund of the Ontario Ministry for Health and Long Term Care. We also thank Whynne Caves, Dr. Mesana, and Dr. Labinaz of the University of Ottawa Heart Institute for keeping us on track in understanding cardiac care. We would also like to thank the NCs who participated in this study.

References

Bogner, M. S. (1994). *Human Error in Medicine*. Boca Raton, FL: CRC Press.

Burns, C. M. (2000). Putting it all together: Improving display integration in ecological displays, *Human Factors, 42*, 226–241.

Burns, C. M., & Hajdukiewicz, J. R. (2004). *Ecological Interface Design*. Boca Raton, FL: CRC Press.

Chow, R. W. Y. (2004). *Generalizing Ecological Interface Design to Support Emergency Ambulance Dispatching*. Ph.D. thesis, Department of Mechanical and Industrial Engineering, University of Toronto.

Enomoto, Y. (2005). *Induction of Decision Trees by Soft Computing for Cardiac Phone Consultation*. White paper for SYDE 625. November 9, 2005. available by contacting the first author.

Hajdukiewicz, J. R., Vicente, K. J., Doyle, D. J., Milgram, P., & Burns, C. M. (2001). Modeling a medical environment: An ontology for integrated medical informatics design. *International Journal of Medical Informatics, 62*, 79–99.

Hoffman, R. R., Shadbolt, N., Burton, A. M., & Klein, G. A. (1995). Eliciting knowledge from experts: A methodological analysis. *Organizational Behavior and Human Decision Processes, 62*, 129–158.

Hoffman, R. R., & Woods, D. D. (Guest Editors) (2000). Cognitive task analysis. *Special Issue of Human Factors*, 42(1), 1–95.

Jamieson, G. A. (2003). Comparison of information requirements from task- and system-based work analysis. In R.N. Pikeur, E.A.P. Koningsveld, & P.J.M. Settels (Eds.), *Proceedings of the International Ergonomics Association XVth Triennial Conference*. Amsterdam: Elsevier.

Klein, G., Orasanu, J., Calderwood, R., & Zsambok, C. E. (1993). *Decision making in action: Models and methods*. Norwood, NJ: Ablex Publishing Co.

Lewis, J. A., & Sommers, C. O. (2003). Personal data assistants: Using new technology to enhance nursing practice. *The American Journal of Maternal Child Nursing*, 28(2), 66–73.

Miller, A. (2004). Work domain analysis for intensive care unit patients. *Cognition, Technology, & Work*, 6 (4), 207–222.

Rasmussen, J. (1985). The role of hierarchical knowledge representation in decision making and system management. *IEEE Transactions on Systems Man Cybernetics* 15, 234–243.

Rikers, R. M., Loyens, S. M., & Schmidt, H. G. (2004). The role of encapsulated knowledge in clinical case representations of medical students and family doctors. *Medical Education*, 10, 1035–1043.

Sharp, T. D., & Helmicki, A. J. (1998). The application of the ecological interface design approach to neonatal intensive care medicine, *Proceedings of the 42nd Annual Meeting of the Human Factors and Ergonomics Society*, 350–354.

Tooey, M. J., & Mayo, A. (2003). Handheld technologies in a clinical setting: State of the technology and resources. *The American Association of Critical Care Nurses (AACN) Clinical Issues*, 14 (3), 342–349.

Vicente, K. J. (1999). *Cognitive Work Analysis: Toward Safe, Productive, and Healthy Computer-Based Work*, Mahwah, NJ: Erlbaum and Associates.

Vicente, K. J. (2002). Ecological interface design: Progress and challenges. *Human Factors*, 44: 62–78.

Chapter 8

Methods for the Analysis of Social and Organizational Aspects of the Work Domain

---

Jonathan D. Pfautz and Stacy L. Pfautz

Contents

Overview

Analysis of the constraints and requirements of social organization is an important aspect of Cognitive Work Analysis (CWA), for which there are many possible approaches but little guidance on their applicability. This chapter describes techniques and technologies for the analysis of social organization and cooperation (e.g., social network theory, communication analysis), to provide the CWA practitioner insight into the advantages and disadvantages of these techniques and technologies for different types of studies. Social analysis begins with the collection of data on entity-to-entity interaction (e.g., human–human, human–automation, human–organization), a process fraught with challenges that need to be understood. Once data is obtained, a variety of processing methods and analytic techniques can be used to draw some conclusions about the types of relationship (e.g., social, communication, organizational) between entities and thus more richly describe the "social ecology" of a work domain. A key concern across all of these methods and techniques is the degree to which assumptions impact the validity of conclusions drawn by a CWA practitioner. This chapter addresses these and other issues that may arise when applying social and organization analysis techniques to CWA.

Introduction

CWA addresses the "analysis of sociotechnical systems," or "systems containing social, psychological, and technical elements" (Vicente, 1999). Although many CWA practitioners acknowledge the need for deeper study of social aspects and interrelationships among actors in the work domain, to date, most have focused on psychological (pertaining to the human operator) and technical (pertaining to the physical system used by the operator) elements. However, social aspects of the domain are increasingly relevant as technologies enable more and new forms of electronically mediated interactions (e.g., the Internet, cellular phones), and systems are designed to incorporate these interactions as part of the workflow. These technologies enable workers to be geographically dispersed, increasing the number and heterogeneity of perspectives on the work. As problem spaces become larger, the need for teaming and interaction between workers becomes more important. Even without considering current technologies and the challenges and opportunities they provide to a social system, the need to understand the communication between workers in a system is fundamental to performing a complete analysis of a work domain. To further motivate the study of interactions among workers, we assert:

■ Work is situated not only in a physical environment but also in a social environment.

> For example, Jed works writing papers in an office that is kept at ~72°. If it is too hot or too cold, he works more slowly. Jed is interrupted in his writing by people interacting socially, asking him for help in their work, and asking him about his progress.

■ Work can include communication of data, information, and knowledge.

> For example, my computer tells me the current temperature outside so I know if I need to wear a coat. Or, an experienced mason teaches his apprentice better methods as they construct a brick wall.

■ Work is often performed collaboratively and cooperatively.

> For example, David and Rahul are developing different components of a software system at the same time, and communicate via e-mail about how their components will interact. Or, a team reaches consensus about task allocation by workers volunteering for tasks and technical directors assigning tasks.

■ Social interaction self evolves and adapts through work.

> For example, Miriam and Enzo tell each other what color or type of pieces they are currently looking for as they assemble a jigsaw puzzle together.

Even in the simple examples just given, the variety and complexity of the social aspects of work are apparent. Clearly, any analysis that hopes to address a work domain comprehensively must find methods for incorporating the influences of complex social interrelationships among workers, whether those workers are humans, machines, organizations, or societies. In fact, the terms *social* or *organizational* can fail to adequately capture all types of interactions among actors that may require analysis (and could even bias the practitioner to only consider affective relationship types). Here, we use these terms to include *any form of meaningful, directed interaction between entities in the domain*, ranging from simple communication of information (e.g., my system tells me it is 5 o'clock and time for my appointment) through more sophisticated political and affective relationships impacting work performance (e.g., Bob struggled to adapt to the last standard issued by ISO, and is therefore unwilling to contribute to defining the next standard). Figure 8.1 shows examples of how we might define some of the entity types in the domain, and presents possible forms of meaningful, directed relationships or interactions between a single worker and those entities.

This figure includes only some potential entity and relationship types—many more definitions and division of any sociotechnical system are clearly possible—but it illustrates the inherent complexity associated with analyzing the wide range of social and organizational aspects in the work domain that makes this step of CWA particularly challenging. Fortunately, other disciplines (e.g., organizational psychology, social science, management science, cultural anthropology), have developed methods to study social interactions and structures that can be applied to

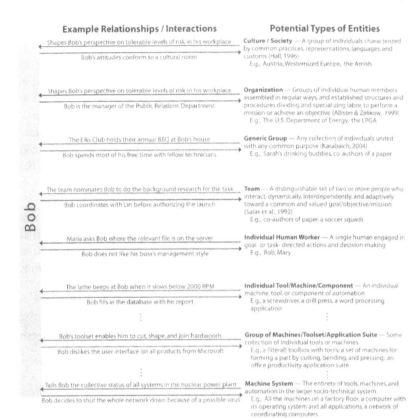

Figure 8.1 **Examples of relationship and entity types in a work domain.**

some, if not all, of the relationship types suggested in Figure 8.1. In this chapter, we present a subset of these methods for collecting and analyzing socioorganizational information, with particular attention to the advantages and disadvantages of each method with respect to their application within CWA. Our goal is to provide the CWA practitioner with tools (and knowledge of the limitations of those tools) to better understand the social and organizational aspects of work.

This chapter is organized around four main sections. The first section introduces basic assumptions used in the chapter and suggests the range of social and organizational analyses that could be relevant to CWA, particularly as we consider entity and relationship types that may be of interest in the work domain. The second section presents methods for capturing data about various relationship types, whereas the third section covers methods for processing that data into a form suitable for deeper analysis. The fourth section discusses specific analytic methods and their ability to support meaningful conclusions about the work domain.

Social and Organizational Analyses in Cognitive Work Analysis

Social and organizational analysis techniques can support CWA in many ways. To scope the discussion here, we must make certain assumptions. First, we assume that the practitioner has invested some time in performing earlier phases of the CWA process and has developed more than a cursory understanding of the work domain. This understanding is necessary because analysis of the work domain should lead to clear knowledge of which specific entities are of interest in the domain (and which are not), and which relationship types are likely to impact work performance. As recognized in the social sciences by Wellman and Berkowitz (1988),

> 'The Community'… is largely a matter of how analysts define ties, where they draw boundaries, and how they raise the level of analytic magnification in order to take into account internal links within clusters. (p.126)

To acknowledge the domain dependency of what constitutes entities and relationships, we also make the assumption that the practitioner is (or will be) comfortable with the inherently "soft" nature of some social and organization analysis techniques—that is, although these techniques may be founded in rigorous experimental or mathematical principles, they do not account for imprecision or lack of rigor in defining what constitutes a meaningful entity or relationship type (e.g., "Bob knows Juan" does not define precisely what "knows" means, yet this statement and statements like it could be used as the basis for a large and complex analysis of social structure). Therefore, the burden of defining meaningful and valid entities and relationships must fall upon the CWA practitioner as a function of the work domain being studied. In the following text, we describe an abstract set of entity types and relationships to drive our discussion of collection, processing, and analysis methods.

Entity Types

In Figure 8.1, we illustrate a number of potential entity types that may be of interest in the work domain, including individual human or machine entities, teams, groups, organizations, societies, and cultures that, together could constitute an entire sociotechnical system. We use *entity* in this chapter to refer to any one of these units of analysis (as well as others we have not stated explicitly, such as combined human–machine teams). With the exception of the individual human, these are rather broad characterizations that require some further refinement, particularly as we consider multiple entities. The specific entity types to consider in a particular CWA must be a function of the work domain and the scope of study.

For example, a "machine" entity could take many forms. We could define this entity as "automation" and refer to levels 2 through 10 of Parasuraman et al. (2000) as our definition. However, referring to levels of automation could restrict an analysis that may also be interested in using social and organizational analysis techniques to study relationships between entities and specific tools (e.g., many users may use a single screwdriver throughout the day, and the frequency of use and coordination of use of that decidedly nonautomated tool may be of critical interest in a CWA). Similarly, we could apply the five levels of system decomposition in the abstraction–decomposition hierarchy to specific levels for groups of machine entities (Vicente, 1999), but many more levels of abstraction may be needed. An analysis may study how various hierarchical software applications are used on different computers on different networks in different corporate locations by different user types. Should the analysis focus on one or all of these levels as "entities"? A CWA focused on a single user of an application suite may want to apply particular social and organizational analysis techniques at this level of analysis, whereas a CWA focused on coordination across sites may need to apply these same techniques at a much higher level.

Similarly, identifying a meaningful definition for more than one human in the domain is challenging. The literature on team performance defines a *team* as a "distinguishable set of two or more people who interact, dynamically, interdependently, and adaptively toward a common and valued goal/objective/mission" (Salas, Dickinson, Converse, & Tannenbaum, 1992). However, this taskcentric definition may miss other important components of a group, such as affective, social, cultural, or political dimensions that may also influence work performance. For example, a group may be defined by a set of entities sharing a common set of beliefs but different, independently executed tasks. And, their beliefs can impact work performance (e.g., a group of neo-Luddities who believe in rejection of computer tools may refuse to use those tools in their jobs—jobs that may include very different tasks, ranging from directing traffic to performing surgery). Other definitions of groups or social and organizational forms abound in the social and related sciences (Hudlicka, Karabaich, Pfautz, Jones, & Zacharias, 2004; Gupta & Hanges, 2004; Kontopoulos, 1993; Miles & Snow, 1986; McKelvey, 1982; Wellman & Leighton, 1979; Sherif, 1963; Bavelas, 1948). However, these definitions may be difficult to translate into practice (Karabaich, 2004) in a particular work domain because they are not specified relative to particular forms of work (e.g., defining a group as "nationalist" may not be instructive in understanding how an individual in the group uses a calculator). Some efforts, though, have been made to define broader cultural influences in terms that may support CWA (Klein & McHugh, 2005; Klein, 2005; Ross, 2004). Broad cultural or societal influences may be critical to consider in a particular domain (e.g., the design of signage on a defibrillator at an international airport), and therefore should be defined. We might need to consider societal and cultural attitudes toward work motivation and behavior (Bond & Smith, 1996), safety and risk, (Renn & Rohrmann, 2000), leadership (House,

Hanges, Javidan, Dorfman, & Gupta, 2004; Ilgen, LePine, & Hollenbock, 1997) and work-related values (Bond, 2002; Hofstede, 2001; Inglehart & Baker, 2000).

Hopefully, these examples show that the definition of the entity types in the work domain is a critical first step when performing social and organizational analyses. The burden is on the CWA practitioner to identify not only the particular entity types of interest (e.g., organization X) but also the assumptions made when identifying these entities (e.g., organization X is considered as a whole, although most interactions with the organization occur with the facilities department). Ideally, the earlier phases in the CWA process and the associated tools, forms, or representations will provide the insight necessary to identify the key entities in the analysis of social and organizational impacts on the work domain. Examples of identifying entities of interest from CWA tools are shown in Table 8.1.

Table 8.1   Examples of identifying entities of interest from CWA tools

CWA tool	Example entities that could be identified
Abstraction–Decomposition Hierarchy (ADH)	Vicente (1999) (pp. 257–265) suggests that different entities will interact with different parts of the work domain as represented in the ADH. The ADH does not explicitly contain a representation of the different entities, but as a trajectory through the ADH is described, each vector can be mapped onto the entity or entities involved (e.g., in a diagnosis, an expert may narrow the search for a problem as related to a particular function such as "the starter motor will not turn over," which a more junior technician could use to focus his or her subsequent strategy). Note also that an entity could also represent an intervention or support tool that provides similar guidance as to particular strategies or trajectories through ADH.
Decision Ladder (DL)	Vicente (1999) (pp. 265–269) also presents methods for mapping particular entities onto a DL. However, these methods do not support the identification of entities involved in a decision until after the CWA practitioner attempts to assign the elements of a decision to the entities in the work domain (e.g., these three people are involved in evaluating options for possible actions). As an artifact, the DL is useful for describing the decision paths, but is unlikely to be useful for identifying participating entities unless this is performed explicitly by the CWA practitioner (e.g., sharing the DL with a worker may reveal, "We don't do any of these deliberations; that is done by a person at a higher level in the organization.").

Relationship Types

As illustrated in Figure 8.1, the relationships among entities in a work domain can be characterized in a variety of ways. The identification of relevant relationship types in the work domain is, as in defining entity types, a challenge that depends on the understanding of the work domain obtained during earlier phases in the CWA process. For example, a characterization of key relationships will vary depending on whether the CWA is considering an individual operating a system, a team interacting to solve a problem, or an individual turning to an organization to request support. Specific tools or representations used in the CWA process could help to establish where the relationships of interest might come from, as shown in Table 8.2.

These examples, following Vicente (1999), are focused on the analysis of socio-organizational elements in CWA with respect to "content" and "form." "Content" refers to communication related to the coordination of work, whereas "form" refers to the organizational structure supporting that communication (pp. 254–256). Other characterizations, categorizations, ontologies, and taxonomies of relationship types abound, such as Fox, Barbuceanu, Gruninger, & Lin, (1998); Fiedler, Grover, & Teng, (1996); Knoke & Kulinski, (1982); Barnes, (1972); Gittler, (1951); Shirky, (2005); Monge & Contractor, (2003); Fleishman & Zaccaro, (1992). We use a categorization of relationship types here that we believe best supports a CWA

Table 8.2 Examples of identifying relationships of interest from CWA tools

CWA Tool	Example relationships that could be identified
Abstraction–Decomposition Hierarchy (ADH)	If we follow Vicente's examples, the traces of an activity in the work domain that cross boundaries between actors represent possible relationships of interest. As in the example in Table 8.1, relationships between entities could be defined as "informs about problem," "seeks guidance to narrow hypotheses," "provides guidance."
Decision Ladder (DL)	The DL provides an opportunity to identify particular classes of relationships among entities involved in a decision process. Each step in the decision ladder could represent a point of information sharing, collaboration, or consensus seeking (e.g., "Did we all see a gorilla?" or "We agree to execute plan B"), and therefore is potentially very useful for identifying these relationship types.
Information Flow Diagram (IFD)	Like the ADH, the IFD is useful for identifying relationships once the actors in the domain and their roles in executing a strategy have been defined (e.g., in this strategy, the inexperienced pilots attempt to execute the maneuver before handing over control to the experienced pilots).

practitioner approaching the wide variety of tools and techniques for the collection, processing, and analysis of social and organizational data. These categories are as follows:

- *Communication* or *Information Transfer Relationships* are interactions between entities that involve the passing of data, information, or knowledge influencing a specific goal-directed task (e.g., Bill told Rakesh that the number of people coming to dinner was "7").
- *Organizational* or *Coordination/Consensus Relationships* are interactions between entities that influence allocation and timing of tasks, define the structure and nature of communications between entities, and support group decision making (e.g., Ada and Bob agreed that they would count down "3, 2, 1" before lifting the heavy box together).
- *Ecological Relationships* are all other interactions between entities that influence performance in the work domain. They include affective (e.g., Aaron dislikes Pat and avoids communicating with him about work), cultural (e.g., train operators in country X have a higher tolerance for risky behavior), political, (e.g., organization X tries to suppress efforts to do similar work by organization Y), familial (e.g., Susan's mother-in-law runs the firm in which Susan is a department head), and even physical relationships (e.g., Elwood talks about his work most frequently with those who sit near him). Note that none of the tools referred to in Table 8.1 or Table 8.2 appear to explicitly support identifying these kinds of relationships.

These categories are not intended to be prescriptive but rather to support our discussion of the advantages and disadvantages of different methods for collecting, processing, and analyzing the potentially high number of types of interentity relationships. In Table 8.3, we provide examples of how questions that might arise in the course of CWA would be categorized according to this scheme.

As described in the following text, the tools and techniques available for collecting, processing, and analyzing data on relationships in a work domain are necessarily generic. They can be used for a wide variety of analyses, of which information flow and organizational structure should be considered only a subset. These techniques could be used creatively to study other features of the work domain or even within the tools used in the CWA process itself.

Methods for Capturing Data for Social and Organizational Analysis

Before any analysis of the social aspects of a work domain can be performed, data must be gathered about the nature of the interrelationships between actors in the

Table 8.3 Categorizing questions about social and organizational aspects of the work domain

Type of relationship	Example question
Communication or Information Flow Relationships	• Who communicates with whom in the work domain?
	• How often do communications occur? That are related to the work? That are organizational or social?
	• What is the pattern of communications relative to other events? To the introduction of a new process or technology?
	• What general type of data or information is communicated to assist in the completion of tasks?
	• How does the content of the message influence its communication?
	• What types of communication media are used?
	• How does the medium influence communication?
Organizational or Coordination/ Consensus Relationships	• Who coordinates with whom in the work domain?
	• Who participates in the work?
	• Who depends on whom to support their work?
	• Who is responsible for a task being completed?
	• What is the structure of the group responsible for a task?
	• What are the perceived organizational risks/ consequences associated with the work?
	• What specific information is communicated as part of trajectories through abstraction–decomposition space?
	• How does a group's structure impact intragroup communication?
	• Who distributes tasks?
	• Who coordinates activities of a group?
	• Who works to build consensus among entities?
	• What structure has emerged to cope with the work? With a new technology or process?

(continued)

Table 8.3   Categorizing questions about social and organizational aspects of the work domain (continued)

Type of relationship	Example question
Ecological Relationships	• What are the affective, familial, political, or cultural relationships likely to impact work?
	• Which of these relationship types impair, enhance, or otherwise transform processes for communication, coordination, cooperation, and/or consensus?
	• What organizational structures are likely to influence these relationship types?
	• Who shares similar beliefs and attitudes about work and about the organizational structures needed to accomplish particular tasks?
	• What personal affinities are present between workers in the domain?
	• What are likely sources of heterogeneity in perspectives about work and organizational structure?
	• What are the physical relationships that impact work performance?
	• Which group members are distributed versus colocated?
	• In what time zones are the collaborating entities?

domain. Like any data collection, the capture of data about relationships is subject to questions regarding completeness, objectivity, accuracy, precision, repeatability, reliability, and validity (perhaps even more so, depending on the relationship types to be defined). Furthermore, because we are considering relationships broadly (i.e., meaningful relationships among actors in a work domain), the applicability of particular methods will vary with the type of relationships to be captured (e.g., communication patterns, organizational reporting structures, interorganizational influence), as well as the type of entities participating in the relationship (e.g., individual humans, automation, organizations). Here, we discuss collection with respect to the three categories of relationships just described. We divide our discussion into methods that are based on the CWA practitioner and his or her associates performing the collection, and methods that are enabled by a variety of technologies.

Human-Based Data Capture

CWA fundamentally relies on observation and interviewing methods to develop an understanding of the work domain. These and other techniques are similarly used in the social sciences to capture data on social and organizational relationships. For example, the French national statistical institute conducted a study of types of interpersonal contacts among 5900 households, which used human-based interviewing techniques and questionnaires (Heran, 1987). One advantage of these approaches is that they can be implemented more rapidly than some technology-based approaches and may be easier for a less technology-savvy CWA practitioner. However, these methods tend to be labor-intensive when studying large and/or distributed organizations and can be subject to inherent biases in the humans involved in the capture of data (particularly when trying to capture data about affective, cultural, political, and other "softer" types of ecological relationships).

Observation

Observation of work is a comparatively straightforward method for capturing key problem-solving activities as part of CWA. Observation of work by a practitioner can be performed with or without knowledge of the subjects and is particularly useful for capturing work done in context. These forms of observation can include detailing the relationships between entities (e.g., advice from an expert is sought when the two primary diagnosis methods have failed), and supports, to some degree, the capture of information flow, and organizational and ecological relationships.

Observation is inherently subject to the limitations of the observer (e.g., errors in capturing a relationship, biases toward capturing a particular type of relationship), and depending on the domain, may be simply impossible (e.g., how would a single observer capture communications among geographically distributed workers? With only limited time, how can an observer capture all long-term affective relationships?). Webster (1994) noted this problem with observation-based data collection of social information, indicating that relationships of interest must be observable and of low-enough frequency to permit recording. However, from a practical point of view, observation may fit well with existing methods for analyzing the work domain. A CWA practitioner may already be sensitized to some of the nuances of the domain, and the effort required to specify and capture the key relationships may not be overwhelming. However, as the size or spatial distribution of the system being analyzed grows, more observers may be required. Furthermore, when deeper, more qualitative analysis is required (e.g., not simply counting number of communications between participants but studying the content of each message), the workload for the observer can easily become too much. While it may have the least validity because of observer bias, it may be most practical in time and cost to simply have an experienced CWA practitioner be aware of all relationships of interest (from simple information flow to large-scale sociocultural effects) and note these in the analysis. This has the added benefit of supporting other phases of the CWA process simultaneously. Again, this approach will not scale to large, distributed groups of entities, and the confidence in the results of any such observation should necessarily be viewed as being less than rigorous until substantiated with other observers' analyses, analytic methods, or evaluation.

Designing the metrics for the study beforehand with particular attention to the ability to accurately and reliably observe particular relationships is critical to using observational techniques. This may only be practical after some initial analysis of the work domain to anticipate likely relationship types that may be observed. For example, in a study of emergency department communications, researchers planned to capture communication events, along with the role of the communication partner, the mode, the location of the event, and any interruptions in the event (Fairbanks, Bisantz, & Sunm, 2008). In another study, flight crew communications were coded into uncertainty statements, action statements, acknowledgments, responses, planning statements, factual statements, and nontask-related statements (Bowers, Jentsch, Salas, & Braun, 1998). Entin (1999) suggests nine general categories to code team coordination, but these nine—information requests, action requests, action requests using a specific resource, coordination requests, information transfers, action transfers, action transfers using specific resources, coordination transfers, acknowledgments—may not be as appropriate to all work domains, and their lack of specificity may not serve the analysis well unless adapted (e.g., if the CWA practitioner is interested in a analyzing specific actions performed by an individual human using a number of different automated resources).

Observational approaches are also sensitive to the time, duration, and location of the collection and the participants being observed. For example, an observation of a community of windsurfers was conducted manually by researchers over the same 2-hour window every day for 31 days (Freeman, Freeman, & Michaelson, 1988). Although they did observe patterns in communication, it is probable that their observations were affected by the specific choices they made. They may have captured individuals' routine actions for the 2-hour window, but they ignored patterns of behavior that occur off the setting of the beach, and at other times. CWA methods for observation may include methods that ask a subject to talk aloud through a simulated situation (Pfautz & Roth, 2006; Roth, Gualtieri, Elm, & Potter, 2002). These methods can fail to capture many ecological relationships that a subject may be unable or uncomfortable expressing (e.g., "my boss makes my job much harder because he is quick to anger"). Ideal observation would maximize the realism of the observed situation while minimizing the impact of the observer on the situation. Using a human observer, on site in a real (not simulated) work environment, may result in different behaviors from the observed entities and could make the capture of sociopolitical, cultural, and affective relationships impossible.

Interviews and Questionnaires

The use of interviewing strategies and questionnaires should be familiar to the CWA practitioner, along with their advantages and disadvantages (e.g., they do not require direct observation where difficult/impossible, but rely on an entity's unknown ability to describe their own problem-solving activities). Tactics for the design of interviews and questionnaires for eliciting data specifically about social and organizational relationships, however, may be less familiar. Fortunately, the Social Network Analysis (SNA) community has been active in designing methods for the capture of certain relationship types, methods that can be extended to capture organizational and communication relationships and/or their impacts on work.

For example, the SNA literature describes methods for interviewing or designing questionnaires that involve indicating relationships (e.g., "talk frequently with"). These methods require specification of the relationships of interest (e.g., "work with," "talk on the phone with," "get help from"), and involve interviewing individuals to indicate (via free recall or by consulting a roster) with whom they have a relationship. An overview of these methods can be found in Marsden (2005) or Wasserman and Faust (1994). SNA methods are focused on the structure of relationships among entities, so these methods would require some effort to ensure that they are tied to specific aspects of how work is performed (e.g., a relationship might be defined as "consult with X to get information on problem Y"). Like any other effort to collect data via interviewing methods, the effects of the interview context (Bailey & Marsden, 1999) and the interviewer (Straits, 2000; Marsden, 1990; Groves & Magilavy, 1986) in collecting social and organizational relationship data require systematic control.

Another method from the SNA community is the use of structured diaries or surveys, in which individuals note the time, duration, and, in some cases, the type of content of their interactions with others. The validity of all of these methods is subject to the reliability with which the individual reports his or her relationships (Casciaro, 1998; Corman & Bradford, 1993). In our own (unpublished) experience collecting data in this manner, we found similar effects as described in the literature (Brewer, 2007; Brewer & Webster, 1999), with substantial changes in reported relationships after a 3-month period (a group of ~50 participants had a low level of recall of previous responses). Using roster-based methods can help reduce this variability somewhat, as can allowing a fixed number of responses and ranking/ordering of responses (e.g., "the top 3 people with whom I get confirmation on my flight plan are ...").

As we have stated throughout this chapter, defining the relationships that are meaningful in the work domain is critical to the application of any of these methods. SNA data capture methods are useful in that they provide a formalized approach to capturing relationship data. However, they do not provide a framework for relationship types. The interviewing and questionnaire methods mentioned here are also centered on the individual human and relationships among individuals as the unit of analysis. The application of these methods to understanding relationships between individuals and organizations, or intergroup relationships, is less well documented. Another issue with these methods is that they aim to define an atomic link between entities (e.g., "Person A knows Person B") that ignores the qualities or meta-information (Pfautz, Roth, Bisantz, Llinas, & Fouse, 2006) associated with the link (e.g., duration, strength, frequency). This meta-information may be useful to aggregate complex data (e.g., years of telephone records could yield "talks to" links only for entities who contact each other more than 10 times for more than 10 minutes each), but critical information about these qualities may get lost. For example, these methods would be less useful for understanding the specific time and duration (relative to a task) of communications between copilots and air traffic management personnel, but would be useful for identifying that "communication occurs" between these entities. This suggests that they are more applicable to identifying general organizational and consensus/cooperation relationships than specific communication or information flow relationships. They are generic enough to be applied to ecological relationships as well, although instruments for measuring affective and cultural relationships would need to be identified (e.g., Hofstede's (2001) approach to culture, the World Values Survey methodology (House et al., 2004), or see Plutchik and Kellerman's (1989) survey of methods for measuring affect).

Analysis of Existing Records

A final approach to the capture of relationship data involves the human examination of previously generated artifacts from interentity relationships (see the next section for technology-enabled approaches), such as shared whiteboards, group membership rosters, transcripts of radio communications, e-mail threads, and trial transcripts. This approach has parallels in CWA approaches where a practitioner will study prior documents regarding training, processes, and procedures in a particular work domain (e.g., an Army field manual describing steps in diagnosing engine problems). This approach has similar limitations—the availability of materials on past relationships may be inconsistent, difficult to obtain, and may only refer to one type of entity (e.g., a list of individuals at a company does not help to understand the company's interaction with other organizations). The interpretation of the information is subject to the biases of the human who processes and aggregates the information. When the volume of available information is large, human processing of the artifacts to extract relevant information becomes impractical, although some researchers have developed methods to allow practitioners to develop an initial classification of relationships that can be applied to the rest of the data with reasonably high reliability (Entin & Weil, 2006).

The ability of this approach to capture all relationship types at the desired level of resolution (i.e., individual, small group, large organization) varies fundamentally with the source documents. For example, in some domains, detailed studies of relationships impacting work performance may be available (e.g., cultural impacts on aircraft pilots (Mumaw & Holder, 2002), financial relationships between investment banks

(Podolny, 1993)); in others, these may have been impossible or impractical to record in any meaningful form. Some creativity may be required to find and interpret archival data (e.g., reading trial transcripts from a civil liability suit to understand different perceptions of a system failure), which makes this approach useful mostly in *a posteriori* analyses of a work domain. It is more desirable for a CWA practitioner to execute a structured study that will generate specific artifacts for a specific domain that are more amenable to analysis (e.g., using experts to provide subjective ratings during execution of a task, the relative degree of cooperation between two individuals).

Technology-Enabled Data Capture

Technology-enabled data capture methods have the potential to overcome the labor-intensive human-based approaches by reducing the cost of collecting data. Theoretically, significantly more data can be collected more consistently and systematically than with human-based methods, meaning that analyses of relationships based on these data could have substantially more scientific rigor. Technology-based approaches may mean that the data can be processed and analyzed on the fly, so that the CWA practitioner can observe patterns and trends of relationships in the work domain without hours of additional analysis. However, as we will see in the following text, technology-enabled data collection is not without its own costs (e.g., purchasing and setting up sensors, obtaining permissions to monitor communications), and the potential overhead of using such data collection methods may outweigh the benefits for some CWA practitioners.

Electronically Mediated Communication

As increasing amounts of communication among human and machine workers are carried through electronic media, they become easier to capture. E-mail, chat, and voice chat (telephone, VoIP) can be collected with the use of cameras, keystroke loggers, screen capture programs, and other software applications (e.g., server-based e-mail capture tools). In some cases, simply requesting logs of e-mail, chat, and even voice interactions is as simple as receiving a file from a system administrator whose tools auto-generate such data, complete with formatted data and metadata fields (e.g., e-mail content, subject line, author, addressees, time sent, time received). Even interactions such as swiping an ID card to gain entry into a conference room could be a rich source of data about the work domain. Although various tools and applications are available for collecting, processing, storing, and securing such information, the practitioner should be aware of the benefits and drawbacks of using these techniques.

Privacy concerns are among the most cited issues surrounding automated data capture methods. It is important to note current changes in legislation (e.g., the Federal Trade Commission is the U.S. government agency charged with protecting the privacy of individuals), as the use of data mining technology can be significantly impacted by legal issues related to privacy. Most privacy advocates want to ensure that large databases of information are properly and legitimately used (Markoff, 2002). Existing software programs and algorithms can help database administrators maintain anonymity and encrypt potentially sensitive information. Although researchers should have a cursory understanding of these laws and requirements, in most cases it is necessary to oversee the protection and use of data (e.g., understand risks and alternatives, remain transparent, and clearly state how the data will be used to those people whose personal information may be captured within a database), as per Institutional Review Board regulations.

Observability is also a factor to consider with automated data capture. Although collecting electronically mediated communication may quickly uncover trends and patterns regarding information transfer, it is limited in its ability to describe more complex ecological aspects. For example, we may know that Joe sent an e-mail to Steve, but there is no way to know what other tasks they were performing at the same time by just looking at the e-mail data. Similarly, if Joe walked to Steve's office to discuss a task and achieve some consensus on an approach to solving it, this would not be observed. Obviously, the degree to which electronic communication is used in the domain fundamentally affects the value of capturing this data. This suggests that, in some domains, electronic communications should be used in conjunction with other methods of data capture. Electronic communications do have the benefit of providing potentially high-resolution data in which a human observer may classify an interaction as "Alice knows Billy"; having the full text of the interaction may allow for additional subtleties to be analyzed (e.g., "Alice knows Billy from a brief meeting at a conference, which was followed up by an exchange of 5 increasingly friendly and casual e-mails to set up another meeting").

Although there may be a large amount of data available from capturing electronically mediated communication, further processing may be needed depending on the type of relationships being studied. That is, this data on its own is unlikely to provide the kinds of insight that could be achieved with a CWA practitioner observing, in situ, a work domain. It may not be correlated in any way with the tasks performed in the domain, meaning that additional work must be done to establish the correlation, either via human or technological solutions.

Commercially available data capture tools may require license fees, system installation and maintenance expertise, as well as training. It is highly likely that additional processing steps will be needed before any substantial analysis can be done on the data (see the following section on methods for processing social and organizational data), particularly for cases in which communications data are being analyzed not only for information flow properties (e.g., who communicated with whom, when, for how long) but also for content that would illuminate consensus and coordination strategies (e.g., Aaron's e-mail requests confirmation from Bob that the coordinates are correct before they launch the weather balloon).

The ability of this form of data capture to support the relationship types described depends largely on the observability of the desired relationships in electronic communications. Some relationship types may fail to appear in the data (Kiesler, Siegal, & McGuire, 1988), and others may require processing steps to extract those relationships. Although the underlying data would have high reliability and validity for a particular work domain, the processes used to transform this data into units for analysis may use heuristics that reduce the validity of the data (e.g., this tool assumes that any conversation that exceeds a certain level decibel indicates an "angry" interaction).

Human Activity Monitoring

A key issue with the aforementioned methods for capturing data on interactions among entities is that not all interactions of interest are electronically mediated. Depending on the domain, a great deal of interaction between humans may occur face to face (e.g., Worker A talked with Worker B about the current heavy workload and perceptions of Worker C's productivity). Similarly, it may not be possible to capture some physical relationships in the work domain (e.g., Worker A stood at the workbench for an hour to assemble couplers, then took a break by walking to Worker B's workbench). Both of these relationship types would be difficult to capture without a human observer present to take note and characterize the interactions between the entities.

Advances in human activity monitoring technologies enable capture of this kind of data without a human observer. In work domains where computer use is of interest, open-source and commercial applications are available that can collect the user's activity at various levels of resolution (e.g., the user used PowerPoint, the user browsed technical manual 54.2-9 from 8:32 am to 9:14 am). These applications are geared towards business owners who want (ethically or not) to monitor their employee's activity, although other products are available for human factors purposes (e.g., TechSmith's Morae software or other open-source system-specific alternatives). This activity monitoring differs from examining logs of electronically mediated communication (as described earlier) in that the relationships of interest here are not human–human but rather human–machine. However, this approach is only useful for work domains where the machines are instrumented to report how they are used (i.e., it is useful for computer-based activities or with tools that are instrumented to record use as with some factory workstations).

Another approach is to use computer vision techniques to identify patterns of activity in a work environment (Gavrila, 1999; Essa, 1999). Computer vision techniques, especially in highly constrained, known environments (e.g., a factory floor with specific workstations) can segment and track motion in a video stream and identify humans versus other moving objects (e.g., Madabhushi & Aggarwal

(1999) identified nine simple activities like sitting down, standing up, hugging, and walking in an office environment; others have done similar work (Bodor, Jackson, Masoud, & Papanikolopoulos, 2003; Masoud & Papanikolopoulos, 2003; Ayers & Shah, 2001). More sophisticated techniques for learning behavioral patterns have been used for detecting, meeting and greeting behaviors for pedestrians crossing a square (Oliver, Rosario, & Pentland, 2000) and certain office behaviors (Oliver & Horvitz, 2005). The use of computer vision techniques can be useful for collecting or inferring certain relationship types, given a situation where a video camera can be placed with a clear view of the work environment and with consent and awareness of privacy concerns. However, because these methods are still largely in the research domain with no known commercial products available, they may represent an approach to capturing certain relationship types that is still too challenging to implement.

Yet another approach relies on the use of different forms of sensors. In the computer-vision-based techniques, an emplaced electrooptical (i.e., video) sensor provides data on human activity. More simple sensors can be used (e.g., ultrasonic radar systems; Nishida, Murakami, Hori, & Mizoguchi, 2004), by which an individual's position in the room could be correlated with the known location of tools such as computer workstations or other workers' desks. There are a large number of techniques that use body-worn sensors of various forms (e.g., ID and RFID tags Smith et al., 2005; Want, Hopper, Falcao, & Gibbons, 1992), accelerometers (Kern, Schiele, & Schmidt, 2003), microphones (Lukowicz et al., 2004), or sensor platforms that combine all of these (Farry, 2006; Choudhury & Pentland, 2004; Choudhury & Pentland, 2002). These systems can identify certain characteristics of human activity (e.g., two people are talking in a conference room, a group is being lectured by an individual) that can be used to identify patterns of relationships without observer biases or observee sensitivities. Unfortunately, like the computer vision techniques, these methods can be quite technically sophisticated. More practical approaches could involve very simple sensors capturing data at known points in a constrained environment (e.g., use a microphone to detect how often a conference room gets used), and may be more useful for capturing information on interentity communication (particularly human–human) than organizational relationships.

Although these methods have the potential to transform the capture of data for studying human interactions with each other and supporting technologies, they currently require a high degree of technical sophistication to customize for a domain, and few off-the-shelf products are available. For a CWA practitioner, it may not be practical (for both time and cost reasons) to employ such techniques. However, in work domains where existing sensors may already be emplaced and accessible, there may be value in using some of these techniques as they can capture a large amount of data on relationships with little human labor.

Modeling and Simulation

An alternative approach to capturing data on social and organizational relation-ships is to use modeling and simulation (M&S) technologies. By creating (or using existing) models of human and machine behavior, a large amount of data can be generated to analyze the impacts of alternative organizational structures, new tech-nologies, and decision-making models, especially at very large scales. For example, many human-in-the-loop and agent-based simulations support research into the concepts of organizational congruence, complex communication tasks, and situ-ation monitoring by individuals and teams (e.g., Kleinman, Levchuk, Hutchins, & Kemple, 2003; Diedrich et al., 2002). The results of such experiments can often provide data that can be used to identify key relationships and parameters between human–machine systems. Simulations are generally abstractions of the real world, using quantitative models implemented on a computer to represent sociotechni-cal systems with varying degrees of fidelity. There are many types of models and simulations that could be used to generate data about social and organizational relationships, including human-in-the-loop simulations, models of human-system performance, cognitive models, and virtual environments. A key benefit of M&S is that work performance can be simulated (or performed by a small set of humans in specific roles), allowing a CWA practitioner to immediately correlate work perfor-mance with interactions between entities.

Human-in-the-loop simulations, such as the Dynamic Distributed Decision-making environment (Kleinman, Luh, Pattipati, & Serfaty, 1992), can generate data about both human performance and its relationship to interactions with technology, operations, and organizations. They have the advantage of providing real human data in addition to the simulated data (e.g., roles could be modeled to simulate interaction with the role of interest, which would be filled by a human). An alternative to human-in-the-loop simulation uses synthetic agents to replace all human entities in experiments. Synthetic agents can range from task-network models (Allender, 2000) to full cognitive models like ACT-R (Anderson et al., 2004), SAMPLE (Harper, Mulgund, Zacharias, & Menke, 2000), and SOAR (Laird, 1987), which simulate the sensory, perceptual, cognitive, or motor behavior of humans. The interested reader should see Pew and Mavor (1998) or Gluck and Pew (2005) for more details on these techniques.

Both of these approaches—human-in-the-loop and synthetic agent simula-tions—have limitations. Large-scale experiments with human teams are expensive and time consuming. Therefore, a very limited number are conducted, mainly test-ing incremental changes rather than revolutionary ones. Current synthetic simula-tions capture human behavior at various levels of fidelity, so their results may also vary in validity. These simulations can be used to study team decision making, and the results can be quite beneficial (Baker et al., 2004; Kang, Waisel, & Wal-lace, 1998; Serfaty & Entin, 1995). However, it is difficult to predict if a limited number of human-in-the-loop experiments or simplified synthetic environments can generate the data required to study more complex social and organizational relationships.

Virtual environments can support massively multiplayer (many humans-in-the-loop) simulations, creating situations that exercise the same skills of critical thinking, planning, communication, and coordination that are required for successful performance in real operational environments. Although their physical models can vary, some environments possess adequate "cognitive fidelity" to provide useful experiences that can be used to study complex human interactions (Haimson & Lovell, 2006; Freeman et al., 2006; Singley & Anderson, 1989). It is difficult to maintain a constant global awareness of all activities that occur during a simulation; therefore, evaluating individuals or organizations can pose a considerable challenge. Luckily, virtual environments often maintain comprehensive records of events that take place inside their simulated worlds. This makes it possible to use system data to support data processing and analysis. Raw lists of actions executed during a scenario are unlikely to be very useful in this regard, however. Searching through a simulator data stream or log to identify important events for review can be an exceedingly difficult and time-consuming process. The high volume of data and low level of granularity at which these data are presented make it difficult to identify critical incidents (e.g., complex social interactions, important communications relating to work tasks) and to understand their implications through unaided human inspection alone. Automated mechanisms can help to sift through logged data to synthesize it into a more useable form. Intelligent behavior recognition technology, for example, can extract higher-level interpretations from clusters of lower level actions and events captured within a simulator data stream (Jackson & Moulinier, 2002; Allen, 1995).

The clear problem with using any M&S technique to generate data for studying social and organizational relationships is validity. Although M&S can be used to generate large amounts of data on information flow, consensus/cooperation, and many ecological relationships, the fidelity and validity of the models used dictate the degree to which any data collected has merit. M&S also allows for work performance to be simulated and correlated to various relationships and patterns of relationships that, while a benefit, still suffers from questions of validity (e.g., How well does the model represent when an entity will request a particular kind of information?). Given that a CWA practitioner studying a domain is likely to research existing models of human performance in that domain, the degree of validity of those models may be assessed, and the use of M&S for generating data may be explored.

Conclusions and Recommendations on Capturing Social and Organizational Data

We have presented a number of methods for capturing data on social and organizational relationships. Each of these has various advantages and disadvantages.

We suggest that a CWA practitioner should consider these based on the following criteria:

- How much time is available for studying the relationships in the domain?
- What level of funding is present to fund the study?
- How many types of entities and relationships need to be considered?
- Which types of entities and relationships need to be analyzed?
  - How observable are those entities and relationships?
  - How reliably do those relationships and their attributes need to be captured?
  - At what level of resolution does data need to be captured?
- How many total entities are being considered? How many relationships are there among those entities?
- To what degree has analysis already been performed? Are there existing data sets or models?
- How viable is the particular collection method given issues of permission and privacy?
- What is the desired level of validity of the analysis?
- How technology-friendly is the CWA practitioner?
- What form of analysis is planned for the data?

Answering these questions while considering the work domain and the just-described collection methods should provide the practitioner with an understanding of the trade-offs inherent in trying to capture social and organizational data.

Methods for Processing Social and Organizational Data

Data processing is often necessary to encode information into a form that can be analyzed. Clearly, the manual processing of data is time consuming and tedious when processing anything but the simplest data sets, and any data processing is necessarily a function of the existing form of the data and the type of analysis to be performed on the processed data. Information processing often begins with the structuring of data—normally accomplished using file structures, databases, and/or the incorporation of meta-data about the files and databases (e.g., pedigree: "these are all messages that came from the chain of command at Agency X"; provenance: "these are all the e-mails from James"; quality: "these phone records are noisy and need to be cleaned up," or "these chat records contain large amounts of misspellings"; reliability: "these data have large holes in time when our human observer went home sick"). Several tools and methods focus on automatic and user-assisted annotation of structured (e.g., a tab-delimited spreadsheet with labeled rows and columns), semistructured (e.g., a log of a chat session that has structures showing who has messaged when, but no structures for the content of the message), and unstructured data (e.g., the prose content of an e-mail message).

The simplest processing methods directly map existing data types into formats required for analysis (e.g., transpose a table showing the time a communication occurred between two entities). These are less interesting in that most CWA practitioners are likely to be familiar with the need for this type of data manipulation. More interesting yet still relatively simple processing methods include search and excerpt extraction. Advanced search is the process of applying a selection criteria (e.g., finding all communications from managers to subordinates) to data to identify relevant information. The result is a set of data ranked by its relevance to the selection criteria. Excerpt extraction is the process of identifying a subset of the data that directly relates to the topic of interest. For example, a blog may contain numerous topics; some of these topics will be relevant to the topic of interest, some may not, and there are varying degrees of relevance (Llora, Yasui, Weige, & Goldberg, 2006). Similarly, data can be organized and classified according to the entities involved in the interaction (a necessary step for some analysis techniques we discuss in the following text). For example, such processing could include queries and searches to "find all verb relationships between Bob and Jane." It may also be useful to use tools to automatically extract entities from the data if these are not known beforehand (e.g., a tool could identify "James" or "Server X" as an entity automatically in a data set, without knowing a priori that "James" or "Server" is an entity that is of interest). Multiple tools are available both as commercial and open-source products (e.g., ANNIE, Balie, Named Entity Tagger, Inxight's SmartDiscovery server, Basis Technology's Rosette). This may be a particularly useful technique when roles are known but individuals are not, and the analysis is focused on understanding how entities in particular roles interact with other entities (e.g., How do intelligence analysts distribute requests for information to the humans and systems responsible for collecting that information?).

More complex processing methods have their roots in the field of Natural Language Processing (NLP) (Allen, 1995). NLP is commonly used to describe the function of software or hardware components in a computer system that analyzes or synthesizes spoken or written language and aims to automate the generation and understanding of natural human language (Feinman, 2006; Dzindolet & Pierce, 2006; Feinman & Alterman, 2003). Practitioners should be aware that different work domains have slight variations in how they categorize the diverse computational linguistic methods. In general, data mining, text analytics, and latent semantic analysis are methods for processing social and organizational data that have gained popularity among researchers trying to study and predict human behavior at individual and group levels by examining a wide variety of largely unstructured data (e.g., Group A's intergroup chat records have key word indicators that a high level of consensus was achieved [Yang, 1999; Joachims, 1998; Lewis, 1991]).

Data mining involves finding "interesting" structure in data (either by following some preconceived notion of what is "interesting" or simply by finding patterns)

(Han & Kamber, 2007). In a broad sense, interesting structure includes statistical patterns, predictive models, and/or hidden relationships, and data that are unusual, previously unknown, and/or result in specific entity actions or behaviors. Data mining supports exploratory analysis of large data sets. This is particularly useful when the types of entities or relationships of interest in the work domain may be unknown or only partially known by the CWA practitioner. Ideally, data-mining tools are used in conjunction with analysis techniques that dictate the form of the required output data. Often, these exploratory data analysis techniques incorporate tools to allow an analyst to interact with visual representations of the data, although these "visual analytic" (Voinea & Telea, 2007; May & Baddeley, 2006) methods work best for relatively small, low-dimensional data sets. Again, these tools might be useful in early phases of a CWA when a practitioner may be struggling to identify the key entities in the work domain and the types of relationships requiring deeper analysis.

Text analytics methods (Karanikas & Theodoulidis, 2002) support the automatic extraction of structured or semistructured information from unstructured data (e.g., text analytic methods could be used to pull all references to dates, in various formats, from an e-mail; Boschee, Weischedel, & Zamanian, 2005). Text analytics include specific algorithms for common classification and retrieval problems such as entity recognition (ER), coreference, and relationship extraction. ER is useful for classifying elements in text into categories, such as people, places, and organizations (Moens, 2006; Grishman, 1997). ER algorithms normally create tagged instances of data from unstructured data (as shown in Table 8.4).

Coreference algorithms identify chains of nouns that refer to the same entity or object (Gaizauskas & Humphreys, 1999). In "I saw Lewis yesterday. He was at the Red Sox game," *Lewis* and *he* probably refer to the same person. Finally, relationship extraction identifies named relationships between entities in text. For example, *Steve Jobs* (a person) *is the CEO of* (a relationship) *Apple* (an organization). Clearly, these methods are extremely relevant to processing unstructured data, particularly data obtained from electronically mediated communication between enti-

Table 8.4 Unstructured data processed into structured data by an ER algorithm

Unstructured data	*Structured data*
David from Widgets, Inc. met with Sally on October 3 in their Boston, U.S., office	`<person>` David `</person>`
	`<organization>` Widgets, Inc. `</organization>`
	`<person>` Sally `</person>`
	`<date>` October 3 `</date>`
	`<city>` Boston `</city>`
	`<country>` U.S. `</country>`

ties. Latent Semantic Analysis (LSA) is another NLP technique. LSA focuses on the frequency and use of terms within a large corpus of text. It can be used to obtain estimates of the contextual usage of words and similarities in meaning or as a model of the computational processes and representation of underlying knowledge acquisition and utilization (Landauer, 1998).

This is just a small sample of the vast amount of text analytics research currently in progress. Even the simplest of these processing methods have particular shortcomings. For example, human transcription of voice records is likely to result in errors even if methods to simplify the process into identifying classes of interentity relationships have been instituted, or methods for extrapolating from these classifications have been used (Entin & Weil, 2006). Using any of the described techniques to transform raw data on a relationship (or to infer or aggregate relationships) introduces assumptions and therefore impacts the validity of an analysis. We would argue that processing to identify and characterize patterns of communication is least likely to introduce problems into an analysis, whereas using methods to characterize the content of interaction is more likely to be fraught with problems. Some ecological relationships could be easily inferred from raw data (e.g., distributed workers are more likely to use electronically mediated communication methods than colocated workers), whereas others should only be inferred with careful documentation of assumptions (e.g., using LSA to determine the "tone" of a conversation in which a group reaches consensus on a decision). The evaluation of the effectiveness of these processing methods is an important research goal to raise confidence for operational use.

Methods for Analyzing Social and Organizational Data

The CWA practitioner is likely to be familiar with statistical methods to summarize and analyze a data set—methods such as sampling, standard deviation, probability distribution, or frequency analysis. Standard methods, such as estimation, correlation, regression, and analysis of variance, can all be used to describe patterns in data. Although these statistical methods are undoubtedly applicable to aspects of analyzing social and organizational data, we focus here on methods that are less likely to be familiar to a CWA practitioner and come from research communities focused on understanding, describing, and analyzing interentity relationships. Specifically, we present methods used by social scientists, organizational psychologists, mathematical psychologists, and others interested in the study of human social and organizational behavior, namely *Social Network Analysis* and *Link Analysis*. We consider these methods somewhat generically while providing example uses, as the work domain and scope of the CWA determine which methods are used and at what level of sophistication.

Social Network Analysis

In the earlier section on interviewing and questionnaires, we introduced methods from the SNA community for capturing data about relationships. SNA is more broadly focused on the analysis of those relationships to draw conclusions about the nature of social structure (for a full history of SNA see Freeman's *The Development of Social Network Analysis* Freeman, 2004). SNA balances two competing perspectives of social and organizational structure—*individualism*, where individual entities act and achieve goals according to their interests, and *holism*, where individual entities act strictly according to the social structure in which they exist. However, at the core of SNA studies is a focus on the structure of interactions between entities, which allows us to avoid this philosophical issue and instead understand how analytic techniques used in SNA, regardless of beliefs about individualism or holism, may be applied to better understanding a work domain as part of CWA.

Social Network Analysis Methods

SNA relies on graph-theoretic methods for analyzing a network of nodes and links (i.e., entities and relationships). Graph theory is a branch of mathematics that supports the description of network properties in a highly formal manner. Social scientists use these graph-theoretic methods and apply them to relationships among entities (typically individuals) to draw conclusions about social structure. It has been applied to a wide variety of domains, including the study of interpersonal (e.g., social, familial) relationships, communication networks, and organizational structure (Monge et al., 2003; Scott, 2000; Degenne & Forse, 1999; Wasserman et al., 1994).

In the following text, we describe the main methods used, along with the potential pitfalls of applying each method in CWA, to understand information flow, and organizational and ecological relationships. A critical aspect of applying these methods, one that gets particularly little attention in the SNA literature, is the importance of carefully defining what type of entity and what type of relationship is being considered. Particular methods are described with terms (e.g., "prestige") that imply a meaning that may be very different from what is actually described by the type of relationship. All of the methods presented in the following text are straightforward to replicate with the aid of one of the better formal SNA texts (Mokken, 1979) and therefore we have omitted relevant formulae for brevity.

First, a CWA practitioner should understand the basic components of SNA. SNA typically begins with one entity type and one relationship type, described in a *sociomatrix* (or *adjacency matrix* in Graph Theory terminology), as illustrated in Figure 8.2a. This matrix describes a graph where a link is present for each "1" entry in the sociomatrix, as shown in Figure 8.2b. A sociomatrix could be generated for each type of relationship of interest in the domain, and across the set of entities in

	Alice	Bob	Carlos	Dot	Emile
Alice	0	1	0	0	0
Bob	1	0	0	0	0
Carlos	1	0	0	1	1
Dot	0	1	1	0	0
Emile	0	0	1	1	0

A

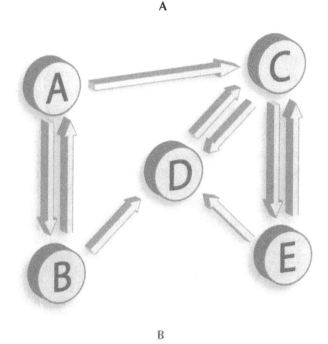

B

**Figure 8.2   An example sociomatrix defining (A) which entities have a "commu-
nicates with" relationship and (B) the corresponding social network.**

the domain. A valued sociomatrix can be created to provide values for the links
(e.g., to represent "strength" or "frequency").

Once data is in the form of a sociomatrix, a number of techniques can be
applied. The most trivial of these is *Degree.* Degree describes the number of links
to an entity. *In-degree* refers to the number of incoming links, whereas *out-degree*
refers to the number of outgoing links. Simply counting the number of links could
be a useful analytic technique in relatively simple cases (e.g., Alfred interacts with

more people in performing his managerial task than Roberto). Another simple technique is *reachability*, which states that two entities are reachable if there is a path between them. This may be useful for the analysis of expected communication pathways (e.g., does Ralph actually consult governing organization X before filing a flight plan?), or for determining the number of steps necessary for two entities to connect (e.g., How many people have to pass along a request for information before it gets to the desired expert?).

However, these are comparatively simple techniques. A more complex concept in SNA is *centrality*, which attempts to describe the "importance" or "prominence" (or, conversely, the "isolation") of an entity within a network. Of course, the semantics of such terms can seem inverted depending on the type of link being considered (e.g., if the links are "dislikes," the most "important" person is the most disliked!). Some of these methods work with undirected graphs, whereas some require directed graphs. Similarly, many of these methods can be applied to a group of nodes to infer group centrality, rather than node centrality (although a node could be defined as a group or an individual, depending on the types of entities chosen for analysis). In Table 8.5, we present the main types of centrality measures with a brief description.

The following figures illustrate these methods using an example network shown in Figure 8.3.

Figure 8.4 shows the application of two centrality measures to the sample network shown in Figure 8.3. Figure 8.4a illustrates how the degree centrality method finds the nodes with the greatest number of linkages (the white node near the

Table 8.5   Centrality and related SNA methods

Method	Description	Examples (with Relationship Type underlined)
Degree Centrality	Measures the number of immediately adjacent entities relative to the total number of entities	In the command center, Joe talks to more people than anyone else. The Elks club has the most financial ties to other organizations.
Closeness Centrality	Measures proximity (in number of hops) of all other entities	Jack can communicate with the most people with the fewest intermediaries, meaning his tasks often get spread through the group first. The Organization X sends data to the largest number of other organizations using the fewest hops. Critical information should be sent to Organization X first to disseminate it most quickly.

(continued)

Table 8.5   Centrality and related SNA methods

Method	Description	Examples (with Relationship Type underlined)
Betweenness Centrality	Measures the proportion of paths on which a particular entity lies	Fabian <u>interacts with</u> people in both departments, meaning that he often carries interdepartmental news. Classified information always has to <u>pass through</u> Agency Q before being exchanged between organizations, so Agency Q may have unique insight into the difference between organizations. Agency Q can also act as a bottleneck on all interorganization information flow.
Eigenvector Centrality	Measures centrality by considering the relative centrality of adjacent entities	All the department heads <u>receive tasking</u> from Ronaldo and pass that tasking to their departments. The department heads are very central, but Ronaldo has higher eigenvector centrality because he is only close to others who are central. Because the City Hospital <u>sends patient information</u> with other branch hospitals that themselves send that information to local and rural hospitals, it could disseminate the most patient information the quickest if a disease outbreak was suspected.
Status or Prestige	In directed graphs, *degree prestige* measures the number of entities with links pointing to an entity In directed graphs, *rank prestige* measures the number of entities with links pointing to an entity but incorporates the corresponding rank of those entities	Joanie is considered the most popular girl in school because the most students <u>like</u> her. Ralphie is considered the most popular boy in school because all the other popular kids <u>like</u> him. Charity Y <u>receives</u> the most <u>funding</u>. Server Z is most likely to have the information because it <u>is sent information</u> from the largest number of other servers that also receive lots of information.

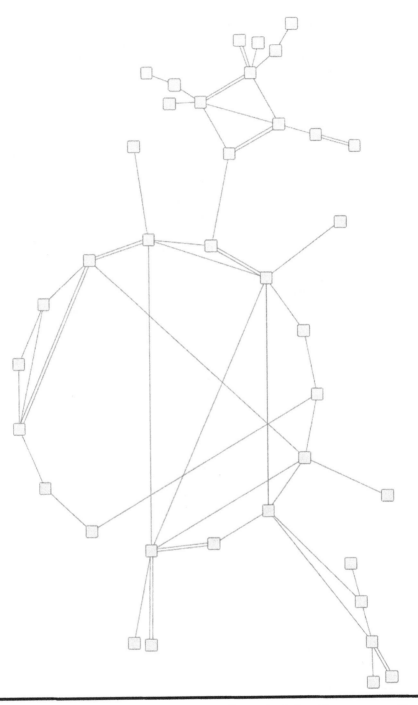

Figure 8.3   An example network showing undirected relationships among entities.

bottom of the figure has eight links). This method could be used to find the "most linked" entity (e.g., this organization uses most of the communication bandwidth, or this entity talks to the largest number of other people and might be a worthwhile starting point for interview-based analysis). Figure 8.4b shows how closeness centrality identifies the entities that are the shortest distances to other entities. For example, this method could be used to locate an organization most likely to find out about an event or most able to disseminate information quickly.

Figure 8.5 shows the application of two centrality measures to the sample network shown in Figure 8.3. Figure 8.5a shows how betweenness centrality identifies the entity that lies on the most paths between other entities. The whitest node connects the smaller upper graph with the larger lower graph, meaning that it is most likely to lie on the most intermediate paths. This form of centrality can help identify "critical links" in a network (e.g., potential bottlenecks in a communication system, the potential for information overload). Figure 8.5b illustrates eigenvector centrality, where the relative centrality of adjacent entities is taken into account. The larger and whiter entities are those that are connected to the largest number of other well-connected nodes. This form of centrality could help identify "key players" in a network (e.g., the members of a team who communicate the most while considering alternative decisions). In each example shown, a different entity is selected as the "most central," even though the underlying network remains the same, which suggests that some effort is required to understand the details and intended use of these methods before application in analysis.

Figure 8.6a shows a version of the network shown in Figure 8.3 with directional links (e.g., while a "communicates" link may be bidirectional, a "likes" link is unidirectional). This directional graph allows us to illustrate, in Figure 8.6b, the computation of status. In this example, the largest and whitest node is the one with the most incoming links coming from nodes that themselves have many incoming links. This method can be useful for identifying hierarchies in a network (e.g., many people turn to their department heads for advice, and the department heads tend to consult each other and their branch chief for advice). As noted earlier, some care should be taken with the use of the status method, as the relationship type can confuse the meaning of "status" (e.g., many people dislike their bosses, who in turn dislike senior management and the CEO; the CEO would have the greatest "status" in this network).

Another set of SNA methods describe the *balance* of a network, particularly as it relates to shared attributes among entities. Balance measures the degree to which subgroups of entities in the network have similarly valued links (e.g., entities are linked by likes/dislike links, or all entities are linked by "believes work should be done collaboratively," with the positive or negative values associated with those links). If a subgroup of entities shares a set of values (e.g., they all dislike all the people in that group), then the graph is said to be balanced. This method was designed to try to measure "psychological tension," but could be applied to identify potential affective/political issues within a group.

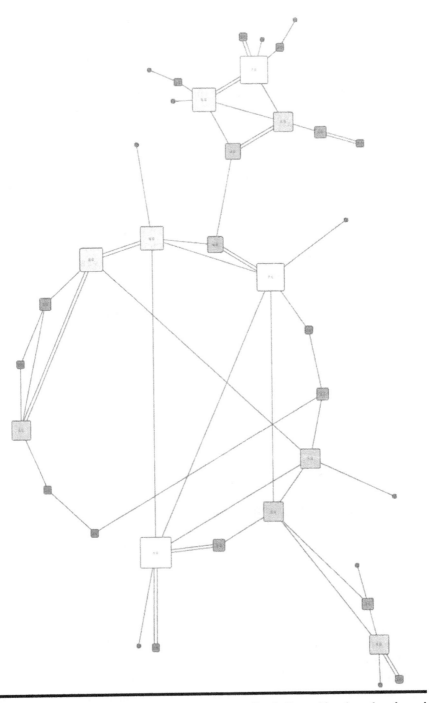

Figure 8.4a Example network with degree centrality indicated by size of node and color (whiter and larger indicates higher centrality).

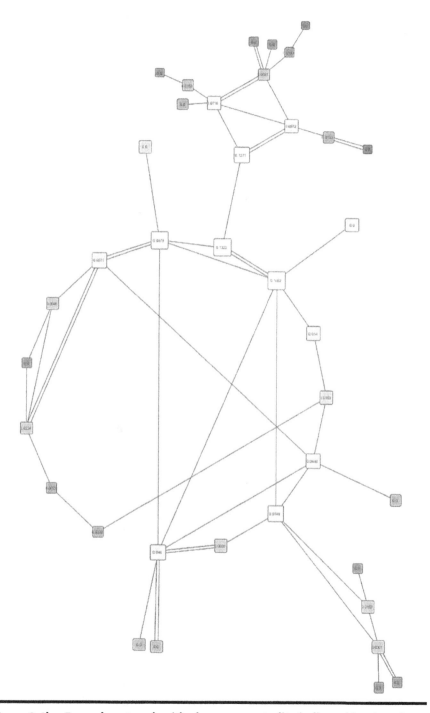

Figure 8.4b   Example network with closeness centrality indicated by size of node and color (whiter and larger indicates higher centrality).

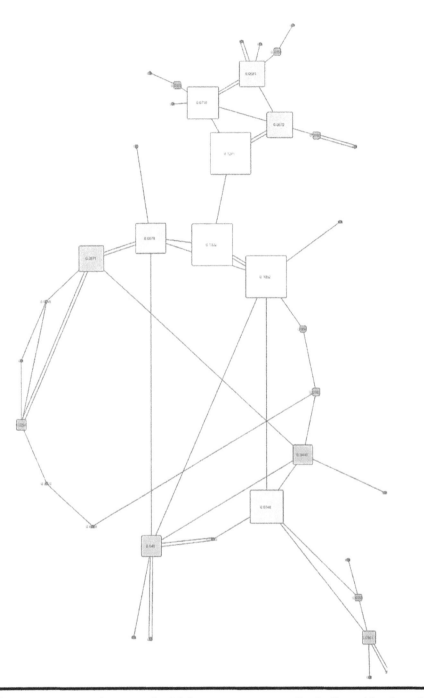

Figure 8.5a **Example network with betweenness centrality indicated by size of node and color (whiter and larger indicates higher centrality).**

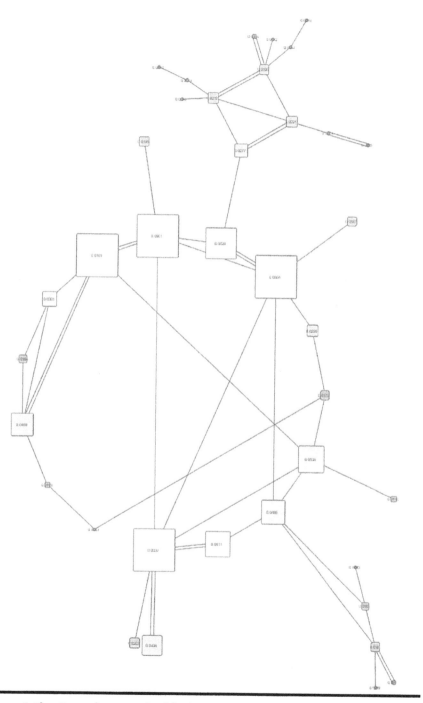

Figure 8.5b   Example network with eigenvector centrality indicated by size of node and color (white and larger indicates higher centrality).

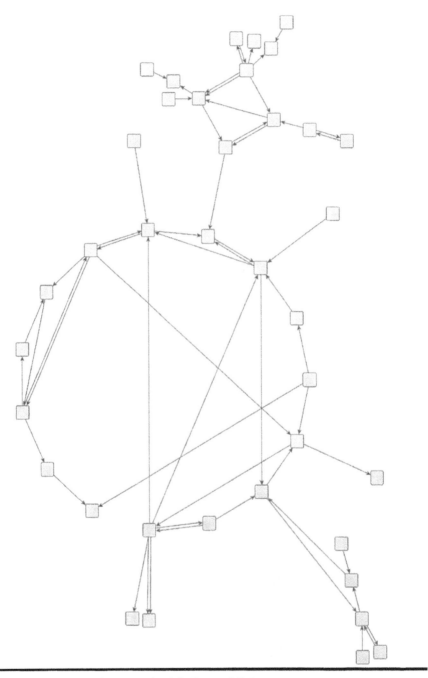

Figure 8.6a   Example network with directed links.

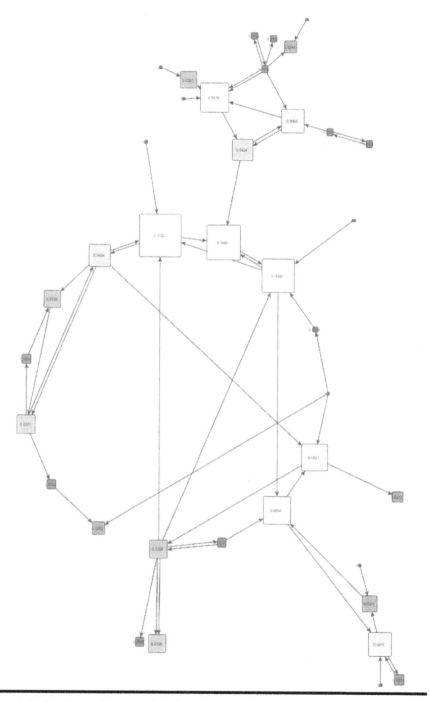

Figure 8.6b   Example network indicating status by size of node and color (whiter and larger indicates higher centrality).

Cohesion, in the terms of SNA, describes methods for finding subgroups within a network that are particularly interconnected. These subgroups are said to have "strong" or "frequent" interconnections (or sometimes, "positive" or "intense" interconnections). Identifying subgroups that are highly linked is a goal of SNA methods that identify *cliques* and *clans*. Cliques (or *n-cliques*) describe a group in which the minimum distance between nodes is *n*. A clan (or *n-clan*) is an n-clique in which the diameter of the group (the maximal distance from any one node to any other) is *n*. These methods could be useful for identifying particular patterns of communication (e.g., although these people are in different departments, they interact frequently; perhaps this is because they work on similar problems), or particular affinities among entities (e.g., this group uses lots of different computer resources; perhaps they are less adverse to automated processes). There are a number of other related methods for identifying cohesive subgroups within the SNA literature (e.g., *n-clubs* (Seidman, 1983), *k-cores* (Seidman & Foster, 1978), *k-plexes* (Borgatti & Everett, 1992; Burt, 1976; White, Boorman, & Breiger, 1976; Lorrain & White, 1971)), in part because of the range of definitions of what constitutes a meaningful "group" in any particular analysis within the social sciences.

SNA methods for *position analysis* focus on (unlike the other methods just mentioned) data across multiple types of relationships with the goal of identifying entities with similar patterns of relationships. For example, positional analysis could be used to identify two air traffic controllers who use similar strategies for resolving emergencies (e.g., "contact craft 1," "contact craft 2," "call supervisor," "contact craft 2," "contact craft 1"). Methods for defining *structural equivalence* are used to identify these patterns and degrees of similarity. Once position has been determined, *role analysis* can be used to study patterns of relationships among positions (e.g., air traffic controllers using similar strategies tend to be located at adjacent workstations and tend to consult each other for advice more often). Role analysis methods can be used to describe the overall structure of positions in a network (i.e., identify the role of each entity in the network), or to define local and individual roles (i.e., identify the role of a particular entity relative to adjacent entities). For example, a global role analysis could reveal a number of entities in a "task allocation" position and the structure between these positions and the "task executor" position. An individual role analysis might identify operators who are in positions indicating high skill levels and are actively mentoring less experienced operators. It should be noted that what can be inferred from role and position analysis (and, in fact, the use of these terms), is somewhat confusing, if not controversial in the SNA community (Monge & Contractor, 2003; Scott, 2000; Degenne & Forse, 1999; Wasserman & Faust, 1994).

Another aspect of a network that could be of interest to a CWA practitioner, particularly in formative analysis, is the *dynamics* of the network, or how it changes over time. These SNA methods have their roots in the study of the spread of innovation (Tarde, 1895), and focus on how particular network structures and types of interpersonal relationships support the adoption of new ideas, technologies, or beliefs (Snijders, 2005; Katz & Lazarsfeld, 1955). These methods tend to focus on the statistical analysis of other, static measures over time (e.g., modeling popularity as a Markov process) and computational modeling approaches to simulating network change according to particular theories (Carley, 2003; Carley, 1999). In CWA, these methods would be critical for in-depth before-and-after studies of new technologies. Although they might be useful to examine baseline variations in a network (e.g., for computer users, how much variation occurs in the main tools they use over time, given the same tasks?), the study of dynamic social networks has the potential to be very useful in evaluating slower changes in the use of and attitudes towards particular technologies, using multiple samples over time (e.g., visiting a site monthly for a year to administer questionnaires and conduct interviews).

The foregoing represents a short summary of a vast literature on SNA methods. We have not given treatment to many other SNA methods—particularly statistical methods for analyzing social networks (e.g., to determine degree-of-fit of a network's structure with a particular theory of social structure). The interested reader is encouraged to refer to Wasserman's 1994 text for a starting point on these and other methods.

Applying Social Network Analysis Methods

The practical application of these methods by a CWA practitioner should start with one of the texts mentioned earlier, or one of the many free, open-source, or commercial software tools. (A set of currently available tools is listed in Table 8.6.) These tools vary in their accessibility to an SNA neophyte and range in feature sets. Of particular interest may be the ability of the tool to provide methods for visualizing the network (e.g., layouts for visualizing the computed attributes of nodes, as in Figure 8.6).

Even as SNA methods could and should appeal to the CWA practitioner trying to study the relationships in a work domain, their advantages and disadvantages should be clearly understood, and critical assumptions should be documented. Because SNA methods are based on graph theory, they are largely generic and can be applied to a variety of relationship and entity types. Each of these methods may have been associated with a particular intent in the social sciences (e.g., identify "psychological tension" in a group), they could be used for the broader aims of CWA (e.g., identify individuals struggling with using particular technologies to improve communication). However, because many of these methods were originally intended for a particular dimension of social science, varying their application from that intent should be done with care (e.g., What does centrality really mean when the entities are departments at a university and the links are "sends more than 50 interdepartmental memos per month"? A department with high centrality may simply have an overeager administrator, and not be more or less important to the functioning of the university).

Table 8.6 Tools for Social Network Analysis

Free and/or open-source tools		Commercial tools	
Huminty	http://www.huminity.com	Anacubis	http://www.anacubis.com/
ORA	http://www.casos.cs.cmu.edu/projects/ora	CopLink	http://www.coplinkconnect.com/
Pajek	http://vlado.fmf.uni-lj.si/pub/networks/pajek	InFlow	http://www.orgnet.com
StOCNET	http://stat.gamma.rug.nl/stocnet/	LinkaLyzer and VisuaLyzer	http://mdlogix.com/linkalyzer.htm
TecFlow	http://www.ickn.org/ickndemo	MetaSight	http://www.morphix.com
UCINET	http://www.analytictech.com/ucinet.htm	NetMiner	http://netminer.com
Visone	http://visone.info	VisiblePath	http://visiblepath.com

Also, SNA methods have been traditionally applied to the static analysis of a network of uniform atomic link and node types (in part because the graph-theoretic nature of the methods are well suited to this simplification). Although there are some methods for the analysis of networks of valued links (e.g., for identifying cohesive subgroups or structural equivalence), such as dynamic networks (Carley, 1999; Carley & Lee, 1998), and varied relationship types (such as position and role analysis), they dramatically complicate the analysis, and appear less in common practice in SNA. There may be more value to the CWA practitioner in using SNA in normative or descriptive analyses rather than in formative analysis (e.g., to describe key entities and their typical relationships in the work domain, rather than to robustly evaluate the impact of introducing a new technology), but again, this depends on the types of entities and relationships studied and the ease with which reliable data can be collected.

An additional issue with the use of SNA methods is the need to correlate particular networks with specific CWA tools. Although Vicente suggests approaches for mapping CWA tools and representations to different entities in the domain, they may fail to capture some of the ecological relationships of interest in the domain (e.g., it might help identify the person responsible for allocating tasks but not the fact that his or her management style causes individuals to avoid reporting status on those tasks, meaning that the work is less coordinated). Even as we have suggested some potential mappings in this chapter (see Tables 8.1 and 8.2), the specific mappings of interest will depend on the domain of interest, and there is no clear CWA tool for explicit representation of ecological relationships that influence work and decision making (e.g., attitudes about work performance).

Perhaps the most significant challenge in applying SNA methods to the analysis of a work domain lies in selecting how to use methods for relatively coarse-grained descriptions of entity interrelationships when the domain is fundamentally rich and complex. For example, simply using SNA methods based on "communicates with" relationships could miss a great deal of subtlety in the domain, as entities communicate with some frequency and in some medium, and the content of the communication is almost infinitely variable. Although a practitioner could define links and use more advanced SNA methods to try and capture this richness, there are other methods, namely, Link Analysis (see the next section), that may be better suited for analyzing a domain at this resolution, depending on the ability to collect data at the desired level of resolution.

Clearly, the particular value of SNA to CWA depends on the work domain and the goals and scope of the particular CWA. The definition of entity and relationship types is critical if these methods are to be useful. Generally speaking, SNA methods are applicable to defining information flow relationships (some, such as betweenness centrality, are particularly well suited for this purpose), organizational relationships, and many ecological relationships. Because the descriptiveness of SNA formalisms may be limited, these methods may serve well as intermediate representations to help identify patterns of relationships before deeper, more fine-grained analysis.

Link Analysis

Link analysis is a subset of many related theoretic fields such graph theory, information theory, and network analysis. Link analysis and discovery aims to identify networks of relationships and patterns of behavior based on the strength and occurrence of associations among people, objects, events, or any other entity of interest. Link analysis tools and techniques can be used to explore associations between objects or to infer relevant information from relational data by considering the commonalities among connections between individual data elements in structured, heterogeneous data sets. Generally referred to as association analysis or "connecting the dots," detecting patterns of interest can be helpful for many types of analysis that are typically difficult to capture with standard statistical models (Getoor, 2003). The task of identifying known, complex, multirelational patterns that indicate potentially interesting activities in large amounts of data is normally referred to as link discovery. With the explosion in the amount of data available, manual link analysis and discovery are generally not feasible. Therefore, link discovery and analysis tools are necessary on medium and large data sets to help automate the detection of patterns, trends, associations, and hidden networks. These hidden networks and structures are usually displayed as a graph of linked objects to facilitate understanding and to expose interactions among disparate types of data. Visual correlations between attributes help drive further investigation when solving problems in diverse application fields. Link diagrams are also commonly referred to as entity-relationship diagrams, connected networks, nodes-and-links, and directed graphs (Westphal & Blaxton, 1998). Examples of applications include:

- *Analysis of communication patterns*: Link analysis graphs are useful for revealing network bottlenecks. Other applications have included the analysis of Enron e-mail logs. The raw data set contains over 600,000 messages from over 150 users. Researchers have cleaned this corpus and analyzed distribution of e-mails per user, characteristics of e-mail threads, number of e-mails between specific users, etc. (Klimt & Yang, 2004).
- *Law enforcement applications*: Link analysis methods have been applied to trace money laundering and illegal goods transfer patterns. It also plays a large role in current counter-terrorism research (Badia & Kantardzic, 2005).
- *Fraud detection*: Analysts can uncover suspicious banking activities, or flag insurance-related activities that demonstrate unusually high correlations and patterns.
- *Consumer behavior*: Link analysis is used in database marketing to identify frequent patterns in purchaser behavior and the characteristics of ideal customers.

Link analysis can be applied by data mining (e.g., investigating hypotheses based on common characteristics that can be linked together in the data), targeted analysis (e.g., filtering data sets based on an entity of interest, such as finding close associates, where the goal is to determine a strong link between two different individuals such as shared address, phone, or credit card), or confidants (where the goal might be to determine if two individuals are working together, possibly indirectly), or as a confirmation tool (e.g., filtering a specific data set based on specific parameters to highlight associations of specific entities). However, unlike SNA, the types of associations and entities are not limited to social structures. Indeed, one strength of link analysis is the ability to include diverse, multidimensional data types within networks. As shown in Figure 8.7, types of entities may include people, places, things (e.g., vehicles, weapons), communications, events, organizations (government, commercial), documents (public records), or money (bank transactions).

Practitioners need to be aware of data concerns such as granularity, aggregation, disambiguation, consolidation, context (how the data was derived), and quality (difference between electronic data in a database versus handwritten data from a century ago). Research activities in this area include mining data from text (as discussed earlier) and from the Web (Henzinger, 2000), the scalability of approaches in large or noisy databases, visualization, and the integration of link analysis with spatial, temporal, or geospatial databases. In the following text we illustrate a brief command-and-control example to show how link analysis methods can support CWA.

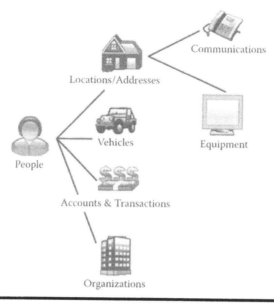

Figure 8.7 Example of a link analysis diagram.

One approach to command center design used link analysis to examine the effects of physical layout on an operator's ability to perform a mission. According to research, the optimal design of the command center may be dependent on many factors, such as communication patterns, physical object location, and environmental considerations. These factors can include human-to-human communications—duration, importance, frequency, or type of communication (chat, e-mail, telephone, face-to-face); auditory interference—time spent talking, reading, or typing; force structure—command hierarchy, team design, and collaboration; visual displays—location of wall displays relative to operator positions, size and type of information on wall displays; and physical locations visited, such as copy machines, server or secondary workstations, and briefing rooms. The results of the analysis provide insights into how information flows through a team—specifically where decision makers receive and send information. The design of an optimal layout was studied by applying traditional CWA techniques (e.g., examining cognitive tasks and decisions required to complete a mission), augmented by insights gained from link analysis (e.g., information flow). In this case, the link analysis found that ratings were high for connections between the Senior Director and the Senior Technician, which suggests that they should be located more closely. Similarly, results show that although the two Operations Technicians communicated with each other, they communicated more frequently, for longer periods, and with more importance, with the Senior Director (Figure 8.8), meaning they should be located near the Senior Director.

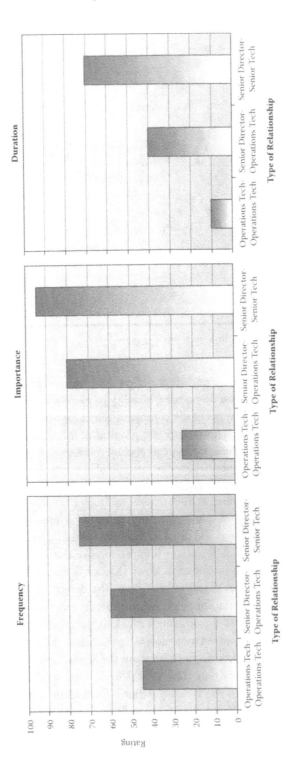

Figure 8.8 Example link analysis ratings between command center operators.

Many of the tools in Table 8.6 can also be applied to link analysis. Still, there are others that explicitly deal with data processing methods (entity extraction, text analytics) and relationship extraction for link analysis. A set of free, open-source, or commercial software tools are listed in Table 8.7.

Although link analysis is useful for understanding how disparate events and entities can be related, and there are tools that support the analysis and representation of large quantities of data, there are still limitations that should be considered. In addition to the aforementioned data concerns, there is a need for human interpretation to define the significance of links, and, without the right contextual knowledge, it is difficult to define significant entities or behaviors. These benefits and limitations should be evaluated before choosing to include link analysis techniques in a CWA.

Conclusions and Recommendations on Analyzing Social and Organizational Data

As with the data collection and data processing methods described earlier, the application of the analytic methods described in this chapter requires a CWA practitioner who clearly understands what particular methods reveal about the work domain. Because many of these methods were originally developed with a specific intent (e.g., the method for computing "status" was developed to understand social roles), their application to dimensions of a work domain require understanding the methods at a deeper level. However, even with this warning, these methods have the potential for enabling the robust and systematic evaluation of many forms of interentity relationships in a domain, providing information about the structures and strategies that relate humans, machines, and organizations and how new technologies and processes might be tailored to leverage those structures and support emergent work strategies.

Conclusions and Recommendations

The goal of this chapter has been to survey methods for the collection, processing, and analysis of social and organizational data. We have focused on particular methods from the social sciences and have attempted to describe their applicability to studying aspects of the work domain as part of CWA. Hopefully, the reader has gained a deeper understanding of how dependent these methods are on the work domain and the unit of analysis (e.g., individuals performing air traffic management, groups managing a chip manufacturing process), how these factors influence the entity and relationship types of interest, and how types influence the application of particular approaches to analyzing these relationships. The tools used in CWA themselves may represent a barrier to understanding the work domain

Table 8.7 Tools for Link Analysis

Tool	Description
Analyst's Notebook (AN)	AN can show the connections between people, accounts, organizations, or other elements.
Annotation Graph Toolkit (AGTK)	The University of Pennsylvania's Linguistic Data Consortium (LDC) developed a customized version of the AGTK for automatically annotating (tagging) pertinent information in source documents to support other tools designed for processing text.
AxisPro	AxisPro includes capabilities for link, temporal, pattern, and geospatial analysis, as well as automated entity and relationship extraction from text documents.
CADRE	Continuous Analysis and Discovery from Relational Evidence (CADRE) is designed to use entity and relation information from other tools (such as knowledge management systems) to detect and predict patterns.
Java Universal Network/Graph Framework (JUNG)	JUNG is a software library for the modeling, analysis, and visualization of data that can be represented as a graph or network. It supports representations of entities and their relations, such as directed and undirected graphs, multimodal graphs, graphs with parallel edges, and hypergraphs, and provides a mechanism for annotating graphs, entities, and relations with metadata.
Link Analysis Workbench (LAW)	LAW provides augmented link analysis using hierarchical and temporal patterns. Analysis is facilitated through a graphical user interface that supports direct graphical browsing and editing of patterns, search strategies, and summaries and details of resulting matches.
NetOwl	NetOwl (developed by SRA International, Inc.) provides a range of advanced text-mining capabilities for dealing with large volumes of unstructured text.
OrionMagic	OrionMagic can organize, collect, and tag note cards to generate link diagrams. Link diagrams are used to graphically analyze the structure of organizations, interrelationships between people, and roles of people and organizations in events. Entities and relationships are represented as the nodes and edges of a link diagram.
Sentinel	Sentinel provides 2-D and 3-D network link chart visualizations that can automatically build network charts. The system automatically organizes data into a coherent and visually clear network when it is queried.

beyond interentity communication into the broader social ecology, and perhaps future work is needed to develop more explicit representations of these ecological relationships in the CWA process.

Even as we have focused on particular aspects of the social sciences, there are many methods from within the human factors community (particularly, those on team performance, such as Arthur, Edwards, Bell, Willado, & Bennett, 2005) that we have mentioned only in passing. We have omitted these perspectives not because these methods lack merit or validity, but rather to provide the practitioner who may be uninitiated in some analytic social science methods a set of interesting and different starting points for their research. The reader is encouraged to pursue the cited papers and texts for more detail on the particular methods that may be appropriate to a given domain of study.

Acknowledgments

The authors would like to acknowledge the contributions of colleagues at Charles River Analytics, MITRE, Aptima, and elsewhere. Specifically, Michael Farry and David Koelle have contributed, over time, to the collection of literature presented here. In addition, we would like to acknowledge our gratitude to the editors of the book, Ann Bisantz and Cathy Burns, and those who have helped review this chapter for their kind and thoughtful guidance. Finally, we would like to thank Karen DeSimone and Yvonne Fuller for their assistance in preparing this material.

References

Allen, J. (1995). *Natural Language Processing.* (second ed.). New York: Benjamin Cummins.

Allender, L. (2000). Modeling human performance: Impacting system design, Performance and Cost. In *Proceedings of the Military, Government and Aerospace Simulation Symposium, 2000 Advanced Simulation Technologies Conference,* M. Chinni (Ed.), (pp. 139–144). Washington, D.C. San Diego, CA: Society for Computer Simulation.

Allison, G., & Zelikow, P. (1999). *Essence of Decision: Explaining the Cuban Missile Crisis.* New York: Longman.

Anderson, J. R., Bothell, D., Byrne, M. D., Douglass, S., Lebiere, C., & Qin, Y. (2004). An integrated theory of the mind. *Psychological Review, 111*(4), 1036–1060.

Arthur, W., Edwards, B., Bell, S., Willado, A., & Bennett, W. (2005). Team task analysis: Identifying tasks and jobs that are team-based. *Human Factors, 47*(3), 654–669.

Ayers, D., & Shah, M. (2001). Monitoring human vehavior from video taken in an office environment. *Image and Vision Computing, 12*833–846.

Badia, A., & Kantardzic, M. (2005). Link analysis tools for intelligence and counterterrorism. In *Proceedings of IEEE International Conference on Intelligence and Security Informatics (ISI '05).* Atlanta, GA.

Bailey, S., & Marsden, P. (1999). Interpretation and interview context: Examining the general social survey name generator using cognitive methods. *Social Networks, 21,* 287–309.

Baker, K., Entin, E., See, K., Gildea, K., Baker, B., Downes-Martin, S. et al. (2004). Organizational structure, information load, and communication in Navy teams. In *Proceedings of Human Factors and Ergonomics Society Annual Meeting.* Santa Monica, CA: HFES.

Barnes, J. (1972). Social networks. *Addison-Wesley Module in Anthropology, 26,* 1–29.

Bavelas, A. (1948). A mathematical model for group structure. *Applied Anthropology, 7,* 16–30.

Bodor, R., Jackson, B., Masoud, O., & Papanikolopoulos, N. (2003). *Image-based reconstruction for view-independent human motion recognition.* In *Proceedings of the International Conference on Intelligent Robots and Systems (IROS 2003),* vol. 2, 1548–1553. Las Vegas, CA.

Bond, M. (2002). Culture's consequences: Something old and something new. *Human Relations, 55,* 119–135.

Bond, M., & Smith, P. (1996). Cross-cultural social and organizational psychology. *Annual Review of Psychology, 47,* 205–235.

Borgatti, S., & Everett, M. (1992). The notion of position in social network analysis. In P. Marsden (Ed.), *Sociological Methodology.* London: Basil Blackwell.

Boschee, E., Weischedel, R., & Zamanian, A. (2005). Automatic information extraction. In *Proceedings of International Conference on Intelligence Analysis.* McLean, VA.

Bowers, C. A., Jentsch, F., Salas, E., & Braun, C. C. (1998). Analyzing communication sequences for team training needs assessment. *Human Factors, 40*(4), 672–679.

Brewer, D. (2007). Forgetting in the recall-based elicitation of personal networks. *Social Networks, 22,* 29–43.

Brewer, D. D., & Webster, C. M. (1999). Forgetting of friends and its effects on measuring friendship networks. *Social Networks, 21,* 361–373.

Burt, R. (1976). Positions in networks. *Social Forces, 55,* 93–122.

Carley, K. (2003). Dynamic network analysis. In R. Breiger & K. Carley (Eds.), *Summary of the NRC Workshop on Social Network Modeling and Analysis.* Washington, D.C.: The National Research Council.

Carley, K. M. (1999). On the rvolution of docial and organizational networks. In S. B. Andrews & D. Knoke (Eds.), *Vol. 16: Special Issue of Research in the Sociology of Organizations: "Networks In and Around Organizations."* (pp. 3–30). Stamford, CT: JAI Press, Inc.

Carley, K. M., & Lee, J. (1998). Dynamic organizations: Organizational adaptation in a changing environment. In J. Baum (Ed.), *Advances in Strategic Management, Vol. 15: Disciplinary Roots of Strategic Management Research* (pp. 269–297). JAI Press.

Casciaro, T. (1998). Seeing things clearly: Social structure, personality, and accuracy in social network perception. *Social Networks, 20,* 331–351.

Choudhury, T., & Pentland, A. (2004). Characterizing social interactions using the sociometer. In *Proceedings of North American Association of Computational Social and Organizational Science (NAACSOS).* June 17–19. Pittsburgh, PA: Kluwer.

Choudhury, T., & Pentland, A. (2002). The sociometer: A wearable device for understanding human networks. In *Proceedings of Conference on Computer Supported Cooperative Work. Workshop on Ad hoc Communciations and Collaboration in Ubiquitous Computing Environments*. November 16–20, New Orleans, LA.: ACM SIGCHI.

Corman, S., & Bradford, L. (1993). Situational effects on the accuracy of self-reported communication behavior. *Communications Research,* 20, 822–840.

Degenne, A., & Forse, M. (1999). *Introducing Social Networks.* London: Sage.

Diedrich, F., Hocevar, S., Entin, E., Hutchins, S., Kemple, W., & Kleinman, D. (2002). Adaptive architectures for command and control: Toward an empirical evaluation of organizational congruence and adaptation. In *Proceedings of Command and Control Research and Technology Symposium (CCRTS '02).* Monterey, CA.

Dzindolet, M., & Pierce, L. (2006). Using linguistic analysis to identify high performing teams. In *Proceedings of Command and Control Research and Technology Symposium (CCRTS '06).*

Entin, E. (1999). Optimized command and control architectures for improved process and performance. In *Proceedings of Command and Control Research and Technology Symposium (CCRTS '99).* Newport, RI.

Entin, E., & Weil, S. (2006). A methodology to predict specific communication themes from overall communication volume for individuals and teams. In *Proceedings of Command and Control Research and Technology Symposium (CCRTS '06).*

Essa, I. (1999). Computers Seeing People. *AI Magazine,* 20(1), 69–82.

Fairbanks, R., Bisantz, A., & Sunm, M. (2008). Department communication links and patterns. *Annals of Emergency Medicine,* 50(4), 396–406.

Farry, M. (2006). *Sensor Networks for Social Networks.* Master's Dissertation. Massachusetts Institute of Technology.

Feinman, A. (2006). *From Discourse Analysis to Groupware Design.* Doctoral Dissertation. Brandeis University.

Feinman, A., & Alterman, R. (2003). Discourse analysis techniques for modeling interaction. In *Proceedings of Third International Conference on User Modeling.* Pittsburgh, PA: Springer-Verlag.

Fiedler, K., Grover, V., & Teng, J. (1996). An empirically derived taxonomy of information technology structure and its relationship to organizational structure. *Journal of Management Information Systems,* 13(1), 9–34.

Fleishman, E. A., & Zaccaro, S. J. (1992). Toward a taxonomy of team performance functions. In R. W. Swezey & E. Salas (Eds.), *Teams: Their Training and Performance* (pp. 31–56). Norwood, NJ: Ablec Publishing.

Fox, M. S., Barbuceanu, M., Gruninger, M., & Lin, J. (1998). An organizational ontology for enterprise modeling. In M. Prietula, K. Carley, & L. Gasser (Eds.), *Simulating Organizations: Computational Models of Institutions and Groups* (pp. 131–152). Cambridge, MA: AAAI Press/MIT Press.

Freeman, J., MacMillan, J., Haimson, C., Weil, S., Stacy, W., & Diedrich, F. (2006). Strategies and studies in game-based training. In *Proceedings of the Society for Applied Learning Technologies.* Orlando, FL.

Freeman, L. (2004). *The Development of Social Network Analysis: A Study in the Sociology of Science.* Vancouver, British Columbia: Empirical Press.

Freeman, L. C., Freeman, S. C., & Michaelson, A. G. (1988). On human social intelligence. *Journal of Social and Biological Structures,* 11, 415–425.

Gaizauskas, R., & Humphreys, K. (1999). Quantitative evaluation of coreference algorithms in an information extraction system. In S. Botley & A. McEnery (Eds.), *Corpus-Based and Computational Approaches to Discourse Anaphora.* Philadelphia, PA: John Benjamins.

Gavrila, D. (1999). The visual analysis of human movement: A survey. *Computer Vision and Image Understanding,* 73(1), 82–98.

Getoor, L. (2003). Link mining: a new data mining challenge. *ACM SIGKDD Explorations,* 5(1), 84–89.

Gittler, J. (1951). Social ontology and the criteria for definitions in sociology. *Sociometry,* 14(4), 355–365.

Gluck, K., & Pew, R. (2005). *Modeling Human Behavior With Integrated Cognitive Architectures.* Mahweh, NJ: Lawrence Erlbaum.

Grishman, R. (1997). Information extraction: Techniques and challenges. In *Proceedings of SCIE '97: International Summer School on Information Extraction,* (pp. 10–27). London: Springer-Verlag.

Groves, R., & Magilavy, L. (1986). Measuring and explaining interviewer effects in centralized telephone surveys. *Public Opinion Quarterly,* 50, 251–266.

Gupta, V., & Hanges, P. (2004). Regional and climate clustering of societal cultures. In R. House, P. Hanges, M. Javidan, & V. Gupta (Eds.), *Culture, Leadership, and Organizations* (pp. 178–218). London: Sage.

Haimson, C., & Lovell, S. (2006). Pattern recognition for cognitive performance modeling. In *Proceedings of AAAI's 2006 Fall Symposium on Capturing and Using Patterns for Evidence Detection.* Washington, D.C.

Hall, S. (1986). Gramsci's relevance for the study of race and ethnicity. *Journal of Communication Inquiry,* 10(2), 5-27.

Han, J., & Kamber, M. (2007). *Data Mining: Concepts and Techniques.* San Diego, CA: Academic Press.

Harper, K. A., Mulgund, S. S., Zacharias, G. L., & Menke, T. (2000). SAMPLE: Situation Awareness Model for Pilot-in-the-Loop Evaluation. In *Proceedings of 9th Conference on Computer Generated Forces and Behavioral Representation.* Orlando, FL.

Henzinger, M. (2000). Link analysis in Web information retrieval. *IEEE Data Engineering,* 23(3), 3–8.

Heran, F. (1987). Comment les Francais voisinent. *Economie et Statistique* (195), 43–60.

Hofstede, G. (2001). *Culture's Consequences: International Differences in Work-Related Values.* Thousand Oaks, CA: Sage Publishers.

House, R., Hanges, P., Javidan, M., Dorfman, P., & Gupta, V. (2004). *Culture, Leadership, and Organizations.* London: Sage.

Hudlicka, E., Karabaich, B., Pfautz, J., Jones, K., & Zacharias, G. (2004). Predicting group behavior from profiles and stereotypes. In *Proceedings of Behavior Representation in Modeling and Simulation (BRIMS).* May 17–20, Arlington, VA. Simulation interoperability and standards organization,

Ilgen, D., LePine, J., & Hollenbock, J. (1997). Effective decision making in multinational teams. In P. Earley & M. Erez (Eds.), *New Perspectives in Industrial-Organizational Psychology* (pp. 377–409). New York: John Wiley.

Inglehart, R., & Baker, W. (2000). Modernization, cultural change, and the persistence of traditional values. *American Sociological Review, 65,* 19–51.

Jackson, P., & Moulinier, I. (2002). *Natural Language Processing for Online Applications: Text Retrieval, Extraction, and Categorization.* Philadelphia: John Benjamins.

Joachims, T. (1998). Text categorization with support vector machines. In *Proceedings of 10th European Conference on Machine Learning.* July 5–9, Paris. Cultures Anglophones et Technologies de l'Information (CATI).

Kang, M., Waisel, L., & Wallace, W. (1998). Team-Soar: A model for team decision-making. In M. Prietula, K. Carley, & L. Gasser (Eds.), *Simulating Organizations: Computational Models of Institutions and Groups* (pp. 23–46). Cambridge, MA: AAAI Press/ MIT Press.

Karabaich, B. (2004). Towards a working taxonomy of groups. In *Proceedings of 13th Annual BRIMS Conference.* Arlington, VA.

Karanikas, H., & Theodoulidis, B. (2002). *Knowledge Discovery in Text and Text Mining Software.* Manchester, UK: Manchester University Centre for Research in Information Management.

Katz, E., & Lazarsfeld, P. (1955). *Personal Influence: The Part Played by People in the Flow of Mass Communication.* Glencoe, IL: Free Press.

Kern, N., Schiele, B., & Schmidt, A. (2003). Multi-sensor activity context detection for wearable computing. In *Proceedings of European Symposium on Ambient Intelligence (EUSAI '03),* (pp. 220–232).

Kiesler, S., Siegal, J., & McGuire, T. (1988). Social psychological aspects of computer-mediated communication. In I. Greif (Ed.), *Computer-Supported Cooperative Work: a Book of Readings* (pp. 657-682). San Francisco, CA: Morgan Kaufman.

Klein, H. (2005). Cultural differences in cognition: Barriers in multinational collaborations. In H. Montgomery, R. Lipshitz, & B. Brehmer (Eds.), *How Professionals Make Decisions.* Mahwah, NJ: Lawrence Erlbaum Associates.

Klein, H., & McHugh, A. (2005). National differences in teamwork. In W. Rouse & K. Boff (Eds.), *Organizational Simulation.* New York: John Wiley.

Kleinman, D., Levchuk, G., Hutchins, S., & Kemple, W. (2003). Scenario design for the empirical testing of organizational congruence. In *Proceedings of International Command and Control Research and Technology Symposium (ICCRTS '03).* Washington, D.C..

Kleinman, D., Luh, P., Pattipati, K., & Serfaty, D. (1992). Mathematical models of team distributed decisionmaking. In R. Swezey & E. Salas (Eds.), *Teams: Their Training and Performance* (pp. 177–218). New York: Ablex.

Klimt, B., & Yang, Y. (2004). Introducing the Enron corpus. In *Proceedings of 1st Conference on Email and Anti-Spam (CEAS).* Mountain View, CA.

Knoke, D., & Kulinski, J. (1982). *Network Analysis.* Newbury Park, CA: Sage.

Kontopoulos, K. (1993). *The Logic of Social Structure.* New York: Cambridge University Press.

Laird, J. E. (1987). Soar: An architecture for general intelligence. *Artificial Intelligence, 33*(1), 1–64.

Landauer, T. K. (1998). Introduction to latent semantic analysis. *Discourse Processes, 25,* 259–284.

Lewis, D. (1991). Evaluating text categorization. In *Proceedings of Speech and Natural Language Workshop,* (pp. 312–318). San Mateo, CA: Morgan Kaufman.

Llora, X., Yasui, I., Weige, M., & Goldberg, D. (2006). Human-centered analysis and visualization tools for the blogosphere. In *Proceedings of Digital Humanities 2007 Conference*, July 5–9, Paris.

Lorrain, F., & White, H. (1971). Structural Equivalence of Individuals in Social Networks. *Journal of Mathematical Sociology*, 1, 49-80.

Lukowicz, P., Ward, J. A., Junker, H., Stager, M., Troster, G., Atrash, A. et al. (2004). Recognizing workshop activity using body-worn microphones and accelerometers. In *Proceedings of 2nd International Conference, PERVASIVE 2004* (pp. 18–32). Linz/Vienna, Austria.

Madabhushi, A., & Aggarwal, J. (1999). A bayesian approach to human activity recognition. In *Proceedings of 2nd IEEE Workshop on Visual Surveillance*, (pp. 25–32).

Markoff, J. (2002). Pentagon plans a computer system that would peek at personal data of Americans. *New York Times*. November 9, 2002.

Marsden, P. (2005). Recent developments in network measurement. In P. Carrington, J. Scott, & S. Wasserman (Eds.), *Models and Methods in Social Network Analysis* (pp. 8–30). New York: Cambridge University Press.

Marsden, P. V. (1990). Network data and measurement. *Annual Review of Sociology*, 16, 435–463.

Masoud, O., & Papanikolopoulos, N. (2003). A method for human action recognition. *Image and Vision Computing*, 8, 729–743.

May, R., & Baddeley, B. (2006). Visual analytics: large and small display environments. In *Proceedings of CHI*. Montreal, Canada: ACM.

McKelvey, B. (1982). *Organizational systematics: Taxonomy, evolution, and classification*. Berkeley, CA: University of Berkeley Press.

Miles, R., & Snow, C. (1986). Organizations: New concepts for new forms. *California Management Review*, 34, 62–73.

Moens, M. (2006). *Information Extraction: Algorithms and Prospects in a Retrieval Context*. New York: Springer.

Mokken, R. (1979). Cliques, clubs, and clans. *Quality and Quantity*, 13, 161–173.

Monge, P., & Contractor, N. (2003). *Theories of Communication Networks*. New York: Oxford University Press.

Mumaw, R. J., & Holder, B. E. (2002). What do cultural dimensions reveal about flight deck operations? In *Proceedings of Human Factors and Ergonomics Society 46th Annual Meeting*. Baltimore, MD.

Nishida, Y., Murakami, S., Hori, T., & Mizoguchi, H. (2004). Minimally privacy-violative human location sensor by ultrasonic radar embedded on ceiling. In *Proceedings of IEEE International Conference on Systems, Man and Cybernetics (SMC '04)*, (pp. 1549-1554).

Oliver, N., & Horvitz, E. (2005). A Comparison of HMMs and dynamic bayesian networks for recognizing office activities. *LNAI 3538*, 205–215.

Oliver, N., Rosario, B., & Pentland, A. (2000). A bayesian computer bision system for modeling human interactions. *IEEE Transactions on Pattern Analysis and Machine Intelligence*, 22(8).

Parasuraman, R., Sheridan, T. B., & Wickens, C. D. (2000). A model for types and levels of human interaction with automation. *IEEE Transactions on Systems, Man, and Cybernetics*, 30(3), 286–297.

Pew, R. W., & Mavor, A. S. (1998). *Modeling Human and Organizational Behavior: Application to Military Simulations*. National Research Council.

Pfautz, J., & Roth, E. (2006). Using cognitive engineering for system design and evaluation: A visualization aid for stability and support operations. *International Journal of Industrial Engineering*, 36(5), 389–407.

Pfautz, J., Roth, E., Bisantz, A., Llinas, J., & Fouse, A. (2006). The role of meta-information in C2 decision-support systems. In *Proceedings of Command and Control Research and Technology Symposium*. San Diego, CA.

Plutchik, R. & Kellerman, H. (1989). In *Emotion: Theory, Research, and Experience, Vol. 4: The Measurement of Emotions*. San Diego, CA: Academic Press.

Podolny, J. (1993). A status-based model of market competition. *American Journal of Sociology*, 98, 829–872.

Renn, O., & Rohrmann, B. (2000). *Cross-Cultural Risk Perception—A Survey of Empirical Studies*. Berlin: Springer.

Ross, N. (2004). *Culture and Cognition: Implications for Theory and Method*. London: Sage.

Roth, E. M., Gualtieri, J. W., Elm, W. C., & Potter, S. S. (2002). Scenario development for decision support system evaluation. In *Proceedings of Human Factors and Ergonomics Society 46th Annual Meeting*, (pp. 357–361). Santa Monica, CA: Human Factors and Ergonomics Society.

Salas, E., Dickinson, T. L., Converse, S. A., & Tannenbaum, S. I. (1992). Toward an understanding of team performance and training. In R. W. Swezey & E. Salas (Eds.), *Teams: Their Training and Performance*. Norwood, NJ: Ablex Publishing.

Scott, J. (2000). *Social Network Analysis*. London: Sage.

Seidman, S. (1983). Network structure and minimum degree. *Social Networks*, 5, 269–287.

Seidman, S., & Foster, B. (1978). A note on the potential for genuine cross-fertilization between anthropology and mathematics. *Social Networks*, 1, 65–72.

Serfaty, D., & Entin, E. (1995). Shared mental models and adaptive team coordination. In *Proceedings of 1st International Symposium on Command Control Research and Technology*. National Defense University, Washington, DC.

Sherif, M. (1963). Intergroup relations and leadership: Approaches and research in industrial, ethnic, cultural, and political areas. *American Sociological Review*, 28(5), 828–830.

Shirky, C. (2005). Ontology is overrated: Categories, links and tags. http://www.shirky.com/writings/ontology_overrated.html [On-line].

Singley, M., & Anderson, J. (1989). *The Transfer of Cognitive Skill*. Cambridge, MA: Harvard University Press.

Smith, J., Fishkin, K., Jiang, B., Mamishev, A., Philopse, M., Rea, A. et al. (2005). RFID-based techniques for human activity detection. *Communications of the ACM*, 48(9), 39–44.

Snijders, T. (2005). Methods for longitudinal network data. In P. Carrington, J. Scott, & S. Wasserman (Eds.), *Models and Methods in Social Network Analysis* (pp. 215–247). New York: Cambridge University Press.

Straits, B. (2000). Ego's important discussants or significant people: An experiment in barying the wording of personal network name generators. *Social Networks*, 22, 123–140.

Tarde, G. (1895). *Les Lois De L'Imitation.* (1979 Reprint ed.). Geneva, Switzerland: Slatkine Reprints.

Vicente, K. J. (1999). *Cognitive Work Analysis: Towards Safe, Productive, and Healthy Computer-Based Work.* Mahwah, NJ: Lawrence Erlbaum Associates.

Voinea, L., & Telea, A. (2007). Visual analytics: Visual data mining and analysis of software repositories. *Computers and Graphics,* 31(3), 410–428.

Want, R., Hopper, A., Falcao, V., & Gibbons, J. (1992). The active badge location system. *ACM Transactions on Information Systems,* 10(1), 91–102.

Wasserman, S., & Faust, K. (1994). *Social Network Analysis: Methods and Applications.* New York: Cambridge University Press.

Webster, C. (1994). Data Type: A comparison of observational and cognitive measures. *Journal of Quantitative Anthropology,* 4313–328.

Wellman, B., & Berkowitz, S. (1988). Part II: Communities. In B. Wellman & S. Berkowitz (Eds.), *Social Structures: A Network Approach* (pp. 123–129). New York: Cambridge University Press.

Wellman, B., & Leighton, B. (1979). Networks, neighborhoods, and communities: approaches to the study of the community question. *Urban Affairs Quarterly,* 15363–390.

Westphal, C., & Blaxton, T. (1998). *Data Mining Solutions: Methods and Tools for Solving Real-World Problems.* Hoboken, NJ: Wiley.

White, H., Boorman, S., & Breiger, R. (1976). Social structure from multiple networks. I. Blockmodels of roles and positions. *American Journal of Sociology,* 81730–779.

Yang, Y. (1999). An evaluation of statistical approaches to text categorization. *Journal of Information Retrieval,* 1(1/2), 67–88.

Chapter 9

Task Analysis and Cognitive Work Analysis: Requirements for Worker Competencies and Beyond

Colin G. Drury

Contents

Overview

Task analysis (TA) and cognitive work analysis (CWA) are human factors tools with a considerable history and widespread current usage. But, how do they relate to each other? TA enumerates the task demands and allows the human factors engineer to link them directly to the human capabilities literature. Hierarchical task analysis (HTA) is the current industry standard, most useful because it is a progressive redescription of goals, functions, and tasks down to whatever level is needed to solve the problem at hand. CWA is a five-phase process that concentrates on the logical constraints inherent in the system and the operator's goals within those constraints. CWA's phases gradually introduce more constraints until at the fifth phase the operator's job is quite completely specified. Both tools show goal-directed interactions between people, their systems, and their "products," but differ in philosophy and practical application. At the simplest level, the last of the five phases of CWA appears to be the only place that TA can be used within this context. Some attempts have been made to show how they fit together (e.g., Hajdukiewicz & Vicente, 2004; Stanton, 2006) reaching different conclusions about the role of TA in CWA. This chapter provides another view where these techniques can be complementary in human factors applications.

In this chapter we explore how and where the two traditions of TA (particularly HTA) and CWA can fit together to achieve the design objectives of human factors. Many practitioners (e.g., Hajdukiewicz & Vicente, 2004) take the obvious view that HTA can fit into CWA only at the fifth phase of Worker Competencies Analysis, but we consider how a broader role may be appropriate. We show that both techniques arose from a common systems basis so that a closer integration may at least be possible.

The Two Traditions: Antecedents of HTA and CWA

Human factors engineering/ergonomics (called HFE in this chapter) has used some form of TA continuously almost from its inception. TA is part of the systems design methodology, most famously conceptualized by Singleton (1974) as a sequence of

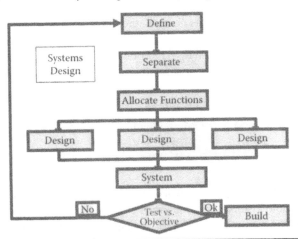

Figure 9.1   Systems design process.

steps (Figure 9.1) leading from the requirements of the system to a physical system design with all HFE issues at least considered, if not always resolved in the operator's favor. TA started by understanding the demands of tasks that need to be performed by an operator, so that for each task the key human capabilities and limitations could be derived. This formal comparison of task demands with human capabilities has been a keynote of the HFE methodology ever since. The concept of systems design exemplified by Singleton is, of course, not the only way in which the joint requirements of the operator and the system can be met. From almost the same time that systems design evolved TA, a parallel body of knowledge was being forged under the term *socio-technical systems design* or STS. Coming from origins closer to civilian industrial psychology than systems design's military experimental psychology, STS evolved ways of considering both the operator's social needs and the system's quality needs. Breakdown of the tasks was not the issue, but rather the functions that the system was meant to accomplish. For each function, the key variances could be derived so that decisions could be taken on how each variance could be controlled to meet system objectives. The overall STS process can be typified by an equivalent diagram to Figure 9.1, shown here in Taylor and Felten's (1993) version as Figure 9.2.

Eventually, from the systems design/TA tradition, HTA emerged as the currently definitive technique. In parallel, STS was one of the parent disciplines from which CWA emerged. Hence, a comparison of HTA and CWA requires a little history as well as snapshots of their current status. Because the rest of this book

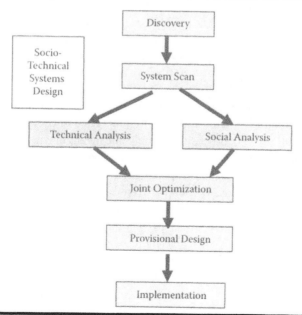

Figure 9.2 Socio-technical systems (STS) design.

provides much on CWA, in this chapter we only explore it as much as is needed to look for potential integration(s) of the two techniques.

The Task Analysis Tradition

Despite links to the distant past of scientific management and occupational job specifications, the origin of TA as we know it today can be traced back to R. B. Miller's 1953 report, *A Method for Man–Machine Task Analysis*. This technique started from the mission and function requirements and used task description as a first step towards understanding how the human element in the system achieved the requirements of the task. These were typically stated in terms of controlling variables within specific limits and with specific response speeds. Since Miller's time, many authors have felt the need to review where we are in TA, for example, Meister (1971, pp. 76–88), Drury (1983), Drury, Paramore, Van Cott, Grey, and Corlett (1987), Luczak (1997), Stammers and Shepherd (2000). Perhaps the definitive compendium of techniques is by Kirwan and Ainsworth (1992), the result of deliberations by the Task Analysis Working Group in the United Kingdom. For a more recent analysis of HTA in particular, see Stanton (2006).

Although TA is often cast as describing what people are *required to do* in a system (e.g., Vicente, 1999, p. 63), this is not exactly how it is seen by users and practitioners. From the systems perspective of Figure 9.1, task description is rather a set of necessary operations to achieve a function or the overall systems goal. The fact that people perform these operations as opposed to machines is typically left for a later decision during human–machine function allocation. Note that I use *operations* as shorthand for "achieving goals" although no overt action may be visible to an outside observer (see Annett, 2000). The distinction between overt behavior, perception, and cognition is historically less recognized in Europe and not seen as useful in HTA (see Chipman, Schraagen, & Shalin, 2000, p. 4). For existing systems, one method of data collection is to observe what people actually do in the current system, but as many authors point out (Kirwan & Ainsworth, 1992, p. 114; Stanton, 2006, p. 63), observation is only one means of data collection among many, for example, manuals, training materials, subject matter experts, and simulations. It would be a rather lazy HFE practitioner who merely observed the current system. There will always be a tension between using what *should be* done to guide the TA and using what *is* done, as they may differ appreciably. In such cases, it is useful to ask why the operator thinks it is a good idea to perform *their* way rather than the way specified, for example, by engineers or ISO 9000 documentation. The answers can be most revealing, either about the spread of industrial mythology or about the role of the operator in innovation.

But the objective of TA is the analysis, not the description. As Stanton (2006) notes: "[T]he HTA representation is the starting point for the analysis rather than the end point." Analysis means anything from a formal balancing of the task demands with human capabilities for each dimension of demand (almost never possible quantitatively) to a simple listing of HFE issues raised at each operation in the description. Analysis was what the original paper by Miller was about. He used an information transaction framework for analysis: what information the person received from the system and what he (usually) did to change the system. Thus, the cues for action, display reading, and control actions are analyzed for HFE implications as is the availability and adequacy of feedback. These same headings for analysis resurface throughout the TA literature, for example, in Drury, Prabhu, and Gramopadhye (1990), Pennington (1992), or Gramopadhye and Thaker (1998), perhaps because they provide the dimensions of the HFE issues or at least a set of objects an HFE practitioner can potentially influence.

The Socio-Technical Systems Tradition

STS started from the realization in the 1950s that effective forms of work organization took account of the social needs of the workforce as well as the technical requirements of the work system. The theoretical framework adopted was based on general systems theory (Ropohl, 1999), emphasizing control of variables essential to effective system performance. Van Eijnatten's (1991) comprehensive history of STS shows four phases, all sharing a system viewpoint but differing in opportunities for interventions of different types. Here, we will continue at the practitioner level, however, following Taylor and Felten's (1993) approach.

Modern enterprises live in a complex and dynamic environment of customer requirements, financial markets, government regulations, and workforce needs. STS recognizes that the enterprise as a system is an open system with many external inputs that can change very rapidly. The price for not being able to react effectively and rapidly is loss of control and ultimately failure—the Law of Requisite Variety (Ashby, 1956). This requirement for agility in an open system implies a flexible systems design, and typically one that is constantly being redesigned. A key principle is "incompletion," for example, the deliberate underspecification of tasks so that people within the organization are free to adapt their choice of tasks to new circumstances. This is in direct contrast to the Tayloristic reliance on sticking to a rigidly specified procedure, often exemplified currently by systems such as the ISO 9000, although, to be fair, that is a criticism of the practice of ISO 9000 rather than its design.

STS, as exemplified in Figure 9.2, uses a systems analysis to derive "variances," or characteristics of each function, that must be controlled for the system to achieve its goals. STS recognizes that these variances, particularly the ones labeled key variances, are controlled by the social system of operators within the enterprise. This appears very close to the systems view of Singleton, where the overall mission is subdivided into functions that must be achieved within specified limits—hardly surprising as both derive from systems theory. For example, Figure 9.3 shows a function and key variance table for sardine packing, derived by students in an

Unit Operation	Variances
Sort Fish	1. Fish dimensions/size/weight
	2. Appearance/color
	3. Freshness (data of arrival)
	4. Source of body of water
	5. Preservative water conditions - Salt concentration
	6. Preservative water conditions - Bacteria amount
	7. Preservative water conditions – Oxygen
	8. Preservative water conditions - Volume
	9. Preservative water conditions – Temperature
Prepare Fish	10. Quality of head/tail/gut removal
	11. Quality of cleanliness
Packing Fish	12. Number of defects for cans
	13. Size of sardines
	14. Weight of sardines
	15. Stability of packing machines
Cook Fish	16. Temperature of cooking process
	17. Time for cooking
	18. Time for cooling
	19. Temperature inside the factory
Add Water	20. Temperature of water
	21. Amount of bacteria for water
	22. Quality of water
	23. Amount of water
Close Can	24. Tightness factor when closed
	25. Other fish present
	26. Obstructions in closing
	27. Weight of the can
Wash Can	28. Amount of water used
	29. Quality of the water
	30. Cleanliness of the can
	31. Dents, dimensions of the can
	32. Quantity of cans being washed
Heat Cycle Can	33. Temperature
	34. Amount of bacteria present
Label Can	35. Material quantity/wear
	36. Wrapping process – Speed
	37. Wrapping process – Strength
	38. Wrapping process – Capacity
	39. Denting or other damage to product
Pack and Ship	40. Sorting and placement of can/box filling – Speed
	41. Sorting and placement of can/box filling - Orientation of cans
	42. Sorting and placement of can/box filling - Accuracy of count
	43. Sorting and placement of can/box filling - Damage to can
	44. Seal master box – Quality
	45. Seal master box - Accuracy
	46. Seal master box - Speed
	47. Labeling of box product/destination info - Legibility
	48. Labeling of box product/destination info – Visibility
	49. Labeling of box product/destination info - Material

Figure 9.3   Unit operations and variances for sardine canning.

STS course entirely from Internet searches and general knowledge of industrial methods. It is quite possible to derive these functions and variances without seeing the functioning system, although operators bring out subtleties of understanding denied to a systems level view.

A key to moving from a variance listing such as Figure 9.3 to a systems design (however deliberately incomplete) is an understanding of where, how, and by whom each variance can be controlled. It is here that constraints on action can arise, for example, from the physics and chemistry of a controlled process, or from the design limitations of existing technical equipment. For example, raising the temperature of a food processing batch depends both on the specific heat of the food product and the maximum output of the heating devices. The recognition of these constraints has been a defining characteristic of CWA.

The deliberate underspecification of an operator's strategy and actions for control of variance, and the explicit consideration of the system as open, lead STS into CWA. This was not an instant process, but an evolution over several years of work by Rasmussen and colleagues. Most of the work was brought together in Rasmussen, Pejtersen, and Goodstein (1994, p. 25, Figure 1.8) where Work Domain Analysis was shown as encompassing various levels of activity analysis. The implication is that any form of TA can best be understood in terms of the "cognitive strategies and subjective preferences" of the actors involved, as by that time the degrees of freedom have been progressively reduced. The choice of strategies depends on their cognitive resource demands compared to the "cognitive resource profiles of the actors" (Rasmussen et al., 1994, p. 30). This is almost exactly how TA is usually described, although the authors to not use the term "task analysis." As Vicente (1999, pp. 72–79) shows, treating TA at this instruction-based level limits its applicability to closed systems with specified operator strategies. In contrast, constraint-based analysis, the essence of CWA, does not have these limits imposed. However, HTA practitioners do not seem to see these limitations, and continue to apply it to open and incompletely specified systems. Are they (actually, we) deluded? We will look more closely at HTA before attempting an answer.

Hierarchical Task Analysis

For HTA, the task description consists of a hierarchy of goals to be met, based originally on a theory of human operation in systems as "goal-directed behavior comprising a subgoal hierarchy linked by plans" (Stanton, 2006). It was developed by Annett and Duncan (1967) as a way to go beyond the simple listing of task steps in traditional industrial engineering and military applications. Their aim was to devise a methodology rooted in then-current concepts of psychology, such as control theory or information processing, which could be applied to the more complex tasks seen in the process industries. Their idea of goal-decomposition description has seen widespread parallel use beyond their original training design to include human–computer interaction analysis and design, such as those demonstrated by GOMS model (e.g., Kieras, 1997).

The easiest way to characterize HTA is by working through an example. My example is part of a systematic examination of the technologies of nondestruction inspection (NDI) used in aviation maintenance. Because NDI inspection systems are technically complex, quite difficult to use, and have high consequences of failure, the incorporation of HFE principles can be most beneficial. This project, undertaken for the Federal Aviation Administration, was designed to examine each of eight techniques and derive Best Practices (or at least Good Practices) for each, specifically incorporating HFE.

The methodology that led to the good practices was multifaceted. It used the usual sources of existing documentation on the inspection process, and incorporated the result of many hours spent working with inspectors. However, aircraft structural inspection is something of a special case; it is a function that I have been involved with for many years and know much more deeply than a typical practitioner would know the domain. In particular, I was able to build on a theoretical background of inspection task decomposition (e.g., Drury, 1992) to structure the tasks being analyzed and to plan my work with inspectors. Also, I was knowledgeable enough in the theory and technology of inspection and the regulatory environment to establish trust quickly in my interactions with inspectors. In the more typical "fresh" domains, HFE practitioners have to read avidly well beyond the process documentation, in parallel with learning about the process from the workforce. For example, in a study to reduce errors in commercial jam-making, much insight was gained from technical papers in the research literature that treated the solidification process as one of polymerization. These papers were not used in the plant, where much of the documentation was recipes and checklists that assumed the background knowledge that I had to acquire explicitly. Only after such insights was I able to interact sensibly with the workforce and begin to ask the sort of questions that helped produce insightful HTAs. In the inspection example given here, insights were much easier given my background in the domain.

All inspection tasks appear to fit a generic model at the function level, that is, the level below the overall system goal of "detect defects in aircraft components." Drury (1992) has defined these generic functions as shown in Figure 9.4. Although these work generically, some functions may need considerable expansion to produce a rich enough task description to derive good practices. For example, fluorescent penetrant inspection (FPI) requires considerable preprocessing of each component before its visual inspection under ultraviolet illumination. Thus, the HTA for FPI reduces most of the functions in Figure 9.4 to a single function of "Read Component" where the visual inspection takes place (Drury & Watson, 1999). In contrast, for the "simplest" NDI technique of visual inspection, these functions become the

Function	Visual Inspection Description
1. Initiate	All processes up to accessing the component. Get and read work documentation. Assemble and calibrate required equipment.
2. Access	Locate and access inspection area. Be able to see the area to be inspected at a close enough level to ensure reliable detection.
3. Search	Move field of view across component to ensure adequate coverage. Carefully scan field of view using an appropriate strategy. Stop search if an indication is found.
4. Decision	Identify indication type. Compare indication to standards for that indication type.
5. Respond	If indication confirmed, then record location and details. Complete paperwork procedures. Remove equipment and other job aids from work area and return to storage. If indication not confirmed, continue search (3).

Figure 9.4   Generic functions of inspection tasks.

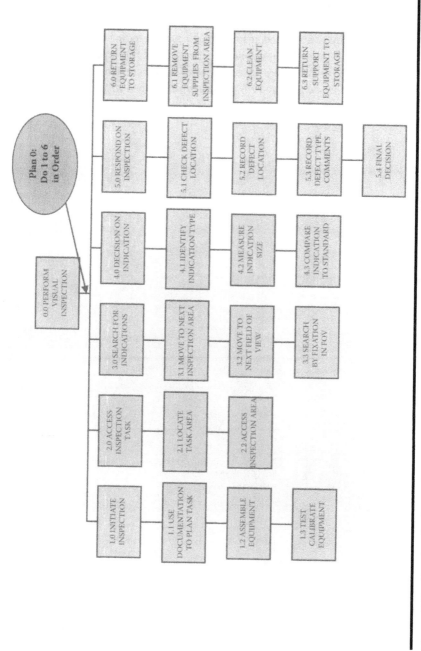

Figure 9.5   Top level of HTA for the example of visual inspection.

first level of the goal hierarchy. Note that others have tackled NDI using HTA techniques (e.g., Kirwan and Ainsworth, 1991, Chapter 14).

Our example here is visual inspection (Drury, 1992), chosen to illustrate some strengths and limitations of HTA. The top level of the HTA is shown in Figure 9.5, with progressively more detailed descriptions in Figures 9.6 and 9.7. There is no defined strategy for visual search, and indeed the concept of a search strategy is seen as quite novel by experienced inspectors. Each inspector has many different strategies, depending upon the component inspected and the expected location of defects such as cracks or corrosion. In experiments with experienced inspectors on actual aircraft (Wenner & Drury, 1997) we have videotaped inspection sequences

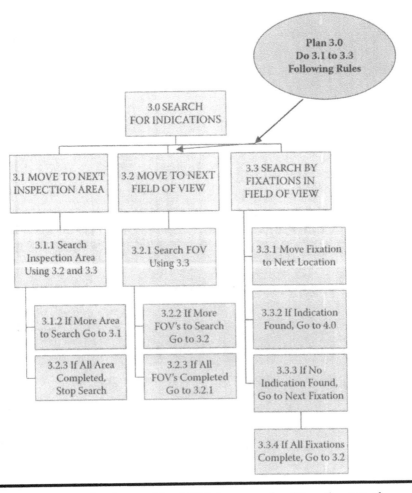

Figure 9.6 **Example of second level HTA for example of Function 3.0 of Figure 9.4.**

3.0 Search for Indications	*Task Description*	*Task Analysis*
3.1 Move to next inspection area	3.1.1 Search inspection area using 3.2 and 3.3 3.1.2 If more areas to search, go to 3.1 3.2.3 If all area completed, stop search	Is area to be inspected remembered by inspector? What path (strategy) is followed by inspector to move FOV's over inspection area? Is search coverage complete? Is sufficient time allowed for reliable search for whole blade?
3.2 Move to next field of view (FOV)	3.2.1 Search FOV using 3.3 3.2.2 If more FOVs to search, go to 3.2 3.2.3 If all FOVs completed, go to 3.2.1	Is FOV movement needed to cover whole inspection area at adequate magnification? Can FOV be moved to all positions needed, e.g. mirror, lighting in correct positions? Can inspector maintain situational awareness as FOV moves? What is scan path followed by inspector? Does scan path cover complete FOV?
3.3 Search by fixations in FOV	3.3.1 Move fixation to next location	Does eye scan path across FOV cover whole FOV? Are fixations close enough together to detect indication if it is in the fixation? Is fixation time sufficient to detect a target? Is inspector expecting all possible indications each time search is performed? Are some indications expected in particular parts of the structure? Do inspector's expectations correspond to reality for this task? Does inspector return to area where possible indication perceived? Does inspector have high peripheral visual acuity? Is contrast between indication and background high? Is indication visible to inspector if an direct line of sight (Fovea)?
	3.3.2 If indication found, go to 5.0	Is there a clear protocol for what is an indication? Is there a clear protocol for remembering how much of search was completed before going to decision?
	3.3.3 If all fixations complete, go to 3.2 3.3.4 If no indication go to next fixation 3.3.1	Does inspector remember whether fixations are complete? Is the policy to scan whole FOV once before stopping? Does inspector try to continue fixations for search while moving FOV?

Figure 9.7   Third level of redescription in HTA for example shown in Figure 9.3.

and recorded this variety. What is shown in Figures 9.6 and 9.7 is a logical model from much theory on extended search (e.g., Morawski, Drury, & Karwan, 1980), omitting the specifics of search strategy. This introduction of search tasks does however bring up the whole issue of the level of CWA implied by a HTA of a task that needs to be flexible.

Note that inspection, even NDI, is not a control task per se, although it is often used as the sensing and feedback loop in a quality control system. Thus, Phase 2 of CWA, control TA, is not strictly applicable. However, strategies analysis certainly is applicable (Phase 3 of CWA). There are many strategies an inspector can use in checking an aircraft component, for example, in starting the search at different points, concentrating on particular areas of the component, or choosing a response criterion. The choice is based on prior knowledge and training as well as specific feed-forward information on the specific component and the aircraft it came from. Thus, for jet-engine fan blades, the inspector may start the inspection with the leading edge where damage is most likely, unless information from others inspecting the same type of blades (e.g., at another facility) indicates that base cracks are being found in these blades. The strategy of searching each blade has elements of knowledge-based, rule-based, and skill-based performance as shown by Drury and Prabhu (1996). Knowledge-based operations are primarily deciding on overall search strategy, for example, where to start or what defects to search for. Rule-based operations comprise, for example, selecting the criterion of how small a defect to report, how far apart to place fixations, and deciding when to stop searching this blade and move to the next blade. Skill-based operations are those generally below the level of conscious choice, for example, control of eye movements, analysis of the visual input during a fixation, use of alternate means of verifying a defect (use of magnifier, use of touch sense). Thus, even at the level of analysis in the HTA example given, the strategies analysis of CWA Phase 3 is implied. With highly trained, skilled, and experienced operators such as aircraft NDI inspectors it is hardly surprising that they are not restricted to a completely prescribed set of actions as implied by Phase 5 of CWA. It is even unlikely that those writing inspection procedures know enough about the task of an experienced inspector to prescribe it at the level of specificity theoretically required for Phase 5 of CWA.

A comment also needs to be made here about Phase 4 of CWA: Social Organization Analysis in relation to HTA. In Singleton's systems design concept (Figure 9.1) the social organization is designed during the systems integration block, after the people, machines, and interfaces have been designed. This typically involves assigning tasks to people, ranging from one-person-one-task (Taylorism) to one-person-all-tasks (craft skills). In the inspection example above, the implication is a craft skill, but note that others in the organization influence the outcome of inspection. Managers may query decisions that would result in a high cost or remove an aircraft from service for too long. Colleagues always discuss in the break room both their findings and what they hear on the industrywide grapevine about the components being inspected. Aircraft that have been through trauma (for example, hard landings) are pointed out to inspectors by managers so that strategies can be chosen appropriately (e.g., special attention to landing gear and high-load pathways in the aircraft structure). Although many of these social and organizational aspects are

Table 9.1 Excerpt of HTA for eddy current inspection

Task description	Task analysis									Observations
	Sub-Systems									
	A	S	P	D	M	C	F	P	O	
3.3 Get the eddy current standard										• No checks on whether thickness of paint on standard and on aircraft is the same. • Uneven wear of painted surface on aircraft cannot be accommodated on standard and could lead to inconsistent inspection.

informal, they do exist and are put into the task analyses and the good practices arising from them (e.g., Drury & Watson, 1999).

This example has concentrated on the ways in which task can be described (i.e., Task Description), with the TA coverage limited to the final column of Table 9.1. This analysis asks the questions one HFE professional sees as important to understanding how task demands are matched to human capabilities. Users of TA, and particularly HTA, recognize that TA is the goal with task description only being a means toward that end. The analysis must be useful to the ultimate user of the HFE effort, who typically never sees the task description, only the final product. As a conclusion on TA, we can take this a stage further to examine the output for the NDI example. The brief was to bring HFE knowledge to bear on improving practices in a set of technologies that are mature in their physics but changing in the ways computer technology can be used to change display, analysis, and storage of information. Here, the users were engineers, managers, and inspectors in industry, and those who design the technology and its associated training programs.

There is no definitive prescriptive way of finding the task demands and human capabilities, although many partial techniques exist and can be useful. As noted earlier, the original task analyses (e.g., Miller, 1953) used a set of human subsystem headings as check-boxes to determine whether a demand existed for each. Drury et al. (1990) used the subsystems below as a way of reminding the analyst of the dimensions along which task demands and human capabilities can vary:

Attention
Sensing
Perception
Decision making
Control
Feedback
Posture
Other (including forces exerted)

Each HTA or subtask is tabulated against these subsystems, with an entry if the analyst deems there to be a potential issue (i.e., if this could become a limiting subsystem and hence have error potential). For example, in eddy-current inspection (Table 9.1), the electronic system must be calibrated against a known standard, usually a component known to have no defects. One issue is that the standard may not accurately represent the aircraft component, so one subtask (3.3 Get eddy current standard) is shown below as an example. The potential problems are deemed either perceptual or decision in nature (i.e., can the inspector judge the paint on the aircraft component versus the standard, and does the inspector reach the correct decision based on this perception?).

In this way, either the task demands can be changed (i.e., a technological solution found for estimating paint thickness on component and standard) or a training requirement specified to allow the inspector to make the correct decision from the evidence available.

The first NDI technique analyzed was FPI, and the report (Drury & Watson, 1999) showed how HTA formed the link between hangar-floor knowledge and HFE knowledge. In subsequent reports, the 20-page HTAs were omitted beyond the top-level view shown in the example of Figure 9.5. Instead, the main presentation of output from the HTA were two sets of good practices. The first, and more detailed, was a table linking the top-level description (Figure 9.5) to the good practices. An example from the search function is shown in Figure 9.8. The second was a more discursive description of system-level good practices such as designing better documentation, personnel and training issues, HFE principles of hardware design (including HCI), environmental design, and shift-work effects. In Figure 9.8, the actual good practices have come from the detailed observation and reading necessary to perform the HTA, industry good practices that made HFE sense, and the application of HFE knowledge (e.g., in improving visual search). A feature appreciated by users was the final column of Figure 9.8 that went beyond the rule-based

Process	Good Practice	Why?
3. Search	Allow enough time for inspection of whole area	1. As shown in section 4.2.1, the time devoted to a search task determines the probability of detection of an indication. It is important for the inspector to allow enough time to complete FOV movement and eye scan over the whole area. When the inspector finds an indication, additional time will be needed for subsequent decision processes. If the indication turns out to be acceptable under the standards, then the remainder of the area must be searched just as diligently if missed indications are to be avoided.
3. Search	Provide clear instructions to inspector of expected intensity of inspection	1. The documentation should give the inspector enough information to provide a consistent choice of inspection intensity. Terms such as "general", "area" and "detailed" may mean different things to different inspectors, despite ATA definitions. Well-understood instructions allow the inspector to make the intended balance between time taken and PoD. If the inspector looks too closely or not closely enough then PoD may not be that intended by the inspection plan.
3. Search	Inspector should take short breaks from continuous visual inspection every 20 - 30 minutes	1. Extended time-on-task in repetitive inspection tasks causes loss of vigilance (Section 4.2.1), which leads to reduced responding by the inspector. Indications are missed more frequently as time on task increases. A good practical time limit is 20-30 minutes. Time away from search need not be long, and can be spent on other non-visually-intensive tasks.
3. Search	If search uses a loupe, ensure that magnification of the loupe in inspection position is sufficient to detect limiting indications.	1. The effective magnification of the loupe depends upon the power of the optical elements and the distance between the lens and the surface being inspected. Choose a loupe magnification and lens-to-surface distance that ensures detection. This may mean moving the lens closer to the surface, thus decreasing the FOV and increasing the time spent on searching. The cost of time is trivial compared to the cost of missing a critical defect.
3. Search	Provide lighting that maximizes contrast between indication(s) and background.	1. The better the target / background contrast, the higher the probability of detection. Contrast is a function of the inherent brightness and color difference between target and background as well as the modeling effect produced by the lighting system. Lighting inside a structure mainly comes from the illumination provided by the personal lighting (flashlight), which is often directed along the line of sight. This reduces any modeling effect, potentially reducing target background contrast, so that lighting must be carefully designed to enhance contrast in other ways.
3. Search	Provide the inspector with approved tools to prevent tools being improvised.	1. Inspectors will improvise tools if the correct one is not available. For example, inspectors use a knife to check elasticity of elastomer seals, or use a rag that catches on frayed control wires to inspect for fraying. While these may be adequate, they have not been tested quantitatively. Wrong indications may result.
3. Search	Use a consistent and systematic FOV scan path	1. A good search strategy ensures complete coverage, preventing missed areas of inspection. 2. A consistent strategy will be better remembered from task to task, reducing memory errors. 3. Searching for all defects in one area then moving to the next (Area-by-Area search) is quicker than the alternative of searching for all areas for each type of defect in turn (Defect-by-Defect search), but the probability of detection is reduced. It may be difficult to help inspectors to work Defect-by-Defect.
3. Search	Use a consistent and systematic eye scan around each FOV	1. A good search strategy ensures complete coverage, preventing missed areas of inspection. 2. A consistent strategy will be better remembered from task to task, reducing memory errors.
3. Search	Do not overlap eye scanning and FOV or blade movement.	1. It is tempting to save inspection time by continuing eye scans while the FOV is being moved. There is no adverse effect if this time is used for re-checking areas already searched. But search performance decreases rapidly when the eyes or FOV are in motion, leading to decreased probability of detection if the area is being searched for the first time, rather than being re-checked.
3. Search	Provide memory aids for the set of defects being searched for.	1. Search performance deteriorates as the number of different indication types searched for is increased. Inspectors need a simple visual reminder of the possible defect types. A single-page laminated sheet can provide a one page visual summary of defect types, readily available to inspectors whenever they take a break from the inspection task.

Figure 9.8   Task analysis of search function, giving good practices and reasons.

specification of good practices (column 2) to a more knowledge-based explanation of reasons for recommending that good practice (column 3). Here, users could get a flavor of the HFE reasoning and become sensitive to new or emerging issues where HFE knowledge might be applied. The good practices were also turned into simple checklists so that users could evaluate their own systems to see which practices were in use and which new ones could be added. The NDI reports have been taken up by the U.S. Army and two engine OEM suppliers, as well as airlines.

Cognitive Work Analysis (briefly)

With the rest of this book devoted to the topic of CWA, there is no need to repeat definitions and examples here. Rather, I will extract those issues which are comparable to, or contrasting with, HTA with the main sources being obviously Vicente (1999) and Hajdukiewicz and Vicente (2004).

CWA has at least one root in the STS tradition, where the principle of incompletion is a design tradition, if not a rule. The idea is that designers, even a team including operators, should not completely specify either the equipment interfaces or procedures, so that future operators have freedom to adapt to the changing demands typical of open systems. This is usually conceived as being a good design objective, potentially reducing stress by increasing operator control as in Karasek and Theorell's (1980) model. Vicente's CWA model is seen as formative (as opposed to normative or descriptive), using physical, chemical, and other constraints to derive what·operations *can* be done rather than specifying what *should* be done (normative) or what is *actually* done (descriptive). Vicente (1999) then goes on to show that CWA is constraint-based in contrast to the instruction-based TA methods. He shows (p. 72) a simple but elegant control theoretic illustration that instruction-based methods only apply to closed systems (i.e., almost no variation from outside the system), whereas constraint-based methods apply to the far more common (perhaps universal?) open systems. This would leave very little room for TA as a tool in HFE. The general impression (Vicente, 1999, p. 63) is that task analytic methods are representative of a class of techniques based on Tayloristic assumptions of "the one best way" of performing a task.

Integrations

This paper has given brief overviews of the HTA and CWA treating them as current endpoints of the TA and socio-technical systems traditions, respectively. Both arise from a framework of general systems theory (e.g., Ropohl, 1999) and attempt to derive a hierarchy of goals and functions for system operation. Neither is seen as leading to a formal document to be archived within the system, but rather as an ever-evolving system description to guide interface design and operator training.

As noted above, CWA was specifically developed to be a formative technique applicable to open systems, and its proponents characterize TA as a prescriptive technique applied to closed (or almost-closed) systems. HTA and CWA have followed quite different evolutionary paths and to some extent have seen their most frequent use in different types of system. TA was originally applied to military tasks which were seen as requiring sets of well-defined procedures. In its later incarnation as HTA, it was specifically designed to be applied to the cognitive tasks found in the process industries (Annett & Duncan, 1967; Stanton, 2006). CWA, in contrast, appears to be applied to more complex control and decision tasks, even though some of these are in domains that appear to be highly constrained (e.g., fast food retail; Bisantz & Ockerman, 2002). Thus, both CWA and HTA grew from process control domains, but have seen wide use beyond that. The contrast between the two techniques is still being debated, for example, in the context of rural car driving (Laberge, Ward, Rakauskas, & Creaser, 2007). They showed that HTA and WDA produced some similar and some complementary decision information requirements in a rural intersection driving situation. WDA produced more requirements overall, although to be fair the authors only present their task description rather than their TA in this short paper.

At least some of the perceived contrast between CWA and HTA arises from their reported uses rather than from fundamental differences. Vicente (1999) quotes Meister (1985) as stating that TA should avoid tasks with branching because it makes for a more complicated task representation. However, many practitioners do apply HTA to branching tasks; indeed any inspection task (such as the examples earlier) must by definition include a branch on whether a defect is or is not found. HTA can handle this quite simply by specifying plans that branch on finding a defect. Once branching is admitted into HTA, any external (open system) inputs can be accommodated. This is only in principle: we cannot hope to specify every possible system disturbance from beyond system boundaries (what about a meteor strike?). We can however specify different tasks, even different goals, contingent upon external events and alternative system states. At this level, one way to reconcile HTA and CWA is possible: HTA specifies task fragments to be launched given different system inputs or disturbances. Thus, the airline pilot relies on prevalidated checklists for different flight phases and events (potentially specified by TA), although a CWA might be a more appropriate description of the overall piloting job. This is essentially how Hajdukiewicz and Vicente (2004, p. 537, Figure 8) see TA as arising at the final phase of CWA when all sources of uncertainty (degrees of freedom in the work domain) have been eliminated. Indeed, Drury (1983) saw such sequences embedded in process control tasks as a legitimate part of task description for process control. Note that use of TA only later in the systems design process follows the restriction on degrees of freedom that take place at the allocation of function level, following Singleton's system analysis model in Figure 9.1.

We should note, however, that practitioners of HTA and its developers Annett and Duncan (1967) do not see any conflict in applying HTA to process control tasks (i.e., tasks with considerable uncertainty and external input). Are they missing something, or just stretching TA beyond its logical framework? Stanton (2006) disagrees with Vicente's (1999) characterization of TA as choosing example and quotations that paint an unnecessarily restrictive picture of HTA practice. Another possible integration is to use the plans in HTA more broadly. There is no reason why the plans element of HTA cannot specify such condition-based operations as "do X and Y while … " as well as simple sequences. There is even no reason in principle why it cannot include the principle of incompletion by specifying "do X, Y, or Z as appropriate," thus leaving the choice of method to the operator in real time. The specific sequences X, Y, and Z may themselves not be completely specified, although their goals must be in HTA. As shown in the inspection example, HTA in practice can provide a description at well above Phase 5 level, going at least to Phase 3, Strategy Analysis. Perhaps HTA practitioners should not have been writing task descriptions at this level of lack of specificity over the years, but they have and the systems analyzed and designed do not appear any the worse for that. This may not be a theoretically appealing way ahead, but it does show why practitioners of either technique seem to use it appropriately across a wide range of system and work types. Perhaps, as Laberge et al. (2007) suggest, the two techniques are indeed complementary in practice, if not in theory.

Conclusions

HFE practitioners have exhibited eclectic tastes when it comes to their techniques, borrowing unashamedly from mathematics, psychology, anthropology, physiology, or even other applied disciplines such as industrial engineering (Wilson & Corlett, 2006). Their motivation is to change, or at least understand and evaluate, systems of humans and technology. But even practitioners are not theory-free. Our discipline appears to reside firmly in what Stokes (1997) characterizes as "Pasteur's Quadrant," that fraction of research with high levels of both practical utility and theoretical understanding. Thus, critiques indicating that such a widely used technique as TA may be only strictly valid under extremely restricted conditions must give us pause. Is HTA indeed restricted to creating short fragments of description/analysis at the fifth level of CWA? Informal discussions with a few users and writers on HTA suggest that this may be a common view, but one that does not prevent them using HTA more widely than their view would warrant. Using the plans function of HTA, a broader interpretation is possible. As shown above, there is clearly a role for HTA as currently used in Phase 3 of CWA, although Phase 5 is still the obvious place to use HTA within a CWA context.

References

Annett, J. (2000). Theoretical and pragmatic influences on task analysis methods. In J. M. Schraagen, S. F. Chipman, & V. L. Shalin (Eds.), *Cognitive Task Analysis (pp. 25–37)*. Mahwah, NJ: Lawrence Erlbaum Associates.

Annett, J., & Duncan, K. D. (1967). Task analysis and training design. *Occupational Psychology, 41*, 211–221.

Ashby, W. R. (1956). *An Introduction to Cybernetics*. London: Chapman & Hall.

Bisantz, A. M., & Ockerman, J. J. (2002). Informing the evaluation and design of technology in intentional work environments through a focus on artifacts and implicit theories. *International Journal of Human–Computer Studies, 56*, 247–265.

Chipman, S. F., Schraagen, J. M., & Shalin, V. L. (2000). Introduction to cognitive task analysis. In J. M. Schraagen, S. F. Chipman, & V.L. Shalin (Eds.), *Cognitive Task Analysis (chap. 1)*. Mahwah, NJ: Lawrence Erlbaum Associates.

Drury, C. G. (1983). Task analysis methods in industry. *Applied Ergonomics, 14*(1), 19–28.

Drury, C. G. (1992). Inspection performance. In G. Salvendy (Ed.), *Handbook of Industrial Engineering, 2nd Ed*. New York: John Wiley.

Drury, C. G., & Prabhu, P. (1996). Information requirements of aircraft inspection: Framework and analysis. *International Journal of Human–Computer Studies, 45*, 679–695.

Drury, C. G., & Watson, J. (1999). *Human factors good practices in borescope inspection*. http://hfskyway.faa.gov/HFAMI

Drury, C. G., Paramore, B., Van Cott, H. P., Grey, S. M., & Corlett, E. M. (1987). Task analysis. In G. Salvendy (Ed.), *Handbook of Human Factors (pp. 370–401)*. New York: John Wiley.

Drury, C. G., Prabhu, P., & Gramopadhye, A. (1990). Task analysis of aircraft inspection activities: Methods and findings. *Proceedings of the Human Factors Society 34th Annual Meeting*, pp. 1181–1185.

Gramopadhye, A., & Thaker, J. (1998). Task analysis. In W. Karwowski & W. S. Marras (Eds.), *The Occupational Ergonomics Handbook (pp. 297–329)*. Boca Raton, FL: CRC Press.

Hajdukiewicz, J. R., & Vicente, K. J. (2004). A theoretical note on the relationship between work domain analysis and task analysis. *Theoretical Issues in Ergonomics Science, 5*, 527–538.

Karasek R. A., & Theorell, T. (1990). *Healthy Work: Stress, Productivity and the Reconstruction of Working Life*. New York: Basic Books.

Kieras, D. E. (1997). A guide to GOMS model usability evaluation using NGOMSL. In M. Helander (Ed.), *Handbook of Human–Computer Interaction* (pp. 391–438). Amsterdam: Elsevier.

Kirwan, B., & Ainsworth, L. K. (Eds.). (1992). *A Guide to Task Analysis*. London: Taylor & Francis.

Laberge, J., Ward, N., Rakauskas, M., & Creaser, J. (2007). A comparison of work domain and task analysis for identifying information requirements: A case study of rural intersection decision support systems. *Proceedings of the Human Factors and Ergonomics Society 51st Annual Meeting, pp. 298–302*.

Luczak, H. (1997). Task analysis. In G. Salvendy (Ed.), *Handbook of Human Factors and Ergonomics (chap. 12)*. New York: John Wiley.

Meister, D. (1971). *Human Factors: Theory and Practice*. New York: John Wiley.

Meister, D. (1985). *Behavioural Analysis and Measurement Methods*. New York: John Wiley.

Miller, R. B. (1953). A method for man–machine task analysis. WADC Technical Report 53-137. Wright-Patterson Air Force Base, Ohio.

Morawski, T., Drury, C. G., & Karwan, M. H. (1980). Predicting search performance for multiple targets. *Human Factors, 22*(6), 707–718.

Pennington, J. (1992). A preliminary communications systems assessment. In B. Kirwan & L. K. Ainsworth (Eds.), *A Guide to Task Analysis* (pp. 252–265). London: Taylor & Francis.

Rasmussen, J., Pejtersen, A. M., & Goodstein, L. P. (1994). *Cognitive system engineering*. Boston: John Wiley.

Ropohl, G. (1999). Philosophy of socio-technical systems. *Society for Philosophy Technology, 4*(3).

Stammers and Shepherd (1997) Task analysis. In J. R. Wilson & E. N. Corlett (Eds.), *Evaluation of Human Work*. London: Taylor & Francis.

Singleton, W. T. (1974). *Man–machine systems*. London: Penguin.

Stanton, N. A. (2006). Hierarchical task analysis: Developments, applications, and extensions. *Applied Ergonomics, 37*, 55–79.

Stokes, D. E. (1997). *Pasteur's Quadrant: Basic Science and Technological Innovation*. Washington, DC: The Brookings Institute.

Taylor, J. C., & Felten, D. F. (1993). *Performance by Design*. Englewood Cliffs, NJ: Prentice-Hall.

Van Eijnatten, F. M. (1991). From autonomous work groups to democratic dialogue and integral organizational renewal: 40 years of development and expansion of the socio-technical systems design paradigm. *Research Memorandum* 91–015. The Netherlands: Maastricht Economic Research Institute on Innovation and Technology, University of Limburg.

Vicente, K. J. (1999). *Cognitive Work Analysis*. Mahwah, NJ: Lawrence Erlbaum Associates.

Wenner, C. L. & Drury, C. G. (1997). Beyond "hits" and "misses": Evaluating inspection performance of regional airline inspectors. *Proceedings of the 41st Annual Human Factors and Ergonomics Society Meeting, Albuquerque*, NM, 579–583.

Wilson, J. R., & Corlett, E. N. (Eds.). (2006). *Evaluation of Human Work*. London: Taylor & Francis.

Chapter 10

Pragmatic Use of Cognitive Work Analysis in System Design: Extending Current Thinking by Adapting the Mapping Principle

William Elm, James Gualtieri, Jim Tittle,
Scott S. Potter, and Brian McKenna

Contents

Overview

Cognitive systems engineering (CSE) provides a means by which the interaction between people and machines can be described as they cope with complexity (Hollnagel & Woods, 2005). Cognitive work analysis (CWA) is useful for the design and testing of effective joint cognitive systems (JCS) only when understood in the larger context of a CSE approach that integrates that analysis with a representational approach to interface design. By adapting Woods' (1991) mapping principle as a framework to integrate the results from a CWA into system design and development for effective decision support, powerful insights become evident to better understand what is needed of a CWA and how system designers can successfully exploit the power of a CWA to deliver uniquely effective computer systems as one component of a JCS. The extended mapping principle is used as the overarching context for attributes of the CWA and also as a link between the CWA and representational design that, when combined with high quality systems engineering practices, offers an end-to-end, traceable, testable design methodology. Just as steam power augmented the physical work that a human could accomplish, the tools of the information age are able to augment the cognitive work that a human can accomplish when those tools are successfully designed using CWA within an overall CSE process.

Introduction

CWA was initially envisioned as an approach for the analysis and modeling of complex systems that has its foundations in the work of Rasmussen and his colleagues (Rasmussen, 1986; Rasmussen, Pejtersen, & Goodstein, 1994). This chapter extends the current thinking on CWA methodologies by utilizing Woods' (1991) mapping principle as a framework to characterize a JCS and to integrate the results from a CWA into system design and development. A JCS is the combination of human problem solver(s) and technologies (including information presentations and any automated agents), which act as coagents to achieve goals and objectives in a complex work domain (Hollnagel & Woods, 2005). In order to achieve their goals, the human problem solver(s) must work with the technology as a coordinated team.

CWA is not viewed as a stand-alone activity, but rather as a means toward the end of powerful, intuitive system design. In this chapter, we examine premises that provide insights into what is needed of a CWA and how the CWA can then be exploited to deliver unique and effective decision support tools for the JCS.

These premises highlight the importance of linking the broader set of CSE-derived CWA artifacts with the perceptual and psychological requirements for good representation design. CSE-based CWA techniques determine the proper information and relationships to map effortlessly to "expert"-equivalent mental models. Then, CSE-based representation design offers the visual form that matches the human perceptual, precognitive, and cognitive processes in order for it to be "internalized" accurately by the operator. When used together in a systems engineering process, both the analysis and design activities must be further adapted to integrate with the other.

By integrating CWA into a larger CSE framework, CWA becomes the key analytic basis for the development of effective decision support. The applied cognitive systems engineering (ACSE; see Elm, Potter, Gualtieri, Easter, & Roth, 2003 of ACSE's precursor, applied cognitive work analysis) methodology is a technique specifically tailored to meet the requirements of a successful CSE process. ACSE's CSE basis creates analytic artifacts that are strongly tied to system design. ACSE is effective because it balances an understanding of the research from several diverse fields (e.g., cognitive psychology, perception, systems engineering, operations research, and decision theory) with the application of that research as part of an engineering process. In striking this balance, ACSE has become a framework within which research findings are organized and integrated into engineering practice.

The JCS concept specifies the scope of the CWA. In systems where automation and technology play a role alongside the human in the accomplishment of goals (which defines a JCS), the JCS is the fundamental unit of analysis (McKenna, Gualtieri, & Elm, 2006). Therefore, the decision making completed by the entire JCS should be the focus of the CWA, not just the decision making of the human agent.

Further, good overall systems engineering practices make CWA part of an end-to-end process in which each visual element (as well as the entire, large-scale design of the workspace) is fully traceable to its CWA requirements and basis. These "design threads" offer insights on the explicit testing of the JCS, utilizing the novel approach of decision-centered testing (DCT) to evaluate the uniquely cognitive aspects of the joint system. DCT is a complementary use of the CWA output, explicitly linked to testing specific artifacts of the analysis output, just as these artifacts were linked to system design elements at the outset.

Underlying It All ... The Mapping Principle

The mapping principle (Woods, 1991) is a depiction of the transformation from the goals and task context within a work domain (i.e., the cognitive work endemic to the work domain, and therefore the cognitive work that the practitioners need to accomplish) to the representation of the information required to support that cognitive work (the encoding mapping) as well as the transformation from the representation to the interpretation and ultimate use of that information by the practitioner (the decoding mapping). It has been constructed as a framework to emphasize both the critical role of a CSE-based analysis in determining the information requirements as well as the fundamental importance of representational design in developing effective decision support systems (DSSs), and, ultimately, effective JCSs. The mapping principle provides several key premises that are discussed below.

Premise 1: The effectiveness of a decision support system (as a component of a joint cognitive system) depends on the effectiveness of the mapping from work domain through the representation to the end user. Effectiveness is not about the visual characteristics of the displays ... it is about how the displays represent the information space of the work domain to support the decision-making requirements.

The relationship between the visual structure established by a particular display element and the constraints and relationships of the domain itself is fundamental to the effectiveness of the visualization. Without an explicit specification of this mapping, this premise implies that it is impossible to determine if the visualizations are supporting users' needs as they intended, or making the "supported" task more difficult. See Figure 10.1 for our adaptation of Woods' original concept.

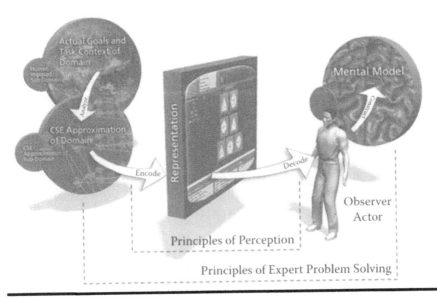

Figure 10.1 The mapping principle (adapted from Woods, 1991) dictates that design of a decision support system must be based on the informational requirements of the work domain—how they are reflected in the representation as well as how they are perceived by the user.

Fundamental to this premise is the representation effect. This refers to the finding that the way in which a problem is represented influences the cognitive work needed to solve that problem (Norman, 1988, 1993; Zhang & Norman, 1994). In other words, the design of artifacts in the DSS either improves or hinders our perceptual and cognitive capabilities, i.e., either improves or hinders the net effectiveness of the JCS.

Therefore, one of the goals of the CWA effort in support of decision support system design is to identify the linkage from the cognitive work requirements (i.e., decisions) being supported, and the form (e.g., visualization technique) for presenting the information to support those demands. The key to accomplishing this goal is to understand the mapping between the information regarding the behavioral state of the domain and the representation of that information within the visualization.

This premise is in part based on cognitive engineering principles that have been called representational aiding (Bennett & Flach, 1992; Woods, 1995; Roth, Malin, & Schreckenghost, 1997) and ecological interface design (Vicente & Rasmussen, 1990, 1992; Reising & Sanderson, 1998; Burns & Hajudukiewicz, 2004).

Premise 2: The mapping principle is actually composed of two mappings—encoding and decoding—that must each be considered in the context of the other. The ability of the problem solver to decode information depends on the design decisions made during encoding. From a design perspective, though, encoding must take place with careful consideration of the way in which the problem solver will decode the information displays.

The mapping principle helps to identify two mappings—encoding and decoding— which define an effective JCS. The first mapping is between the domain referents and perceptible (usually visual) elements in the representation within the computer medium. Ideally, the representation captures, at a basic, intuitive level, what is meaningful about the work domain—an *encoding* or *work domain to representation mapping*. As indicated in Figure 10.1, this encoding mapping actually consists of two partial mappings. Encoding-Partial-Mapping-A (represented by the "analyze" arrow in the figure) is an analytical mapping to determine the essential properties, concepts, abstractions, changes, and relationships in the world that define the information space essential to effective decision support. This partial mapping forms the CSE approximation of the work domain. Encoding-Partial-Mapping-B (represented by the "encode" arrow in the figure) is a design mapping where the analytic CSE model of the work domain is transformed or encoded into the structure and behavior of elements in the computer medium.

The second or *decoding* mapping consists of the relationship between the representation presented on the computer medium and the observers' or practitioners' perceptual internalization of information from it—*the mapping of representation into cognitive work*. As with the encoding mapping, the decoding mapping also consists of two partial mappings. Decoding-Partial-Mapping-A (represented by "decode" in the figure) is the user's perceptual extraction of information from the representation. This perceptual extraction partial mapping depends on the display characteristics of the visual elements as interpreted by the human perceptual system. Decoding-Partial-Mapping-B (represented by "construct" in the figure) is the internalized construction of a mental model of the work domain. Only this internalized mental model is actually employed in the problem solving.

This decoding mapping affects the cognitive work of observers as they carry out tasks in the work domain under the conditions of actual task performance (e.g., demands to shift focus of attention, varying tempo and time pressure of operations, risk of different types of failures in tasks, uncertain evidence about the state of higher order concepts). The decoding mapping is analogous to discussions of the affordances provided by the representation—how the encoding mapping affords the decoding of domain meaningful issues for practitioners (Flach, Hancock, Caird, & Vicente, 1995). Regardless of the model used in establishing the encoding mapping, the decoding mapping remains constant—it is the constant of human perception.

The end-to-end mapping for a decision support system can significantly affect the cognitive work needed to carry out domain activities in various ways. For example, some mappings may place a greater burden on the human component of the JCS, forcing a greater reliance on more deliberative, serial, resource-consuming cognitive processing, whereas other solutions shift the emphasis to technological solutions which could serve as an external memory. Some mappings may trap observers in a narrow keyhole, making it harder to revise focus as situations change; some mappings may restructure problems so that higher order properties are directly perceivable.

Premise 3: In order to effectively design DSSs, constructing a CWA is the essential catalyst for executing the mapping principle. With a CWA as an integral part of the mapping principle (and defining the content of the information to be represented in the DSS), the resulting representations are grounded in the critical concepts and relationships in the referent domain.

With the mapping principle as the foundation, a CWA needs to construct a model of the work domain that is consistent with expert mental models in order to facilitate the effective decoding process. The critical step in employing the mapping principle is the transformation of aspects of the referent domain (i.e., elements from the CWA) into properties of the dynamic computerized representation. Thus, the CWA provides a generative foundation on which the design is built. The encoding mapping depends on analyses of the nature of the work domain (i.e., CWA) to determine the interesting properties and changes of the domain referents, given the goals and task context of the practitioners to provide the key guidance and requirements to define successful encoding.

Having the mapping principle as the critical overarching framework for a CWA means that CWA artifacts cannot be considered in isolation—they must always be considered the means to discover the critical concepts and relationships as well as the critical characteristics and properties of the representation. As such, CWA artifacts must enable this encoding and decoding mapping. It is insufficient to have a hard break in the analysis-to-design effort or to have vague information requirements from the CWA serve as the basis for design. Rather, it is essential to have specific CWA artifacts that explicitly link to both specific content of the information spaces and specific properties of the representation.

In developing an effective mapping, design (built on a base of CWA) provides a setup for cognitive work in context by using the relationships, context, changes, events, and contrasts that give data meaning to organize those data. This process depends on having models of the processes in the work domain. As a result, representations that achieve direct correspondence are consistent with model-based design (Mitchell and Saisi, 1987). Designing effective representations is concerned with managing both encoding and decoding mappings so that the resulting human–computer team operates as an efficient JCS.

Using Problem Solving Research for DSS Design

Because mental models affect cognitive performance, research into expert complex decision making (Newell & Simon, 1976; Johnson-Laird, 1983; Gary & Wood, 2005) should inform the execution of a CWA when designing a DSS. Specifically, this research should inform what the CWA must produce in order to construct a CSE approximation of a work domain. Figure 10.2 illustrates how this link between research on effective mental models and the CSE approximation of a work domain is part of the larger mapping principle. For the purposes of this discussion, Rouse

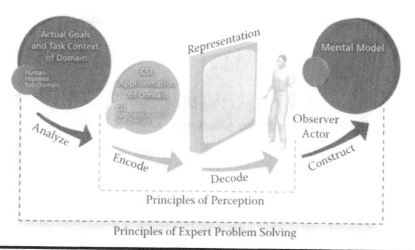

Figure 10.2 To effectively conduct a CWA, it is crucial to understand the characteristics of mental models and use that understanding to define success for a CWA output.

and Morris' (1986) definition of mental models is used: "the mechanism whereby humans are able to generate descriptions of system purpose and form, explanations of system functioning, observe system states, and prediction of future system states." The cognitive approximation that is a mental model effects problem solving in several ways: event anticipation and explanation, inferences about system state, representation effects, and allocation of attentional resources.

In order to advance the state of the art in CWA, it is necessary to develop approximations that provide a picture of functional concepts and their relationships, as they exist in the world (Rasmussen, 1986; Norman 1988). Operators of DSSs develop mental models of how automation works to anticipate system behavior and to explain why the automation reacts as it does. In other words, the CWA should represent the mental model of a given problem domain, which becomes the only means of conveying a representation to the operator. Which concepts and relationships the CWA needs to focus on is highly dependent on the goals and intentions of the decision maker. As Gentner and Nielsen (1996) noted, users cannot help trying to understand the "intestines" of the automation behind the system's interface. This compulsion toward making a mental model of the "internal logic" of the automation demands the application of the CWA within the context of the mapping principle. Adopting this perspective changes how CWAs are conducted; specifically, the approach suggests that the abstraction–decomposition space (Vicente, 1999), the most commonly implemented form of CWA, is insufficient to describe the complexities that exist within the problem domain (mental models are not reducible to a five-by-five grid). This is particularly problematic because some (Linegang et al., 2006; Sanderson, 1998) have argued that the abstraction–decomposition space is the foundation on which CWA rests. The CSE approximation must capture the operator's mental model in order to support the desired performance within the JCS.

Capturing the Internalized World Model to Understand a Problem Domain

The existence of mental models has important implications for JCS design. To interact with the world effectively, human agents must understand the state of the world, as well as the impact of intended action in that world. It has been suggested that users understand and reason about complex systems on the basis of a mental model of the system's internal mechanics (Halasz & Moran, 1983). These internalized mental models are the basis for all aspects of expert problem solving, from directing attention to the most detailed action decisions. Thus, when conducting a CWA, the analyst must consider the goal-oriented functional and abstract cognitive structures (concepts and their relationships) that humans form as their internalized understanding of the domain.

CWA was developed as a conceptual framework by Rasmussen and others (Rasmussen, 1986; Rasmussen et al., 1994; Vicente, 1999; Elm et al., 2003) to analyze nonbehavioral components of a task domain—the cognitive demands necessary to achieve the JCS's goals. CWA examines the work domain in which the JCS is situated. From a CSE perspective, the purpose of CWA is to define the content of the DSS design as part of a JCS solving problems in the work domain (i.e., the encode arrow from Figure 10.2).

CSE considers the humans who interact with the system to be agents involved in their work-related actions. Focusing on decision-relevant behavior on the job, CSE views JCS interactions in the context of work activities. It assumes that, in order to be able to design systems that work harmoniously with humans to form a JCS, the designer has to understand:

■ the goals of the JCS
■ the relevant cognitive work that needs to be accomplished
■ the information needed to accomplish that behavior
■ the larger functional context in which work is completed

Therefore, the product of a good CWA should focus simultaneously on the cognitive tasks agents perform, and the environment in which that performance occurs. *Premise 4: When conducting a cognitive work analysis, the analyst must consider the goal-oriented functional and abstract cognitive structures (concepts and their relationships) that humans form as their internalized understanding of the domain—a mental model.* Through interaction with the world and thoughtful study of it, humans form mental models that capture the essence of the functional relationships in the world as invariant forms, not situation-specific memories (Hawkins, 2004). Situational variability interact with these invariant forms to provide the contextual variability that the JCS experiences. Research has shown that there exists a causal link between the structure of memory and the characteristics of the situation (Anderson & Schooler, 1990), as well as being the key reflection of expert versus novice cognitive performance. These mental models are demand driven; they exist to make information more accessible. Hawkins (2004) goes on to propose that humans make a model of the world and constantly check that model against reality. In fact, several times a second, the human visual cognition system is making "predictions" about what it will perceive next. When that prediction is wrong, the human attentional processes are immediately aroused. (This instinctual arousal can be used to assist operators in redirecting their attention to areas of interest, which can be used as the basis for an effective alarm. The concepts are discussed in detail in the next section.) By understanding the interrelationships and constraints in the world, through the presence of mental structures, human operators are able to process information about the world effectively, and to utilize the information to achieve goals.

Mental models also govern the access of information, through the selection of relevant functionally invariant features for attention (Gibson, 1966). This enables decision makers to sample the environment at a high level of abstraction, without having to perform a bottom-up analysis of individual physical components (Rasmussen, 1986). Thus, mental models map completely to the relevant part of the functional model of the domain (Van der Veer et al., 1990). When inconsistencies between the mental model and the actual functional model of the domain occur, errors and unexpected effects will likely follow. This one-to-one mapping between the mental model and the actual goals and context of the domain implies a positive correlation between the complexity of the task domain and the complexity of the CWA approximation of that domain. For that reason, the results of the CWA must be graded against these requirements rather than a stand-alone bundle of artifacts evaluated against a particular theoretical perspective.

ACSE that adopts this perspective, that is expert mental model approximation as the basis for requirements, means that the human operator simply cannot be considered a conduit transforming data into actions utilizing the artifacts of the JCS. Very complex human responses can be generated with relatively primitive inputs when they are indexed against the operator's internal mental model. It is, therefore, of critical importance that the CWA accurately approximate an expert operator's mental model. This CSE-based approximation of the domain serves as the requirements template and the presentation concept specification for the representational design of decision-aiding tools to develop a JCS.

Achieving Decision Effectiveness

Premise 5: The structure of abstract concepts is informative as a working engineering approximation of a mental model that is not empirically observable. This engineering approximation must be embedded in the DSS design for effective decision performance of the JCS.

To achieve decision effectiveness, an expert mental model, the target of the CWA, once captured must serve as the basis for the design of the information space within a DSS. The structure of abstract concepts is informative as a working engineering approximation of a mental model that is not empirically observable. As noted in the previous section, mental models are centrally important to dynamic decision making and problem solving—therefore, in order to promote effective cognitive performance, a JCS must contain a model of the domain in which it is attempting to achieve goals in order to be effective (Chi & Glaser, 1985; Robertson, 1990). JCSs with a model of the system with which they were interacting have shown significantly better performance than JCSs without that model for novel problems (Halasz & Moran, 1983). Thus, embedding an analytic approximation of an appropriate mental model of the actual goals and context of the domain within the DSS design for the JCS is critical for having an effective and robust JCS capable of achieving goals (Gott, Lajoie, & Lesgold, 1991). Woods, Johannesen, Cook, and Sarter (1994) went further, stating that it is essential to have a thorough understanding of the functional structure of the domain (both in terms of its abstractions and in terms of relationships) and to be able to use this knowledge in operationally effective ways. This leads to robust and reliable decision performance (i.e., at the skill, rule, or knowledge levels of problem-solving) in both ordinary and unusual or novel situations (Rasmussen, 1986).

There is considerable evidence that the core determinant of skilled performance is the knowledge base accumulated in long-term memory stored in the form of a mental model (Simon, 1982; Sweller, Chandler, Tierney, & Cooper, 1990). Research has shown that experts employ "better" mental models than novices—models with richer domain knowledge, more structure and interconnections, and a better basis in the underlying principles of the work domain (Glaser and Chi, 1989; Gualtieri, Folkes, & Ricci, 1997; Woods and Roth, 1988). Mental models include a range of knowledge about concepts, known "facts," perceived causal relationships, and heuristics about a particular task or work domain. In this context, accurate mental models of the domain result in more appropriate and more effective decisions, and therefore better outcomes. A JCS's performance on a task is a function of the accuracy of that JCS's representation of causal relationships between variables in the decision environment. Highly effective JCS will "know" more about the causal relationships between variables in a decision environment. Therefore, the DSS portion of the JCS must reflect an explicit representational encoding of the cognitive system engineer's best approximation of an expert mental model in order to enable effective human performance within the JCS (Chi & Glaser, 1985; Halasz & Moran, 1983; Robertson, 1990). Thus, the decision effectiveness of the resulting JCS should be the criterion by which all CWAs should be measured.

Expanding CWA to Facilitate Decision Effectiveness

Premise 6: The CSE approximation of the functional complexities of the work domain must be integrated with the design of the DSS components and interactions in order to form an effective JCS.

The ACSE technique is an expansion to CWA, which was developed as a process to capture and record a CSE approximation of the functional complexities within the domain and integrate this with the design of the components and interactions of the JCS. ACSE emphasizes a series of parallel processes to reduce the gap between analysis and design to a series of small, logical, engineering problems. These artifacts form a continuous design thread that provides a principled, traceable link from CWA to visualization. The CWA artifacts serve as a post hoc mechanism to record the results of the design thinking and as stepping stones for the subsequent step of the process. At each intermediate point, the resulting decision-centered artifacts create the spans of a design bridge that link the demands of the domain as revealed by the cognitive analysis to the elements of the decision aid. Each intermediate artifact also provides an opportunity to evaluate the completeness and quality of the analysis effort, enabling modifications to be made early in the process. The linkage between the CWA artifacts also ensures an integrative process; changes in one-artifact cascades along the design thread necessitating changes to all. Details concerning each step of the ACSE approach can be found in Elm et al. (2003). ACSE is a structured and principled methodology to systematically transform the problem from an analysis of the demands of a domain to developing visualizations and decision-aiding concepts that will provide effective support. The result is a traceable and pragmatic methodology that provides a repeatable engineering process for the design of effective systems.

Perceptual Decoding and Representational Design

As we stated earlier, practitioners form mental models of the systems and processes under their control and use those models as the basis for problem solving, for example, to interpret new data and unexpected events. In the absence of effective representation design, practitioners may form incomplete or inaccurate models of the processes under their control. This is an example of the representation effect (Norman, 1993; Zhang & Norman, 1994): How a problem is represented influences the cognitive work needed to solve that problem, either improving or degrading performance.

Given the representation effect, representation designers must attempt to develop problem representations to support the cognitive work involved in a particular work domain. These efforts at representation aiding or ecological interface design attempt to connect:

■ the processes and tasks in the work domain
■ the cognitive work of practitioners responsible to achieve goals in that work domain
■ the representations that support the ability of practitioners to understand and act successfully in the work domain

Although it is clear that an important goal of representation design is to aid the practitioner in forming appropriate mental models of the work domain, as shown in Figure 10.3, there is no direct route by which the designer can influence such high-level cognitive processes. Rather, CSE designers only have control of the representation presented to the user's perceptual systems and the selection of what automation to include in the system. Although the choice and use of sophisticated automation is an important component of DSS design, we will focus here on the nature of representations presented to an individual, and the implications of the decoding of those representations by human perception (currently primarily the visual system).

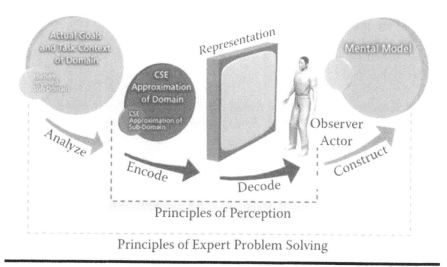

Figure 10.3 **The mapping principle contains the inner encode–decode component of representation design that emphasizes the importance of principles of human perception (decode) on the design decisions within the DSS (encode). Effective encoding is based on a thorough understanding of decoding.**

Effective Decision Support

Premise 7: Effective representation design requires that the CWA explicitly identify the cognitive work and information relationship requirements from the work domain.

Within the discipline of CSE, knowledge about the work domain is obtained through CWA (Rasmussen et al., 1994; Vicente, 1999; Elm et al., 2003); therefore, achieving this goal requires having mechanisms for transferring the insights obtained from our analyses to our representation designs. Described as part of an engineering design process, effective representation design requires a successful transition from analysis artifacts to design artifacts (Potter, Gualtieri, & Elm, 2003). In the ACSE approach, this is accomplished by transforming information relationship requirements (IRR) into an explicit statement of the intent of the display concept, or a representation design requirement (RDR). Working opportunistically backwards and forwards through such design threads provides a mechanism for a design team to constantly calibrate whether the proposed representations effectively support the cognitive work required by practitioners in the work domain and the display representation of the problem in the proposed visualization.

Establishing effective representation design starts from a simple basic principle, derived from the generalization that data become meaningful based on their relationships to other data, to larger frames of reference, and to the goals and expectations of the observer—to build conceptual spaces by depicting relationships within a frame of reference.

This basic principle can be decomposed into four sub-principles (Woods, 1995). Each principle is concerned with how to establish a direct correspondence between the representation and the process represented, given the goals and tasks of practitioners in that field of practice.

- **Choose frames of reference** from the CWA that effectively support the relevant cognitive work. Each frame of reference is like one perspective from which someone extracts meaning from data about the underlying process or activity. The CWA should define the appropriate frames of reference for interpreting information.
- **Put data into context**. Data should be in the context of related values and around important issues in the work domain. One prerequisite is to know what relationships are informative in what contexts in the field of practice. Putting data into context requires modeling the work domain to know what are the related data that give meaning to any particular datum (and how these sets change with context), and what the higher order issues of interest suggest about the collection and integration of data.
- **Highlight change and events**. Representations should reveal the dynamics, evolution, and future paths for the process in question. Events are temporally extended behaviors of the device or process involving some type of change in objects or situations. For representations to reveal the dynamics of the monitored process, they must be based on determining what are operationally interesting changes or sequences of behavior (from the CWA).
- **Highlight contrasts**. Meaning lies in contrasts—some departure from a reference or expected course. Since what is informative depends on the goals and expectations of the observer, mechanisms need to be in place to see actual conditions relative to expected norms for that context. Representing contrast means that one indicates the relation between the contrasting objects, states, or behaviors—how behavior departs or conforms to the contrasting case.

A particularly important technique for achieving some of the information highlighting described above involves taking advantage of the way in which the human visual system processes different types of stimuli to influence the perceived *salience* of particular graphical elements of the representation. A salience map provides a mechanism for relating the most important elements of the display to their perceived visibility or relevance. Figure 10.4 is the design artifact in which the representation designer makes a specific rank ordering of the various elements that will constitute the final representation design. The overall display space can be structured on the basis of decisions about the relative importance of information obtained from the CWA.

The appropriate number of levels in a salience map depends on the complexity of the domain to be represented and the nature of the stimulus elements in the display design. In the salience map in Figure 10.4, we show five salience levels. At the

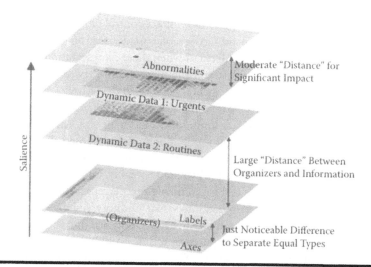

Figure 10.4 **The use of a salience map to structure information in a visual display. Controlling salience allows a representation designer to effectively direct attention and highlight information based on the RDRs obtained from a CWA.**

lowest level of salience are graphical elements that organize and provide structure for the overall display (e.g., plot labels). At the next level of salience are so-called static data or reference marks. These are important values for the system, but they are known in advance and do not change in the course of normal operations. The last three levels of salience are reserved for dynamic data elements. Routine dynamic data should be the least salient of the three levels. Data that are urgent should be more salient than the routine data. Finally, abnormalities or alarms should be the most salient items on the display.

Salience in the design results from specific decisions made by the designer, and is a statement of design intent which results from an assessment of the relative importance of different information as determined by a CWA (e.g., abnormalities should be made most salient on the display in order to immediately draw the attention of the operator). However, perceived salience of the visual display, by a practitioner, is based on the pre-attentive perceptual mechanisms of the human visual system, and results from spatial and temporal contrasts of the individual display elements (Rosenholtz, 1999). In order to effectively direct practitioner attention, create relevant contrasts to highlight information, and group data into meaningful combinations, designers must utilize characteristics of the human's pre-attentive perceptual mechanisms.

Perceptual Processes as Constraints

Premise 8: Perceptual processes are primarily cognitively impenetrable.

Because representation designers only have direct control of the encoding of information, it is crucial that every encoding decision is made based on our understanding of the perceptual decoding that will occur when those representations are observed by human practitioners. To understand the relationship between relatively high-level cognitive representations (e.g., mental models), and relatively low-level perceptual representations (e.g., perception of the spatial structure in a visual display) it is useful to consider an important assumption behind many information processing and computational models of cognition. Simon (1969) proposed that cognition is a "nearly decomposable" process, meaning that it could be recursively decomposed into more basic processes, each of which only weakly interacted. For example, human mental capabilities can be broken down into perceptual and cognitive processes. Perception can be decomposed into visual, auditory, smell, taste, and tactile processes, and visual perception can be further decomposed into subprocesses such as the four perceptual processes proposed by Palmer (1999): image-based, surface-based, object-based, and category-based. Such an assumption has been the basis for much of the work in cognitive psychology, which operates by isolating and studying individual cognitive or perceptual systems independently of others.

Furthermore, researchers such as Fodor (1983) have argued that perceptual processes are primarily bottom-up (i.e., sensory input driven) and "cognitively impenetrable," which means that we can neither be consciously aware of, nor directly influence, low-level perceptual processes (e.g., perceptual grouping or texture segregation). Now this is not to say that cognition can never influence the nature of perceptual representation. Indeed, every introductory psychology textbook has notable examples of such interactions (e.g., the Necker cube or bi-stable pictures with multiple interpretations). However, it is important to note that these examples all involve very carefully constructed stimuli, designed to be ambiguous with multiple interpretations that never occur naturally. The importance of this point for representation designers is that neither training nor practitioner experience can significantly alter the perceptual decoding of the representations we design. Indeed, as studies of the representation effect have shown, deficiencies in representation of a problem must be compensated for by increases in cognitive work from human problem solvers. Thus, it is crucial that we understand the way in which the human visual system decodes visual representations so that our designs may take advantage of natural human capabilities rather than fight against them.

A classic example of the effects of this cognitive impenetrability of perception can be seen in the phenomena of visual search. For example, Triesman and Gelade (1980) found that the human visual system is very adept at finding a target visual element in a field of distracters elements if the target differs from the distracters on a single perceptual feature (e.g., shape or color). In fact, we are so good at this task that the time to conduct a visual search in such conditions can be independent of the number of distracters. However, if the difference between the target visual element and the distracters is based on a conjunction of features such as shape and color (e.g., searching for a red triangle in a field of red and green circles and green triangles) then performance is greatly diminished and search time increases proportionally with the number of distracters. More recent studies suggest that certain feature conjunctions can lead to effective visual search (Triesman and Gormican, 1988), but it remains to be determined if these conjunctions are actually higher-order features. Such a result has an obvious implication for representation designers: despite the advantage of encoding data elements as conjunctions of multiple visual dimensions (i.e., the ability to uniquely encode a very large number of visual elements), doing so can render useless the powerful capability of the human visual system to process and search many visual elements in parallel. In doing so, the fast and accurate perceptual task of pre-attentive visual search is turned into the slow and error prone cognitive task of identifying and categorizing each visual element one at a time.

The visual search paradigm is the primary method used in perception research to identify the many pre-attentive mechanisms that the visual system utilizes to provide powerful processing to those regions of the visual field where we are not concentrating our attention (Wolfe, 1998). The results of such studies have provided evidence for the existence of pre-attentive perceptual mechanisms to process a variety of stimulus properties including: abrupt visual onsets (Yantis & Jonides, 1984), color (D'Zmura, 1991; Nagy & Sanchez, 1990), form (Brown, Weisstein, & May, 1992; Triesman and Gormican, 1988), line segment orientation (Foster & Westland, 1995; Landy & Bergen, 1991), motion (McLeod, Driver, Dienes, & Crisp, 1991; Nakyama & Silverman, 1986), and size (Duncan & Humphreys, 1992; Triesman and Gelade, 1980). Such pre-attentive perceptual mechanisms play an important role in allowing the designer to focus the practitioner's attention to particular regions of the visual display, group similar or related information, and create contrasts to highlight particularly important information. Specifically, the more important aspects of the work domain need to be more easily detected in the display representation, and thus an effective representation design will include specific controls for the relative visibility of different elements in the overall representation design.

Offloading Cognitive Work

Premise 9: Representation designers should seek to offload as much cognitive work to the pre-cognitive perceptual systems as possible.
As we saw from the discussion of visual search, understanding characteristics of the human perceptual systems can be used to aid cognition with a DSS by effectively encoding information. Thus, as a general principle of effective representation design, whenever possible, difficult cognitive tasks should be transformed into relatively easier perceptual tasks.

A particularly elegant example of the value of this approach can be seen in the design and use of pattern view or configural displays (Woods, 1995; Burns

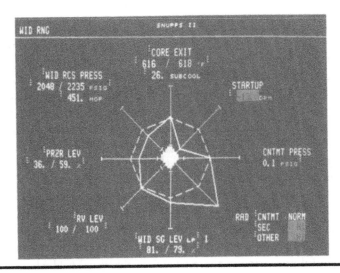

Figure 10.5 Example of a pattern view display that integrates many individual system parameters into a visual form whose deformation exploits the perceptual sensitivity to nonrigid motion and results in emergent patterns (i.e., shapes) that map onto unique system states. Explicitly coding such displays using understanding of visual perception allows pre-attentive situational awareness.

& Hajdukiewicz, 2004), which are integrated spatial representations that interrelate data to provide a coherent view into a process (see Figure 10.5). An effectively designed pattern view display allows practitioners to pick up status at a glance rather than having to read, integrate, and interpret many data elements. As such, configural displays are a clear example of using a core perceptual competency (perception of visual form and perceiving nonrigid motions of that form) to transform a difficult cognitive task into a relatively simple perceptual one.

In this section, we have emphasized the role of perceptual mechanisms and capabilities in guiding representational design. As we have seen in the cases of visual search and pattern view displays, by choosing our representations appropriately the robust capabilities of the human visual system enable the design of decision aids which can effectively and quickly provide observability of the underlying system processes and support the development of a rich mental model of the underlying work domain. However, it is important to note that these same powerful perceptual mechanisms transform the abstract information space, resulting from our CWA, into a perceptual representation from which the practitioner must build his or her mental model of the work domain. If this representation is not well-matched to the cognitive work required by the practitioner, it will have to undergo additional mental operations, with their corresponding costs (e.g., Zhang & Norman, 1994) to bring it into alignment with the user's mental model.

Decision Centered Testing

Premise 10: Tests represent a natural complement to "close the loop" by evaluating the resulting effectiveness of decision aiding tools.
Achieving a high level of decision-making effectiveness by the JCS requires that the operator has an effective mental model of the fundamentals of the work domain. A CWA-based DSS design will help generate this model. However, the display is only the designer's hypothesis about the way the domain works (Woods & Hollnagel, 2006). As is the case with all hypotheses, the display design must be rigorously tested before it is accepted into use. Each display element can be tested to see if it provides decision-making support for the cognitive work for which it was designed. Just as the CWA is used in the creation of a display design, it also must be the foundation on which designs are tested to ensure that cognitive work is being supported.

The test becomes a mechanism for ensuring the visualization was designed in a manner that allows an operator to routinely make correct decisions. Although a CWA is used to capture the functional and information relationships necessary to make these decisions, errors can be made in the analysis itself or in the encoding that is based on the analysis. Testing to identify the decision-making support provided by a DSS, therefore, must be used to ensure that the transformations reflected in the Mapping Principle are executed properly.

Premise 11: By testing against the analysis, the success or failure of the visualization to support decision making can be attributed to how closely the CSE practitioner approximates the domain in an analysis and how well the analysis artifacts are encoded into a display.
It is with this in mind that decision-centered testing (DCT) was created (Rousseau, Easter, Elm, & Potter, 2005). DCT aims to assess the net decision-making effectiveness of a JCS by extracting cognitive work from the analysis and making it the focus of the test. A user is then asked to perform this cognitive work using a visualization created specifically to support that work while under stressors specifically selected for that type of cognitive work. These results can be compared to results produced from users on other DSS display concepts (the results can also be compared against a "gold standard" of performance, i.e., the best performance possible given the state of the world).

Differentiating it from standard system evaluations, DCT requires using measures that go well beyond typical performance metrics (typical methods often include such metrics as the number of subtasks achieved per person per unit of time and the corresponding simple baseline comparisons or workload assessment metrics). Although these standard metrics can facilitate comparisons between systems, they do not address the resilience of the JCS in the face of novel situations, nor do they present solid evidence that the display enhances the decision support provided to the user. A key metric for all DCTs is whether or not the correct decision has been made. This is different from measuring performance results, because, due to the number of variables that go into the end result, a good decision can still lead to bad performance, and a poor decision can still end with good performance. Therefore, DCT seeks only to ensure that the DSS will support good decision making, even when intense cognitive pressure is exerted on the operator.

It is important to single out the cognitive work being supported when testing the decision-making support provided by a visualization. In doing so, it becomes easier to identify whether the correct decision is made or not. The results of the test will either support or refute the hypothesis that was generated by the designer about the particular design supporting that decision (i.e., did a particular DSS element assist the operator in making the correct decision?). Having a CWA-based design, therefore, provides the opportunity to examine what led the designer to make that hypothesis. The designer has a place to focus, and reanalysis or redesign can quickly be completed. This is more beneficial than testing the system as a whole, where the hypothesis under test becomes too broad to determine the impact specific display elements have on decision making.

Finally, through CWAs, analysis patterns can be abstracted across settings and situations (Woods & Hollnagel, 2006). Problem areas in one domain are likely to be problem areas in another domain (e.g., it is always difficult for an operator to balance multiple, competing goals). As more tests are run, more problem areas can be identified. As the system has been tested in the past, designers will be able to identify, a priori, decision-difficult areas within an analysis. In effect, analysis patterns will emerge that carry with them known, inherent risks. Having identified these problem areas in the analysis, effective design solutions can be created early in the system design process (i.e., design patterns follow their analysis pattern counterparts).

Premise 12: DCT tests the decision support provided by a design, and, because each test is based on a CWA, DCT provides insights back into the analysis.

DCT has been designed to provide a much-needed evaluation methodology for decision support systems designed for complex work domains. With DCT, it is possible to evaluate whether or not the new human–computer team truly demonstrates increased decision-making effectiveness in situations specifically designed to stress the potential weaknesses in the JCS. If a new DSS excels in a DCT weakness-centered evaluation, it will ensure a much more robust decision-making team and thus be better able to deal with the complex decision-making demands of the work domain.

Conclusion

The only pragmatic engineering approach to producing an effective JCS is through a rigorous design process that uses a CWA as its starting point—epiphanies are possible but not dependable. *But, although a cognitive work analysis is valuable, its value is not as a stand-alone artifact, but as part of a larger methodological basis for designing systems.* The CWA must define the structure and the content of the information to be encoded in the display space, which must then be decoded by the observer (Premise 1). However, the CWA in itself does not dictate this encoding; a successful transition from analysis to design requires that the designer understands how experts solve problems and how people perceive objects in the world.

The importance of understanding perception and problem solving is best represented by the reflective nature of the mapping principle. *Although designers cannot change the way in which people perceive information or form mental models, many years of research have been conducted in these fields that must be considered during the design phase.* Understanding how people form mental models necessarily influences the way the analysis must be structured. As people make decisions based on a functional representation of the world (Premise 4), then the analysis should represent the work domain's functional relationships (Premise 6). Understanding how people perceive objects in the world will have a large impact on the way that the designer should encode information. As an example, designers must carefully control a display's salience in order to ensure that the operators can redirect their attention to the most important events (Premise 7).

With this in mind, it becomes even more important to have a structured methodology, such as ACSE, that specifies both the structure of the analysis as well as the way to transform this analysis into a design. Because the analysis must provide an approximation of the mental model of a domain expert, it makes sense that the analysis output should also provide key insights for the way that information must be encoded within the display (thereby allowing for the creation of visualizations that allow novices to see the domain as experts do). However, making the conscious choice to encode this information is not enough. A designer must be cautious about how this information is encoded; how the perceptual system works cannot be altered (Premise 8). Even with the correct analytic content, a poor encoding can lead to perceptual ambiguity and faulty mental models being constructed. The ACSE methodology allows a traceable link from the display to the information relationships, which can then be used to determine if the display has done an adequate job of representing the information. Even if the proper relationships between data are identified, if they are not encoded to adhere to the principles of perception, the human is placed at a disadvantage when operating within the JCS. Making a display that adheres to the principles of perception without encoding the proper data relationships places the operator at a similar disadvantage. In order for a design to provide the proper support to the user, it must account for all transformations represented within the mapping principle.

The premises offered in the previous sections are intended to shape and strengthen not only the way that CWAs are used, but the entire design process, as well. *In using a structured methodology, there can be accountability for all actions taken in the design process.* Every action in the process must be done with a purpose, so that it shows the way forward, while allowing a link back to other actions that have been taken. They show the fundamental bases on which decision support systems must be designed, and allow explicit DCT to link back to a specific flaw in a specific analysis or design action.

Although these premises do not dictate what a system design should look like or the exact design process to use, Elm et al. (2003) and Potter et al. (2003) discuss ACWA, a precursor to ACSE—a methodology that was specifically tailored from this approach. They create three core requirements for system design: an analysis must be completed that describes how an expert thinks about the domain; the system must be designed from this analysis; and the way the analysis is encoded in the design must reflect the perceptual abilities of people. Only when all three of these requirements are met can the designer be confident that a truly effective JCS has been created.

References

Anderson, J., & Schooler, P. (1990). *Reflections of the Environment in Memory.* Pittsburgh, PA: Carnegie Mellon University Press.

Bennett, K. B., & Flach, J. M (1992). Graphical displays: Implications for divided attention, focused attention, and problem solving. *Human Factors, 34,* 513–533.

Brown, J. M., Weisstein, N., & May, J. G. (1992). Visual search for simple volumetric shapes. *Perception and Psychophysics, 51*(1), 40–48.

Burns, C. M., & Hajudukiewicz, J. R. (2004). *Ecological Interface Design.* Boca Raton, FL: CRC Press.

Chi, M., & Glaser R. (1985). Problem solving ability. In R. Sternberg (Ed.), *Human Abilities: An Information Processing Approach.* New York: Freeman.

D'Zmura, M. (1991). Color in visual search. *Vision Research, 31*(6), 951–966.

Doyle, J. (1997). The cognitive psychology of systems thinking. *System Dynamics Review, 13*(3), 253–265.

Duncan, J., & Humphreys, G. W. (1992). Beyond the search surface: Visual search and attentional engagement. *Journal of Experimental Psychology: Human Perception and Performance, 18*(2), 578–588.

Elm, W. C., Potter, S. S., Gualtieri, J. W., Roth, E. M., & Easter, J. R. (2003). Applied cognitive work analysis: A pragmatic methodology for designing revolutionary cognitive affordances. In E. Hollnagel (Ed.), *Handbook for Cognitive Task Design.* London: Lawrence Erlbaum Associates.

Flach, J., Hancock, P., Caird, J., & Vicente, K. (Eds.). (1995). *Global Perspectives on the Ecology of Human–Machine Systems.* Hillsdale, NJ: Lawrence Erlbaum Associates.

Fodor, J. (1983). *Modularity of Mind.* Cambridge, MA: MIT Press.

Foster, D. H., & Westland, S. (1995). Orientation contrast vs. orientation in line-target detection. *Vision Research, 35*(6), 733–738.

Gary, M. S., & Wood, R. E. (2005). *Mental models, decision making and performance in complex tasks. Proceedings of the 2005 International System Dynamics Conference.* System Dynamics Society: Albany, NY.

Gentner, D. R., & Nielsen, J. B. (1996): The anti-Mac interface. *Communications of the ACM, 8,* 70–82

Gibson, J. (1966). *The Senses Considered as a Perceptual Systems*. Boston, MA: Houghton Mifflin.

Glaser, R., & Chi, M. (1989). Overview. In M. Chi, R. Glaser, and M. Farr (Eds.), *The Nature of Expertise*. Hillsdale, NJ: Lawrence Erlbaum Associates.

Gott, S. P., Lajoie, S. P., & Lesgold, A. (1991). Problem solving in technical domains: How mental models and metacognition affect performance. In R. F. Dillon (Ed.), *Instruction: Theoretical and Applied Perspectives*. New York: Praeger.

Gualtieri, J., Folkes, J., & Ricci, K. (1997). Measuring individual and team knowledge structures for use in training. *Training Research Journal, 2*, 117–142.

Halasz, F., & Moran, T. (1983). Mental models and problem solving in using a calculator. *Proceedings of the SIGCHI conference on Human Factors in Computing Systems*. Boston, MA.

Hawkins, J. (2004). *On Intelligence*. New York: Times Books.

Hollnagel, E., & Woods, D. D. (2005). *Joint Cognitive Systems: Foundations of Cognitive Systems Engineering*. Boca Raton, FL: CRC Press.

Hubel, D., & Wiesel, T. (1968). Receptive fields and functional architecture of the monkey striate cortex. *Journal of Physiology, 195*, 215–243.

Johnson-Laird, P. N. (1983). *Mental Models: Towards a Cognitive Science of Language, Inference, and Consciousness*. Cambridge, MA: Harvard University Press.

Landy, M., & Bergen, J. (1991). Texture segregation and orientation gradient. *Vision Research, 31*(4), 679–691.

Linegang, M. P., Stoner, H. A., Patterson, M. J., Seppelt, B. D., Hoffman, J. D., Crittendon, Z. B., & Lee, J. D. (2006). Human-automation collaboration in dynamic mission planning: A challenge requiring an ecological approach. *Proceedings of the Human Factors and Ergonomics Society 50th Annual Meeting*. San Francisco, CA.

McKenna, B., Gualtieri, J., & Elm, W. (2006). Joint cognitive systems: Considering the user and technology as one system. *Proceedings of the International Council on Systems Engineering 2006 Annual Symposium*. Orlando, FL.

McLeod, P., Driver, J., Dienes, Z., & Crisp, J. (1991). Filtering by movement in visual search. *Journal of Experimental Psychology: Human Perception and Performance, 17*(1), 55–64.

Mitchell, C., & Saisi, D. (1987). Use of model-based qualitative icons and adaptive windows in workstations for supervisory control systems. *IEEE Transactions on Systems, Man, and Cybernetics, 17*, 594–607.

Nagy, A. L., & Sanchez, R. R. (1990). Critical color differences determined with a visual search task. *Journal of the Optical Society of America–A, 7*(7), 1209–1217.

Nakayama, K., & Silverman, G. H. (1986). Serial and parallel processing of visual feature conjunctions. *Nature, 320*, 264–265.

Newell, A., & Simon, H. (1972). *Human Problem Solving*. Englewood Hills, NJ: Prentice-Hall.

Norman, D. A. (1988). *The Psychology of Everyday Things*. New York: Basic Books.

Norman, D. A. (1993). Cognition in the head and in the world: Introduction to a debate on situated cognition. *Cognitive Science, 17*, 124–138.

Palmer, S. E. (1999). *Vision Science: Photons to Phenomenology*. Cambridge, MA: MIT Press.

Potter, S. S., Gualtieri, J. W., & Elm, W. C. (2003). Case studies: Applied cognitive work analysis in the design of innovative decision support. In E. Hollnagel (Ed.), *Cognitive Task Design*. New York: Lawrence Erlbaum Associates.

Rasmussen, J. (1986). *Information Processing and Human–Machine Interaction: An Approach to Cognitive Engineering*. New York: North-Holland Series in System Science and Engineering, Elsevier Science Publishing.

Rasmussen, J., Pejtersen, A., & Goodstein, L. (1994). *Cognitive Systems Engineering*. New York: John Wiley.

Reising, D. V., & Sanderson, P. (1998). Designing displays under ecological interface design: Towards operationalizing semantic mapping. In *Proceedings of the Human Factors and Ergonomics Society 42nd Annual Meeting*, Santa Monica, CA: Human Factors and Ergonomics Society, pp. 372–376.

Robertson, W. (1990). Detection of cognitive structures with protocol data: Predicting performance on physics transfer problems. *Cognitive Science, 14*, 253–280.

Rosenholtz, R. (1999). A simple salience model predicts a number of motion pop out phenomena. *Vision Research, 39*, 3157–3163.

Roth, E. M., Malin, J. T., & Schreckenghost, D. L. (1997). Paradigms for intelligent interface design. In M. Helander, T. Landauer, and P. Prabhu (Eds.), *Handbook of Human–Computer Interaction,* 2nd Ed. Amsterdam: North-Holland.

Rouse, W. (1983). Models of human problem solving: Detection, diagnosis and compensation for system failures. *Automatica, 19*, 613–625.

Rouse, W., & Morris, N. (1986). On looking into the black box: Prospects and limits in the search for mental models. *Psychological Bulletin, 100*(3), 349–363.

Rousseau, R., Easter, J., Elm, W., & Potter, S. (2005). Decision-centered testing: Evaluating joint human–computer cognitive work. In *Proceedings of the Human Factors and Ergonomics Society 49nd Annual Meeting*, Santa Monica, CA: Human Factors and Ergonomics Society.

Sanderson, P. M. (1998), Cognitive work analysis and the analysis, design, and evaluation of human– computer interactive systems. *Proceedings of OzCHI '98*. Adelaide, Australia.

Shepard, R., & Cooper, L. (1982). *Mental Images and Their Transformations*. Cambridge, MA: MIT Press.

Simon, H. (1969). *The sciences of the Artificial*. Cambridge, MA: MIT Press.

Simon, H. (1982). *Models of Bounded Rationality, Vol. 2: Behavioral Economics and Business Organization*. Cambridge, MA: MIT Press.

Sweller, J., Chandler, P., Tierney, P., & Cooper, M. (1990). Cognitive load as a factor in the structuring of technical material. *Journal of Experimental Psychology, 119*(2), 176–192.

Triesman, A., & Gelade, G. (1980). A Feature-integration theory of attention. *Cognitive Psychology, 12*(1), 97–136.

Triesman, A., & Gormican, S. (1988). Feature analysis in early vision: Evidence from search asymmetries. *Psychological Review, 95*, 15–48.

Van der Veer, G., Guest, S., Haselager, P., Innocent, P., McDaid, E., Oesterreicher, L., Tauber, M., Vos, U., & Waern, Y. (1990). Designing for the mental model: Interdisciplinary approach to the definition of a user interface for electronic mail systems. In D. Ackermann and M. Tauber (Eds.), *Mental Models and Human–Computer Interaction.* Amsterdam: North-Holland.

Vicente, K. (1999). *Cognitive Work Analysis.* Mahwah, NJ: Lawrence Erlbaum Associates.

Vicente, K., & Rasmussen, J. (1990). The ecology of human–machine systems II: Mediating direct perception in complex work domains. *Ecological Psychology, 2,* 207–249.

Vicente, K., & Rasmussen, J. (1992). Ecological interface design: Theoretical foundations. *IEEE Transactions on Systems Man and Cybernetics, 22,* 589–606.

Wolfe, J. M. (1998). Visual search. In H. Pashler (Ed.), *Attention.* London: University College Press.

Woods, D. (1991). The cognitive engineering of problem representations. In G. R. S. Weir and J. L. Alty (Eds.), *Human–Computer Interaction and Complex Systems.* London: Academic Press.

Woods, D. (1995). Toward a theoretical base for representation design in the computer medium. In J. Flach, P. Hancock, J. Caird, and K. Vicente (Eds.), *Global Perspectives on the Ecology of Human–Machine Systems.* Hillsdale, NJ: Lawrence Erlbaum Associates.

Woods, D., Johannesen, L., Cook, R., & Sarter, N. (1994) Behind human error: Cognitive systems, computers, and hindsight. *State-of-the-art report CSERIAC SOAR 94-01.* Wright–Patterson Air Force Base, Ohio.

Woods, D., & Roth, E. M. (1988). Cognitive engineering: Human problem solving with tools. *Human Factors, 30* (4), 415–430.

Woods, D. D., & Hollnagel, E. (2006). *Joint Cognitive Systems: Patterns in Cognitive Systems Engineering.* Boca Raton, FL: CRC Press.

Yantis, S., & Jonides, J. (1984). Abrupt visual onsets and selective attention: Evidence from visual search. *Journal of Experimental Psychology: Human Perception and Performance, 10*(5), 601–621.

Zhang, J., & Norman, D. A. (1994). Representations in distributed cognitive tasks. *Cognitive Science, 18,* 87–122.

Chapter 11

Integrated Constraint Evaluation: A Framework for Continuous Work Analysis

John Flach, Dan Schwartz, April Bennett,
Sheldon Russell, and Tom Hughes

Contents

Overview

Although few doubt the value of work analysis for developing insights relevant to design, there have been concerns about whether the artifacts from work analysis are useful to those who did not participate in their production. This chapter describes a system, integrated constraint evaluation (ICE), for archiving the products of work

275

analysis. ICE is designed to provide easy access to artifacts generated in the process of work analysis. It also is designed to encourage and facilitate revision and additions to the database throughout the lifecycle of a work system. The chapter provides a concrete illustration of how this archive might be developed for a domain related to military command and control.

Introduction and Motivation

Consider the following design scenarios.

Scenario 1. Imagine a young human factors engineer, recently graduated from a psychology program at a small Midwestern university (in Dayton, Ohio). He goes to work for a large engineering corporation and is told that he will be participating in a project to design the next generation of decision support for Combined Air and Space Operations Centers (CAOCs). One of the primary goals for this decision support is to facilitate the ability to prosecute time-sensitive targets (TSTs). Where does he start? Perhaps he locates a resident domain expert and arranges a meeting. He introduces himself and begins, "Can you tell me everything that I need to know about TST?" Where does the domain expert begin? How does he capture and communicate a lifetime of experience in a 1-hour meeting? Perhaps, he provides a detailed work analysis that had previously been done, or perhaps he turns to his bookshelf and says, "Start by reading this shelf!"

Suppose that there had been a previous work analysis, how valuable do you think it would be? Would it reflect the new technologies and software options that will be implemented in the next generation CAOC? Would it accurately reflect the demands of future air operations? Even if the analysis did anticipate future opportunities and demands, would the data be in a form that would be useful to the young human factors engineer? That is, could he learn what he needed to learn about the domain from reading the work analysis report? Would the data in the report be represented in a way so that the important dimensions of the work problems would be salient?

Scenario 2. MG is a retired AF officer who has been working for the last 15 years for a large engineering corporation. He has become the resident expert on TST. In fact, he is often heard to remark that he was "doing TST before it was called TST." He has an uncanny way of getting quickly to the heart of the matter and can skillfully talk with both the domain customers and the design teams, which includes a wide range of different engineering disciplines. He is the "go-to" guy for any design question that requires deep domain knowledge. Next week, MG is retiring.

Is it necessary to find or grow another MG? Is it possible to archive some small part of MG's experience or is this experience lost to the corporation? Is it possible to externalize the experiences of MG in a way that it could be shared or utilized by others?

Scenario 3. DW is a senior scientist at a large engineering consulting company. He currently has his fingers in a dozen different projects. Some of these projects involve areas that DW has studied on and off throughout his 30-year career. A younger colleague is describing a particular problem on one of the projects at lunch. DW immediately recalls reading a tech report that addressed that exact problem. However, it was a number of years ago, and he cannot remember the author or title of the report. He thinks he still may have it, but he has no idea where to even start looking for it. Although he promises his colleague that he will look for it, when he returns to his office he is immediately engaged by more immediate demands and soon forgets his promise.

These three scenarios each highlight different aspects of the data overload problem faced by system designers.

- Scenario 1: How do you consolidate and effectively communicate domain experience to the many different people, from many different domains, that are working on a design project?
- Scenario 2: If you are lucky enough to have a domain expert who has consolidated his own experiences and can effectively communicate with the other disciplines on the design team, how do you preserve that expertise so that it does not walk out the door with the expert when he retires?
- Scenario 3: How do you keep track of all the potentially relevant domain information that you encounter over a working career?

Our experience suggests that the person conducting a work or task analysis typically learns from the process, the insights gained can sometimes be translated into smart design decisions, and that within an individual, expertise can accumulate over years of experience. However, it is rare that the archives of the work analyses are ever used by other people in future evolutions of the system. It is more likely that at the next stage in the life cycle of the system, the work analyses will begin from scratch. The new group of human factors engineers will assemble their own set of materials for table-top analysis; do their own field observations; do their own interviews with operators and other domain experts.

It may be that there is no substitute for the direct experiences of conducting the work analysis. It may be necessary for each new design participant to immerse herself in the work domain and recapitulate the experiences of previous analysts. However, we find it difficult to give up the hope that the products of a work analysis conducted today will have value for future generations. We find it difficult to give up hope that it might be possible to construct a data archive that will evolve with a work domain. An archive that will carry forward and make salient those aspects of work that might be invariant; and at the same time, that will adapt to accommodate new insights, new demands, and new opportunities.

Despite our hopes, we do recognize the enormity of the challenge. In fact, despite the enthusiasm that cognitive systems engineering (CSE) has sparked in those seeking simple answers to complex problems, we believe that the central theme of CSE is that systems are far more complex than has classically been assumed. Further, CSE is not an antidote to cut through the complexity, but rather it is a challenge to embrace the complexity. It is a challenge to scale up our skills of management and control so that we can utilize as many degrees of freedom as possible in order to adapt resiliently to the requisite variety of a complex work domain.

In the next section, we will give a brief perspective on CSE. Then we will outline our vision for a data management tool—ICE—that might:

■ help to make the accumulated products of cognitive work analysis (CWA) accessible as a tool for multidisciplinary collaboration
■ help the discovery and specification of invariant patterns or dimensions of a work domain
■ support an iterative process to elaborate and refine questions and deepen insights

The CSE Challenge

The classical "scientific management" or "Tayloristic" approach of identifying a single "best" way through detailed work analysis and then designing the work methods to insure that the "best" way is followed results in "brittle" solutions when the target work domain either has a large number of degrees of freedom (e.g., nuclear power) or involves rapidly changing contingencies (e.g., flexible manufacturing), or both (e.g., military command and control, health care systems). In such work domains, there is no single, stationary "best way." In such work domains, unanticipated variability (i.e., surprise) is to be expected. Thus, a critical design goal is to provide operators the tools they need to adapt in real-time to novel situations. This realization was a prime motivation leading people to explore for alternative approaches to work analysis and design that have come to be generally identified with a CSE perspective (e.g., Flach & Rasmussen, 2000; Hutchins, 1995; Rasmussen, 1986; Rasmussen, Pejtersen, & Goodstein, 1994; Rochlin, 1997; Vicente, 1999, 2004; Woods & Hollnagel, 2006).

One of the key realizations underlying this new perspective was that a focus on work activities and work methods was *not* sufficient—because the work activities had to be adapted to the changing contexts created by interactions among the many degrees of freedom and changing contingencies (e.g., unanticipated variability). In addition to looking at human work (or problem solving) activities, it was necessary to consider the context of those activities—to consider the constraints—to consider the workspace or problem-space. Another way to say this is to say that work (or cognition) is "situated" (e.g., Suchman, 1987). Thus, work analysis must attend to properties of the situation (Flach, Mulder, & van Paassen, 2004; Vicente, 1999). In many respects, this recognition of situated constraints simply echoes the realization of Simon and others who were studying human problem solving—a clear specification of the problem-space was essential to progress (e.g., Simon, 1981).

The divergent point between CSE and academic research on human problem solving was that the academics had the luxury to focus on puzzles where the problem-space was well specified (tic-tac-toe, go, chess, crypto-arithmetic, etc.). Thus, specification of the problem-space in terms of states, operators, and constraints was simply a matter of clearly articulating the rules of the games being studied. Problem specification was a necessary, but relatively uninteresting dimension of the research agenda. The problem-space was "predefined"; there were no ambiguities, there was no need to discover constraints of the problem. Thus, interest focused on the "constraints" that the problem solver imposed due to limitations, biases, heuristics, or expertise.

However, when it comes to regulating a high degree of freedom system (e.g., a nuclear power plant, an oil-cracking plant, or a soccer team) discovering the "rules" or the "states," "operators," and "constraints" that link the activities to the situation dynamics is not a "given." Experts in these systems will often argue about the "best way" and about the underlying rationality of how or why one way will work or another will not. This was clearly evident in our work on landing displays where we discovered that pilots and aeronautical engineers had diverse opinions about the significance of energy constraints for safe landings (Amelink, Mulder, Paassan, & Flach, 2005; Flach et al., 2003). Thus, in contrast to the academic study of puzzle solving, discovering the "rules of the game" or the work constraints can be the most difficult challenge when evaluating complex systems. It becomes a challenge to differentiate the subjective constraints (limitations, biases, heuristics) from the objective constraints of the work domain.

In many respects, CSE begins with the recognition of this challenge—to identify the demands of the work domain, as an essential feature of work analysis. In other words, CSE recognizes the need for an understanding of situations as a prerequisite to any understanding of awareness (Flach et al., 2004). This is a feature that has not been addressed well by either human factors (with its focus on human limitations) or HCI (with its focus on the attributes of technology). Neither of these approaches provides much motivation or guidance for how to meet the challenge of describing the workspace and problem-space or situation (on the other side of the computer)—that is, what the human is using the technology for. Human factors and HCI have a lot to say about the nature of "awareness," but provide little guidance for dealing with the nature of "situations." Flach and Dominguez (1995) have suggested the term "use-centered" design as a stimulus to get focus off of the "user" or the "device" and onto the work problem that is to be solved.

Recognition of the need to focus on the workspace as a set of constraints that humans must adapt to is the common thread that links most of the people who are searching for new approaches to analysis and design. This will be assumed as a given, noncontroversial aspect of the general field of CWA. The point of this paper will be to challenge another vestige from classical approaches to work analysis and design. That is, the assumption that it is possible to do a complete work analysis prior to design. Although probably unintended by the authors, introductions to CWA (e.g., Vicente, 1999) can lead to the impression that it is possible to carry out a complete work analysis as a prerequisite to any design opportunity. Such assumptions tend to be molded into the formal frameworks for systems engineering and acquisition. Thus, there seems to be a persistent assumption that work analysis can be completed early in the design process. That specification for the design can then be handed over the wall to the system builders. Finally, there is the assumption that if the system is designed to these specifications, a robust, adaptive cognitive system will be the result. This assumption fails to fully recognize the inherently adaptive nature of work in these complex contexts. It assumes a naive "waterfall" model of the design process and fails to appreciate the creative, iterative nature of innovation associated with the evolution of complex work environments (e.g., Royce, 1970).

The premise of this chapter has been chosen for the very same reasons (e.g., high degree of freedom, fast pace of change) that require a focus on work constraints; work analysis will never be complete. In other words, the horizon of any work analysis will always be finite and limited—such that most work systems will outlive any specification based on finite work analysis—regardless of the framework used for that analysis. For a system to be truly adaptive, it will be required that work analysis be integrated as part of the system life cycle. In essence, a capacity for self-awareness (evaluation, mindfulness, and meta-cognition) must be integrated into any system whose survival depends on adaptation to changing or unknown contingencies. Thus, every stage in the design and life of a system provides opportunities for new discoveries about the problem-space. This has led us to consider formal ways to integrate, archive, and represent these discoveries so that the acquired information can support adaptation throughout the life cycle of a system. This chapter will describe the ICE framework that we are exploring as a means to provide structured access to data that have been accumulated in the process of CWA.

Overview of ICE

As a description of a problem-space, we call our framework Integrated Constraint Evaluation (ICE). The term "constraint" is used to reflect the boundaries on possibilities—as it might be used in a state-space representation of a control system. Boundaries may represent anything that might constrain action or preference, such as physical limitations (e.g., brake dynamics); values (speed limits or comfortable g-forces); or information limits (perceptual thresholds). Although we use the term "constraint," our interest is ultimately in the complement of constraint—the possible and desirable states (i.e., the "field of possibilities"). However, in many problem-spaces that include continuous state variables—it is far easier to specify the boundaries (constraints) on this field than the infinite possibilities within the field. In analogy to the types of puzzles studied in academic fields—we want to know what moves are possible and legal (i.e., the operators) and also to know what moves are desirable (i.e., the goals and reward or cost structures). Although in many cases the "activities" of expert workers will be a guide in discovering the "constraints," for work analysis, we are less interested in what these experts *do*; and more interested in what they *could do*. At the design stage we are particularly interested in how new technologies impact this field of possibilities.

Consistent with Rasmussen (1986) and Vicente (1999) we find it useful to organize our thinking about constraints along levels of abstraction and decomposition. And further, we recognize that there is a correlation between the level of decomposition (i.e., amount of detail) and the level of abstraction, such that higher levels of abstraction generally require less detail and lower levels of abstraction generally require more detail. This is typically visualized as the diagonal of a two-dimensional abstraction/decomposition space having particular value for understanding the relations among constraints that are significant to work.

At the highest level, ICE structures data according to levels of abstraction. ICE diverges from the discussions of Rasmussen and Vicente, in how the constraints are mapped across levels of abstractions. In ICE, constraints at higher levels of abstraction will be used to provide categorical distinctions at lower levels of abstractions. Each lower level of abstraction can inherit categorical distinctions from higher levels and these categories can be used to organize the greater detail at the lower levels. This is difficult or impossible to visualize in the abstract. So, the next sections will provide an explicit example to illustrate how this is being applied to the domain of time sensitive targeting (TST).

Survey of ICE-T

ICE-T is a hypertext database that we have been developing to archive data acquired in the process of studying TST in combat air operations. We will use ICE-T as a concrete case to illustrate how a living database might be constructed that becomes a learning tool that can potentially support and facilitate learning and self-organization. Or in other words, ICE-T is intended to serve as a memory-aid to provide a level of mindfulness essential for an evolutionary design process. The overall structure of the nested indexing system is illustrated in Figure 11.1. At the highest level in the database there are five levels to choose from. Within each level, an indexing scheme is constructed from the dimensions of that category or from dimensions inherited from higher levels. The arrows indicate the propagation of categories across levels. After a brief overview of the domain, we will systematically move through the levels and try to explain the source for the indexes at each level and the rationale motivating our choices. Hopefully, the logic of the indexing system will gradually become clear. Following the survey view of the indexing landscape, we will describe some routes through this database to illustrate how we envision the interactions with this archival tool might work.

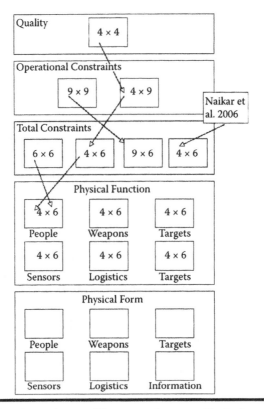

Figure 11.1 This schematic map illustrates the nested indexing relations. The dotted arrows indicate how categories from higher levels are used as indexes at lower levels.

Combined Space and Air Operations and Time Sensitive Targeting

Currently, the U.S. Air Force manages air combat from within a Combined Space and Air Operations Center (CAOC). Typically, an air tasking order (ATO) is generated to cover an 8- to 10-hour period of operations. The operations division is responsible for managing this plan in real time. The Air Operations Division includes a TST Cell. This cell is responsible for coordinating dynamic and time sensitive targets. *Dynamic targets* have been specified within the ATO, but because of the nature of the targets (e.g., mobile missile launchers), the time and location of the attack cannot be specified. Thus, the targets have to be identified, located, and targeted in real time. The TST Cell is responsible for managing this process and coordinating with the operational forces and other Ops Cells within the CAOC to get the right resources to the right place at the right time. *Time sensitive targets* are generally high value targets of opportunity that were not anticipated in the current ATO, but that may have a significant impact on the long-term success of the operations. In addition to coordinating with the operational forces and the other Ops Cells, time sensitive targets require coordination with the highest levels of command to get permission for execution. In essence, these targets are examples of unanticipated variability and an important function of the TST Cell is to adapt to this variability.

Defining the Goals: The Quality of Operations

The top level of Rasmussen's (1986) abstraction hierarchy is typically labeled "Goals and Purposes." In a process control system like a nuclear power plant, things like production, economic, and safety targets might be considered at this level. In a process like Vicente's (1999) Duress micro-world these goals can be well specified in terms of targets for the temperature and volume and rate of the water output. However, in many complex systems including TST, the goals may not be so clearly specified.

In our view, the fundamental issue to be addressed at this level of abstraction is *quality*. In other words, at this level of abstraction we are interested in what determines "goodness" or "rightness." And consistent with Pirsig's (1974; 1991) metaphysics of quality, we believe that quality is a foundation for all other distinctions related to the work experience. That is, quality is the raw material from which all other distinctions within the workspace derive their meaning. This flips conventional wisdom on its head. Conventionally, the objective properties of entities (e.g., physical characteristics) are considered to be primary, whereas value judgments are considered to be derived attributes. In the metaphysics of quality, this is reversed. Quality, value, and goodness are considered primary properties from which more objective properties are derived. In essence, any properties that are not relevant to quality will not be considered in our description of the problem-space. Quality becomes the foundation for "reality" with respect to the workspace. It becomes the context that determines what is meaningful.

In addressing quality, it is important to attend to what the domain experts consider to be "good" or "satisfactory" solutions. What are the distinctions that domain experts use to grade performance? This can be a difficult question. Our experience is that although domain experts can generally recognize quality, they are not always able to specify exactly what dimensions contribute to this judgment. For example, you may have a preference for Van Gogh's paintings, but can you specify the dimensions that are important to your preference? Why do we consider his paintings to be of higher quality than others? If our goal is simply to enjoy the paintings, we may not need to answer this question. However, if our goal is to create quality paintings, then every attempt we make is in essence a struggle to address this question (at least implicitly).

With TST it is common to hear repeated references to *right* asset, *right* target, *right* place, and *right* time. We take this as evidence that the dimensions for assessing quality will include the assets (resource constraints), the targets (intentional constraints), the place (spatial constraints), and the time (temporal constraints). Thus, we use these dimensions to create an index within the highest level of abstraction—which for the domain of TST we will label the strategic level. These four dimensions now provide a structure for explorations into questions about quality and they are implemented in ICE-T as a matrix of virtual file drawers created by crossing these four dimensions as illustrated in Figure 11.2.

The virtual files created by the crossing of right asset, right target, right place, and right time provide locations for filing data relevant to the quality of solutions. For example, one drawer would allow storage of information about the right assets and another would allow storage of information about the relations between right target and right time. The kind of data that would be archived in these drawers could range from analyses of historical military events, to military doctrine, to specific cases elicited from domain experts. For example, Arquilla and Ronfelt (1997) speculated about the nature of cyberwar based on generalizations from the 13th-century Mongols:

What distinguishes the victors is their grasp of information—not only from the mundane standpoint of knowing how to find the enemy while keeping it in the dark, but also in doctrinal and organizational terms. The analogy is rather like a chess game where you see the entire board,

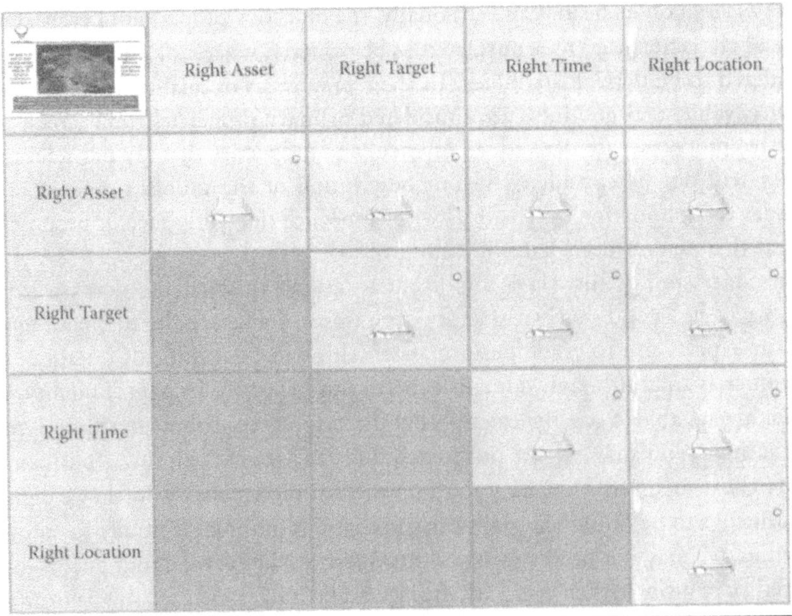

Figure 11.2 A file cabinet is currently being used as the graphical metaphor for creating categories from the crossing of index dimensions.

but your opponent sees only its own pieces—you can win even if he is allowed to start with additional powerful pieces.

We might appear to be extrapolating from the U.S. victory in the Gulf War against Iraq. But our vision is inspired more by the example of the Mongols of the 13th century. Their "hordes" were almost always outnumbered by their opponents. Yet they conquered, and held for over a century, the largest continental empire ever seen. The key to Mongol success was their absolute dominance of battlefield information. **They struck when and where they deemed appropriate**; and their "Arrow Riders" kept field commanders, often separated by hundreds of miles, in daily communication. Even the Great Khan, sometimes thousands of miles away, was aware of developments in the field within days of their occurrence. (pp. 23–24, *emphasis added*)

At the other end of the continuum, we were able to interview an officer who was part of the TST cell during the Iraqi Freedom campaign. He described in great detail a specific TST event including an explanation of why the F-117 was clearly the "right" asset for this particular target. This discussion referred to attributes of the target, the location, and the timing that went into the choice of the F-117. Snook's (2000) case study of a friendly fire accident also could have important lessons for this and other levels within ICE-T. This case study illustrates how factors might lead to convergence on the "wrong" target.

Data might also include more specific details about divisions within the four global dimensions. For example, a distinction is made between four categories of targets: Scheduled Planned Targets, On-Call Planned Targets, Unplanned Immediate Targets (Known), and Unanticipated Immediate Targets (Unknown). Each of these targets creates different demands on the TST process. This will be discussed in more detail later.

Currently, we do not have a firm idea about how information should be subindexed within the file drawers. Our expectation is that a logical substructure will emerge from the data itself. This substructure might reflect the continuum from general to specific, as alluded to above; or it may reflect the nature of the knowledge elicitation process or the form of the representation (e.g., documents, critical incident reports, observations, process charts, and concept maps). We expect that multiple forms of representation might be included at the basic data level. For text-based sources we recommend that specific relevant short passages be presented directly (something that you might highlight when you are reading the original source), along with a reference or when possible a link to the full source. For example, a short passage from the interview with the TST officer would be directly accessible through the Right Asset drawer in both written and audio formats. In addition, there should be a link to the full interview for those who wanted more detail.

Operational Constraints: Principles of War

The second level in Rasmussen's (1986) abstraction hierarchy was originally labeled "Abstract Function," and in Vicente's (1999) Duress example, this level focused on the physical laws related to conservation of mass and energy. Consistent with Vicente's lead, our first hypothesis for thinking about this level of abstraction was to focus on the physical dynamics of TST. Thus, we were thinking about things like aerodynamic, electromagnetic, and other physical sources of constraint. However, these dimensions did not seem to bring the clarity that we were hoping for.

In later work, Rasmussen et al. (1994) added the term *priority measures* to the second level of abstraction. It seems that this was added in recognition that in addition to physical dynamics, information and organizational dynamics might be considered at this level. Although we realized the significance of physical constraints, we have come to the conclusion that TST is fundamentally a problem of military science. Thus, we feel that the physical constraints were best addressed at lower levels of abstraction (e.g., as properties of specific assets). For our second level of abstraction we have turned to military doctrine for thinking about the abstract functional constraints and priority measures. We have concluded that principles of war would provide the dimensions needed to structure explanations at this level. Table 11.1 lists nine principles of war taken from *Air Force Doctrine* (Air Force Doctrine Document 1, 2003).

With the introduction of the principles of war, we have generated two subindexing systems for directing our search and organizing our data at the second level of abstraction (as illustrated in Figure 11.1). The first set of virtual files reflects the crossing of the nine principles of war. The contents of these files will include information relevant to the principles themselves and the interactions among these principles. The second set of virtual files reflects the crossing of the dimensions from the first level of abstraction with the dimensions introduced at the second level. Thus, the categorical classification at this level will cross the four goals (right asset, right target, right time, right location) with the nine principles of war (e.g., visualize a 4 × 9 array of file cabinets). The contents of these files will include information explicitly relating the principles of war to the dimensions of quality associated with the four objectives.

As with the higher level, we expect that various types of data might fit within these categories. Things that we have included so far range from simple definitions of the various dimensions, along with links to the source documents; historical analyses; and excerpts from interviews with domain experts. For example, the story of the Mongol's Arrow Riders might be used to illustrate how managing information can be critical for achieving Unity of Command.

This level of abstraction, in particular, is one that rarely is addressed in more classical approaches to work and task analysis. In our interactions with other researchers who are studying command and control, it is rare to find people who have actually considered basic *Air Force Doctrine*. Similarly, there are many people studying

Table 11.1   Principles of war

Principle	Definition
Objective	Concerned with directing military operations toward a defined and attainable objective that contributes to strategic, operational, and tactical aims.
Offensive	Act rather than react and dictate the time, place, purpose, scope, intensity, and pace of operations.
Mass	Concentrate the effects of combat power at the most advantageous place and time to achieve decisive results.
Economy of force	Judicious employment and distribution of forces. Maximum effort should be devoted to primary objectives. Allocate minimum essential resources to secondary efforts.
Maneuver	The ability to quickly integrate a force and to strike directly at an adversary's strategic or operational centers of gravity.
Unity of command	Coordinate effort toward a commonly understood object.
Security	Never permit the enemy to acquire unexpected advantage. Friendly forces and their operations must be protected from enemy action.
Surprise	Attack the enemy at a time, place, or in a manner for which they are not prepared.
Simplicity	Avoid unnecessary complexity in organizing, preparing, planning, and conducting military operations.

Source: From *Air Force Doctrine* Document 1, United States Air Force, 17 November 2003. http://afpubs.hq.af.mil.

aviation human factors, who do not understand basic principles of aerodynamics. It is also true that operators in these domains often cannot explicitly articulate basic doctrine or the basic physical principles governing the processes they control. And, it is not unusual to find that some of the things they might say about their own work actually are inconsistent with basic principles. For example, some pilots have a relatively naive explicit understanding of the aerodynamic principles that govern flight—even though their skilled performance might suggest an implicit deep understanding of these principles.

However, we find that this level of abstraction provides a very important context for our explorations. We find that knowledge of basic principles helps us to guide our attention when observing operations and talking with operators. Knowledge of these principles is essential in order to distinguish between constraints associated with the workspace and constraints associated with the operators' beliefs about that space. Knowledge of these principles is essential for generalizing from activities to possibilities and for grading the different possibilities with respect to their quality.

Vicente (1999) makes a distinction between correspondence and coherence-driven work domains. Nuclear power plant control and piloting are examples of correspondence-driven domains. In these domains, there are clear principles or laws associated with the physical dynamics that play a significant role in constraining possibilities. For correspondence-driven domains, these principles or laws would be important to consider at this level of abstraction. It is important that operator's actions be guided by a "correspondence" with the physical principles.

Vicente (1999) describes coherence-driven domains as "domains that [do] not impose dynamic environmental constraints on the actions of actors" (p. 6). Pejtersen and Goodstein's (e.g., Goodstein & Pejtersen, 1989; Pejtersen, 1989) work on public libraries is typically used as an example for a coherence-driven domain. The rationale is that there are no well-understood dynamic principles or laws that govern people's unique preferences in terms of reading enjoyment. On the basis of our work with TST, we are beginning to question this distinction. We prefer to think that there always are dynamic principles that govern the dynamics of a work domain. However, the dynamics are not always physical and they are not always well understood. Sometimes the limiting dynamics are physical, but in other domains the limiting dynamics may be psychological, social, organization, political, or economic.

For physical domains, the dynamics will often be well defined as a result of intensive scientific programs of study. For nonphysical domains the dynamics are typically not so well understood. In these domains we will not be able to define the dynamic constraints in relatively simple terms such as mass and energy balance equations. Perhaps, the best that we can do is to specify some of the dimensions that might contribute to the dynamics (such as the nine principles of war). In this case, part of the job of work analysis is the challenge of discovering the dynamic constraints. For example, a key contribution of the book house work was a deeper understanding of the categories that people use to think about fiction. In this respect, work analysis promises to play an important role in basic scientific programs that have the goal of describing psychological and social systems.

Thus, it is important to be clear that the nine principles of war do not allow us to specify the abstract functional constraints for TST in the same way that it is possible to specify the abstract functional constraints for a feed-water control system or an aircraft. In essence, the principles of war simply suggest the dimensionality of the dynamic constraints. It suggests what space we should search in our efforts to discover what the significant dynamic constraints might be.

Also, although we use the term operational to label this level, this is not intended to provide a prescriptive list of operations, processes, or even strategies. This level of abstraction is where possible operations or processes might be evaluated and compared to see how lessons might generalize from past experience. For example, how can the lessons of Vietnam be applied to the current challenges of Afghanistan or Iraq? Again, the principles of war provide a dimensional framework for exploring and debating the relative merits of alternative strategies. As Vicente (1999) suggests, the goal here is to support a "formative" understanding of work. Certainly, in this process descriptive and normative models can be instructive—but the fundamental challenge is to "test" the conventional wisdom of current practices and the assumptions underlying presumed "normative" models.

In this respect, ICE-T is simply a tool for managing the data that we accumulate in our efforts to make sense of the dynamics of military command and control. Also, we should not necessarily expect that specification of psychological or social dynamics will take the same form as the specification of physical dynamics (e.g., differential equations). It may be the case, that a case-based narrative of a prototypical or critical situation may be the most effective way to specify psycho-social dynamics (e.g., Crandal & Getchell-Reiter, 1993).

Comparing Tactics: Practical Operations

The third level in Rasmussen's (1986) abstraction hierarchy is typically labeled "Generalized Functions." At this level, work is typically described in terms of functional stages. For cognitive systems, this would commonly reflect stages of information processing. Descriptions at this level of abstraction are widely used in more classical approaches to work and task analysis. It is common to represent data at this level of abstraction using block diagrams and flow charts.

For TST it is common to characterize the functional flow using the ominously titled "kill chain." The kill chain involves six stages of processing that are typically involved when engaging a target: Find, Fix, Track, Target, Engage, and Assess. For example, this might involve first:

- Detecting (Find) a mobile SCUD missile launcher.
- Locating it in space (Fix), typically involving specifying coordinates (e.g., lat and long).
- Following it (Track) if it begins moving.
- Pairing it with an attacker (Targeting) (e.g., assigning the task to a specific aircraft). This is the primary role of the TST Cell to find the appropriate asset for executing the attack. Key considerations include the requirements to achieve the desired effect (e.g., can the aircraft get there in time with the right ordinance and without unnecessary risks) and the consequences on the overall ATO (e.g., what other tasks the aircraft was scheduled for).
- Engaging it.
- Assessing whether the attack was successful.

These are the primary dimensions that will be added to our indexing system at this level of abstraction. Following the logic introduced for the second level, these dimensions can be combined with the dimensions from the higher levels to generate three different subindexing systems to manage the data at this level of abstraction (see Figure 11.1). The first subindex was formed by crossing the six stages of the kill chain. The virtual file drawers created by this crossing will contain detailed information about each of the specific stages (which might typically also take the form of block diagrams showing the subfunctional flows nested within the more general functions). The interaction drawers (e.g., Find × Fix or Find × Engage) would provide more detail about the links between stages. This could include discussions of how products from one stage are used in another stage as input or feedback. It might also include discussions of precedence constraints between stages in the process.

The second subindex was created by crossing the six elements of the kill chain with the dimensions of quality from the highest level of abstraction. The drawers of this virtual file would include information related to how the functional stages contribute to the overall goals of the system. That is, how the stages can contribute to the goals of getting the right asset, to the right target, at the right time and right place. Or conversely, what is the right asset, right target, right time, and right place with respect to the constraints of a particular functional process. For example, the ideal asset for finding, fixing, and tracking a target, may not have the capability to engage the target (e.g., a UAV or special forces observer).

A third subindex is created by crossing the six elements of the kill chain with the nine principles of war introduced at the second level. The drawers within this file system would include data relative to the most effective ways to manage the individual stages of the TST process. For example, this data might include information about what "mass" means in the context of different stages of the kill chain. In the context of finding, fixing, and tracking a target, "mass" may be relative to intelligence resources (i.e., the number of eyes you can put on the target). In the context of engaging a target, "mass" may refer to the number and types of ordinance available. Another example might include critical events illustrating effective ways to be proactive (offensive) in terms of specific stages of the kill chain. How to manage intelligence assets to anticipate the need for finding and fixing targets? Or how to manage attacking aircraft to insure that they are available to engage a target when it pops up?

We have also included a fourth subindexing system at this level that introduces dimensions that were not primary indexes at higher levels of abstraction. This may suggest that we should reconsider our choices at the higher levels. Or it may simply indicate that no simple propagation of dimensions will cover all the potentially valuable ways to parse a complex work domain. The fourth indexing scheme was inspired by Naikar, Moylan, and Pearce's (2006) contextual activity template. One dimension of this template reflects work functions. In the case of TST, the elements of the kill chain provide this axis. The other dimension of this template reflects work situations. For this dimension, we have chosen the different categories of targets discussed in the previous discussion of data stored at the first level.

The different types of targets place different demands on the TST process. For example, a scheduled planned target will often be in a fixed location (e.g., an enemy air base), so that the process of fixing and tracking may be as simple as looking up the coordinates in a central database. However, such a target is likely to be well defended, so the processes of targeting and engaging might require careful coordination. Unanticipated immediate targets could include support for joint search and rescue to recover a downed pilot. In this case, the stages of the TST process may involve very different information, coordination, and decisions than required to prosecute a fixed target. For example, this will often involve diverting resources from planned targets, requiring considerations of priorities. For planned targets, such considerations would have been addressed in the ATO planning process—prior to the kill chain.

Naikar et al. (2006) suggest that decision ladders (á la Rasmussen, 1986), indicating different information flows, might be an effective way to represent how different situations create different demands on specific functional stages. We agree that this could be a very important form of data that might be included in the virtual files created by this fourth subindex at this level of ICE-T.

In our initial efforts to populate ICE-T, we have found that much of the data at higher levels of abstraction are in narrative forms. As we move down in abstraction, more graphical forms of representation become useful. Thus, at this level of abstraction block diagrams, decision ladders, and other forms of flow charts can be very useful forms of representation. In lower levels, we will see other forms of graphical and pictorial representations become useful.

As you can begin to see, the dimensions are beginning to multiply. This might be a good place to raise the issue of a "stopping rule." For example, in describing functional flows it is possible to generate flow diagrams within flow diagrams, ad infinitum. For example, the engage target stage of the kill chain could be described with its own block diagram that would reiterate most of the stages of the kill chain in terms relevant to the pilot who will have to find, fix, etc. Each of these stages could then be parsed into various stages to describe the information processing behind each stage. Where is the limit to this reduction? How do we know when we are at the right level? Is there a level comparable to Gilbreth's "therbligs" that we might identify as fundamental to the work analysis? How do we know where to stop our search?

For our part, we are approaching these questions pragmatically. That is, we are skeptical about any single level of detail or any single level of abstraction as being privileged with regard to a complete or deep understanding of the workspace. Rather, we believe that multiple perspectives are essential to a full understanding. Each representation may offer unique insights. Thus, the stopping rule has to be derived from the practical aspects of the design problem that motivated the analysis. The "test" of any representation will be the insights relative to design decisions that it offers.

One constraint that we do suggest for the scope of ICE-type databases is to keep the number of dimensions introduced at each level in the range of 7±2. Remember, that the point of the ICE database is to help human designers to manage the information. This limit of about seven dimensions per level will keep the number of chunks within manageable limits. However, by crossing or nesting dimensions at the different levels—a large matrix (7^5) of possible categories can be created. This seems to be compatible with the natural way that people manage complex information—nesting information within meaningful chunks.

The value of a living data archive like ICE-T is not that it will eventually reveal the "atomic" level of description that will answer all our design questions. Rather, it is that the database can grow through the life cycle of the system and it can be adapted to and tested against multiple design challenges that are likely to be faced at different stages of this cycle. The ICE formalism does not offer any answers. It is simply a framework for guiding the search process and for integrating the data accumulated in our search for meaningful ways to think about a work domain. Every design decision is a "test" of our understanding of the work domain and a source of "data" that can be fed back into the ICE database.

Function Allocation: Modalities of Operation

Rasmussen's fourth level of abstraction is typically labeled the "Physical Function Level." At this level, consideration about what type of agent or tools might be involved with the work processes. For example, will the surveillance (finding, fixing, and tracking) be accomplished using manned or unmanned vehicles? Or will the target position be specified using GPS, lasers, or line of sight? We feel that this is the level where the different physical laws will become relevant for TST. For example, the physics of GPS or lasers might have significant implications for their utility in guiding bombs. Or the aerodynamics of an aircraft will constrain the type of situations that it will be suited for. Also, the choice of the type of physical solution to satisfying a functional requirement introduces important constraints on the workspace that might not be recognized based on functional decompositions. For example, having a pilot in a vehicle places limits on g-forces, life support, and weight that might not constrain the functioning of a UAV.

In ICE-T we have tentatively organized the data into six virtual rooms associated with different classes of systems: People, Weapons, Targets, Sensors, Logistical Support, Information Support. Within each room is a virtual file cabinet with drawers created by crossing the general functions from the third level with the goals from the first level. Although the index is identical to that used in the third level, the content of this virtual file cabinet will be specific to the physical category. For example, one virtual file will focus on People, and a drawer within that cabinet would focus on people involved with a specified function in relation to a specific goal. In the Weapons room, a file with a similar categorical structure will archive information about weapons in relation to functional stages relative to different goals.

The kind of data that might appear at this level of analysis might be a diagram showing the positions of operators in the TST Cell with labels and descriptions of their various roles within each stage of the TST process. Also, link diagrams representing the frequency of communications between each of the operators as a function of the type of communications tools available (e.g., chat rooms versus voice-loops) might provide valuable insights.

At this point ICE-T only includes a single virtual file within each of the six rooms identified with different classes of systems. However, all four indexing systems introduced at the third level might be incorporated within each room. By doing this, the general functions become the hinge for connecting the physical form dimensions with dimensions at each of the other levels of abstraction. The point of this would be to help people to explore how the physical form level constraints might interact with constraints at the other levels. For example, in addressing questions about manpower, it might be useful to consider what each person contributes toward the goals within each stage of the process in relation to different situations (types of target). Klein's (Zsambok, Klein, Kyne, & Klinger, 1992) work on identifying information flow and key decision makers would be an example of the type of data that would be considered at this level of abstraction.

Physical Components: The Puzzle Pieces

The lowest level in Rasmussen's abstraction hierarchy is typically identified as the physical form level. As with the other levels, important constraints arise at this level that would not normally be considered at other levels of abstraction. For example, the proximity of electrical and hydraulic lines in a common space might create interactions, such as a leak shorting electrical signals to sensors that would be difficult to anticipate based on a parsing into different physical systems or general functions. In the TST domain, an important physical detail is the type of refueling apparatus that an aircraft and tanker have (boom and receptacle or probe and drogue). This physical detail will determine what tankers might be available for a specific aircraft that an operator might want to redirect to a TST. Other physical details might address the specific capabilities of specific aircraft or the specific formats for displaying information by the decision support software used by the TST chief.

The physical components level is organized around the same top-level categories used in the physical function level (fourth level): People, Weapons, Targets Sensors, Logistical Support, and Information Support. However, the subindexing at the fifth level is so designed that it does not depend on the higher levels of abstraction with respect to TST. Rather, the indexing is designed around the objects themselves. For example, Weapons is currently divided into Manned Aircraft, Unmanned Aircraft, Land Systems, Sea Systems, and Explosives. Explosives are divided into smart (guided) and dumb bombs. Within a subcategory, the data are listed in alphanumerical order. Thus, this level allows people, who have the name of an aircraft or a piece of software; or a partial description of a type of bomb, to quickly find more information about that asset.

The primary data at this level is in the form of "fact sheets" that include detailed descriptions of particular physical systems, along with pictures and appropriate illustrations. For example, the fact sheet for the F-18 contains both photographs and movies of the F-18 in action. It also includes detailed information about the aircraft, including the primary function (fighter/attack), the contractors, the power plant, accommodations (e.g., there are one- and two-seat versions), performance (Mach 1.7), armament (e.g., can carry up to 13,700 lb [6200 kg] of external ordnance), external dimensions, fuel loads, etc. Many of these fact sheets are similar to baseball cards—there is a picture, with tables of statistics about the entity.

Although we have not been able to do this with ICE-T, it might be useful to have links to particular software tools that would allow people to directly interact with the software (perhaps in the context of representative scenarios) so that they can explore the details of how the software functions. Minimally, detailed descriptions of the software and existing operational interfaces should be available—including sample screenshots and perhaps video or animation of sample events.

Source documents are also archived at this level. Documents might be indexed chronologically and alphabetically as a function of author and title. Again, this would help people to access specific sources without having to trace through the functional properties of the domain. Although, one of the values of an ICE-like database may be that it reduces the need to wade through endless stacks of documents for the few golden nuggets of information that are relevant to your design questions.

When we were demonstrating ICE to a TST officer who had provided some of the critical incident reports that we are integrating into the database, we were demonstrating that someone could click on the citation associated with an entry in ICE and immediately link to the source Air Force manual. His response was, "Do you know how many manuals I have read in my career?" Looking at the many sources we had looked at in the course of building ICE-T, we were thinking that it must be a large number. But his answer was, "None!" And, that was why he was enthusiastic about ICE-T, because he could see how ICE-T could let people get to the important information without having to sludge through volumes of documents to find it. On the other hand, as academics, we recognize the value of being able to trace back to original sources.

ICE-T Dynamics: A Route Perspective

In describing ICE-T it is natural to follow a hierarchical path through the five levels of abstraction, as we have done in the preceding sections. However, our intention is to design ICE-T so that it functions as a network, not a hierarchy. That is, ICE-T is designed so that it is possible to enter at any level. For example, the young human factors engineer described in Scenario 1 will be faced with an enormous amount of jargon (e.g., F-18, TST, SCUD, ISR, etc.). He could enter ICE-T at the lowest level (Physical Components) choose the category and then find the term in an alphabetically organized list. Clicking on the term he would get a definition and other information. For the F-18 he would find a detailed fact sheet with pictures and information about the performance characteristics of this aircraft.

In preparation for a visit to Nellis, AFB to observe exercises in CAOC-N, the young engineer may want to enter at the tactical level so that he could get an idea of the process flow that he will be observing. He may also find the Physical Function level to be useful. He can go to the People section and get a list of the people involved with each stage of the process, along with information about their roles and functions.

Once he returns from Nellis, he may be puzzled by some of his observations. Why do they do it that way? To answer this kind of question he may find the higher levels within ICE more useful. It might be useful for him to consider his observations in the larger context of the goals and values (quality level) or the principles of war (strategic level).

ICE-T is designed so that there are multiple connections across levels. For example, in reading a critical incident describing why the F-117 is the ideal asset for a particular situation (Archived at the highest level of abstraction), it would be possible to click on "F-117" in the text and directly link to the picture and fact sheet archived at the lowest level of abstraction.

Also, the same piece of data, for example a specific critical incident report, can be stored in (or linked to) many different places. Our expectation is that the "meaning" or "significance" of a particular item will shift as a function of where it is stored. For example, a critical incident might be perceived one way, when encountered in the context of questions about the right asset. It might be perceived another way, when encountered in the context of a principle of war, such as economy of force. And it may be perceived still another way, when encountered in the context of a particular stage of the kill chain.

Thus, the hope is that people could enter the problem at many different levels and that they could easily explore in many different directions. We hope that the constraints that guide this search process would reflect the domain constraints and the design questions. In other words, we hope that the user-experience would be of navigating through the TST domain—not navigating through a database. We hope that as a result of navigating in ICE-T people would explicitly and implicitly learn about the TST domain. For example, at no point would we make the argument about the need to study *Air Force Doctrine* in order to understand the TST process. But we hope that as people navigated through ICE-T they would begin to realize how the nine principles of war shape the TST processes and how these principles might be used to differentiate preferred solutions.

Most importantly, ICE-T is designed as a living database. That is, it should be possible to continuously add data. In the early stages of development, the focus needs to be on the indexing dimensions at each level of abstraction. This becomes the skeleton for the database. It is important to choose dimensions that should be invariant over the life of the work domain. In the current stage of ICE-T, we hope that the indexing dimensions at each level are fairly stable. Although we are open to the possibilities, there is room for improvement. We like to compare this to Pirsig's (1991) indexing card system described in the book *Lila*. We are open to the possibility that at some point we may find the current indexing system to be too clumsy. For example, we may find that the principles of war are the wrong dimensions for the strategic level or we may find that the matrix inspired by Naikar et al. (2006) is a better indexing system for data at the physical function level than our current choice.

Once a solid indexing skeleton is established, then updating will primarily involve adding data to the virtual file cabinets, and perhaps, reorganizing or recategorizing data within a drawer. As a domain evolves, our expectation is that most of the change will be in terms of the specific physical equipment or types of physical equipment that are employed. However, it is expected that the functional goals and abstract functional constraints will be fairly stable.

The protocol for managing updates will depend critically on the nature of the organization utilizing ICE-T. In a small team, it could be possible that all members could have authority to add data. In essence, the data would be managed as a wiki. For larger organizations, it is likely that a subset of the organization would act as a gatekeeper, to control and monitor changes to the database. All in the organization would be able to access the data, but only a designated subset would be able to add or revise the database.

One final aspect of the ICE framework is that the indexing system derived from crossing the dimensions at each level not only helps to organize the data that has been acquired, but it can also be a guide to the search process. For example, if a drawer in any of the matrices is empty (and in fact, most draws are empty in our prototype, as we still have much to learn about the TST domain), this can stimulate analysts to ask whether this is due to the fact that the category is not interesting, or whether this simply indicates that the right questions have not yet been asked. We imagine this to be analogous to the way that the periodic table has stimulated the discovery of new elements in chemistry. Thus, ICE is not simply an archival tool, but it may also function as a map into yet to be explored territories.

For the young human factors engineer in Scenario 1, the challenge might be to build an ICE like data-base from the start. Thus, he will have to begin by choosing the levels and the indexes within each level. Then hopefully, when he is one of the "old men" described in Scenarios 2 or 3, he will have a logically organized database so that it is easier to recover that relevant article that he read many years ago and so that he can leave this database behind, as a legacy so that the organization can benefit from his expertise.

CSE (Collaborate, Specify, Elaborate) or Die

ICE is our attempt to address the practical problems of learning about complex work domains that is partially illustrated by the opening design scenarios. It is intended to be a common problem-space in which people from many different disciplines can share information that is potentially relevant to understanding the domain, because any complex domain will demand this type of collaboration. It is not intended to teach military doctrine, aerodynamics, cognitive psychology, or electronics. But it is intended to help everyone appreciated how dimensions of these fields might contribute to shaping the work domain. It should help experts in these disciplines to articulate what they know in domain relevant terms. And it should prevent novices in any discipline from completely ignoring constraints that might be fundamental to a domain.

ICE does not specify design solutions. It simply helps to manage the data accumulated in the search for those solutions. The key innovation is the indexing system, which we view as a skeleton constructed to reflect the domain constraints. Thus, we hope that this indexing scheme captures the spirit of Rasmussen's (1986) abstraction hierarchy and Vicente's framework for a formative approach to work analysis. We hope that the organization of the database can:

- facilitate retrieval of relevant information
- facilitate communications across multiple disciplines
- guide the search for missing elements of the work domain
- inspire new insight to guide design innovations

Thus, ICE is viewed as a tool for design **C**ollaboration. It is designed to help people from multiple disciplines to **S**pecify how properties of their domain impact the possibilities and the qualities of work. And ICE allows continuous **E**laboration as our understanding of the work domain grows and as the domain itself changes. In this respect, we offer ICE as a **CSE** Design Tool. Essentially, ICE is a visualization tool for Human System Integration.

References

Air Force Doctrine Document 1, *Air Force Doctrine*, United States Air Force, 17 November 2003. http://afpubs.hq.af.mil.

Arquilla, J., & Ronfeldt, D. (1997). *In Athena's Camp: Preparing for Conflict in the Information Age*. Santa Monica, CA: Rand Corp.

Amelink, H. J. M., Mulder, M., van Paasan, M. M., & Flach, J. M. (2005). Theoretical foundations for total energy-based perspective flight-path displays for aircraft guidance. *International Journal of Aviation Psychology*, 15, 205–231.

Crandall, B., & Getchell-Reiter, K. (1993). Critical decision method: A technique for eliciting concrete assessment indicators from the intuition of NICU nurses. *Advances in Nursing Sciences*, 16(1), 42–51.

Flach, J. M., & Dominguez, C. O. (1995). Use-centered design. *Ergonomics in Design*, July, 19–24.

Flach, J. M., Jacques, P., Patrick, D., Amelink, M., van Paassen, M. M., & Mulder, M. (2003). A search for meaning: A case study of the approach-to-landing. In E. Hollnagel (Ed.), *Handbook of Cognitive Task Design* (pp. 171–191). Mahwah, NJ: Lawrence Erlbaum Associates.

Flach, J., Mulder, M., & van Paassen, M. M. (2004). The concept of the "situation" in psychology. In S. Banbury & S. Tremblay (Eds.), *A Cognitive Approach to Situation Awareness: Theory, Measurement, and Application* (pp. 42– 60). Aldershot, England: Ashgate.

Flach, J. M., & Rasmussen, J. (2000). Cognitive engineering: Designing for situation awareness. In N. Sarter & R. Amalberti (Eds.), *Cognitive Engineering in the Aviation Domain* (pp. 153–179). Mahwah, NJ: Lawrence Erlbaum Associates.

Goodstein, L. P., & Pejtersen, A. M. (1989). *The BOOK HOUSE system: Functionality and evaluation*. Roskilde, Denmark: Ris¾ National Laboratory, Ris¾-M-2793.

Hutchins, E. (1995). *Cognition in the Wild*. Cambridge, MA: MIT Press.

Naikar, N., Moylan, A., & Pearce, B. (2006). Analyzing activity in complex systems with cognitive work analysis: Concepts, guidelines, and case study for control task analysis. *Theoretical Issues in Ergonomic Science*, 7(4), 371–394.

Pejtersen, A. M. (1989). *The BOOK HOUSE: Modeling Users' Needs and Search Strategies as a Basis for System Design*. Roskilde, Denmark: Ris¾ National Laboratory, Ris¾-M-2794.

Pirsig, R. M. (1974). *Zen and the Art of Motorcycle Maintenance*. New York: HarperCollins.

Pirsig, R. (1991). *Lila: An Inquiry into Morals*. New York: Bantam.

Rasmussen, J. (1986). *Information Processing and Human-Machine Interaction*. New York: North-Holland.

Rasmussen, J., Pejtersen, A. M., & Goodstein, L. P. (1994). *Cognitive Systems Engineering*. New York: John Wiley.

Rochlin, G. (1997). *Trapped in the Net*. Princeton, NJ: Princeton University Press.

Royce, W. (1970). Managing the development of large software systems, *Proceedings of IEEE WESCON*, 26 (August), 1–9.

Simon, H. A. (1981). *The Sciences of the Artificial*, 2nd Ed., Cambridge, MA: MIT Press.

Snook, S. A. (2000). *Friendly Fire*. Princeton, NJ: Princeton University Press.

Suchman, L. (1987). *Plans and Situated Actions: The Problem of Human-Machine Communication.* Cambridge: Cambridge University Press.

Vicente, K. J. (1999). *Cognitive Work Analysis.* Mahwah, NJ: Lawrence Erlbaum Associates.

Vicente, K. J. (2004). *The Human Factor: Revolutionizing the Way People Live with Technology.* New York: Routledge.

Woods, D. D., & Hollnagel, E. (2006). *Joint Cognitive Systems: Patterns in Cognitive Systems Engineering.* Boca Raton, FL: CRC Press.

Zsambok, C. E., Klein, G., Kyne, M. M., & Klinger, D. W. (1992). *Advanced Team Decision Making: A Developmental Model.* (Contract MDA903-90-C-0117, U.S. Army Research Institute for the Behavioral and Social Sciences). Fairborn, OH: Klein Associates.

Chapter 12

On an Ontological Foundation for Work Domain Analysis

Eric Little

Contents

Overview

Both realist ontologies and work domain analysis (WDA) models attempt to provide an accurate picture of objects, attributes/properties, processes, and relations as they exist in the world. Many of the relations, entities, and processes developed during work domain modeling may be more precisely defined within a formal ontological framework, because ontologies provide a rigorous theoretical foundation for understanding various sorts of categorical decompositions (e.g., subsumption relations between classes and part-to-whole relations). This chapter explores the theoretical and practical (i.e., design-oriented) interactions between ontological models and work domain models, especially as they apply to the abstraction/decomposition space (ADS). In this context, we will examine the effectiveness of ontologies to better understand some of the kinds of formal relations found in the ADS. The chapter will also provide an analysis of the system of a home, in order to (1) compare and contrast the similarities and differences between ontological and WDA approaches and (2) show how ontologies can, in turn, provide improved theoretical and methodological underpinnings for WDA.

Introduction

Cognitive systems engineering (CSE) utilizes analysis methodologies and models, which describe the complexities and challenges faced by human operators in their control of complex systems. Outputs from CSE methodologies support the development of information display design requirements, information requirements, human–automation function allocation decisions, and operator knowledge and skill requirements based on successive and iterative analyses of system and task constraints. One focus within CSE is the systematic description of aspects of the work domain comprising the environment in which human operators must act and make decisions. Specifically, models and techniques in WDA have been developed that capture the complexities and constraints of the work domain serving to shape and constrain the behavior of domain practitioners. Thus, work domain models serve as purposeful, albeit bounded, descriptions of a portion of reality, that is, the system of interest (Vicente, 1999; Rasmussen, Pejtersen, & Goodstein, 1994; Naikar, Hopcroft, & Moylan, 2005).

Formal ontologies also provide descriptions of reality, but unlike work domain models, they are not limited to work environments, but rather provide a more basic form of categorization that can be applied to any kind of spatial, temporal, or spatio–temporal items (Smith, 2002; Sowa, 2000; Casati & Varzi, 1999; Gomez-Perez, Fernandez-Lopez, & Corcho, 2004; Simons, 1987; Smith & Grenon, 2004; Little, 2003; Smith, 2001). Ontologies are currently used for a number of applications (e.g., informatics, inventory/organization purposes, domain specification/identification, database integration, information fusion, data mining, and information querying) within a wide range of academic and industry-based fields (Smith, 2001; Little, Rogova, & Boury-Brisset, 2005; Kokar et al., 2004; Boury-Brisset, 2003; Nowak, 2003; Kokar & Wang, 2002; Welty & Smith, 2001; Little & Rogova, 2008; Little, Eberle, & Turino, 2008).

The purpose here is to identify ways in which work domain models and ontologies can mutually inform one another within the context of a systems engineering design problem, particularly as it pertains to the ADS, which draws means–ends and part–whole relations between items within a domain of interest. Specifically, this piece shall identify (1) ways in which world descriptions contained within an ontology can map onto, and provide input to, a work domain model, and (2) ways in which the types of concepts and relationships contained within work domain models can augment, enrich, and appropriately constrain an ontological description of the same phenomena, particularly at the subcategory and instance levels of the ontology. To accomplish this goal, I first provide a more detailed description of both work domain and ontological models in isolation, along with a set of conjectures regarding the overlap and potential resonance between these descriptive methodologies. Next, I provide details regarding the kinds of formal relations useful for constructing work domain and ontological models. Finally, based on the outcomes from these analyses, I shall propose a potential methodology in which work domain models and ontologies can be used in a mutually informing way, and discuss the benefits such cooperation can bring in terms of the content of the models and their applicability to systems-design problems, particularly as it relates to an augmentation of the ADS.

What Is Work Domain Analysis (WDA)?

WDA is the initial phase of Cognitive Work Analysis (CWA), which, in turn, is an approach to designing flexible sociotechnical systems that can address the kinds of dynamic cognitive and conceptual complexities associated with work environments (Rasmussen et al., 1994; Vicente, 1999; Vicente, 2000; Bisantz & Vicente, 1994; Kirwan & Ainsworth, 1992; Naikar et al., 2005). These complexities are exacerbated by technological changes (e.g., the ever-increasing use of computerized systems), high levels of integration, the coupling of work-related systems, changing objectives of work-related tasks and requirements, a variety of dispersed resources, and various cognitive features pertaining to a worker's roles and activities within a given work domain (e.g., background beliefs, cognitive/perceptual biases, varied conceptual categories, etc.). WDA is used to model the constraints of both the purposeful (i.e., cognitive, intentional) and physical contexts in which workers operate. It is particularly the latter where one can argue that ontologies can provide beneficial representational models.

WDA models provide categorical structures capable of capturing the kinds of purposes, values and priority measures, functions (both general and object specific), and physical objects for a given domain. In this sense, WDA models provide insight into the general kinds of constraints that a physical environment can impose upon workers, by specifying the kinds of physical objects needed to perform a job and the subsequent functional capabilities and limitations of those objects. In this sense, WDA must be understood as providing abstract categorical models, as opposed to particular instance-based models (i.e., those that capture particular items at particular times, exhibiting particular behaviors). In other words, WDA models provide information about *types* of items (i.e., abstract categories), rather than specific *tokens* (i.e., particular instance-level items). The distinction between types and tokens can be thought of, for example, as the relationship between a universal category, such as "dog," and an instance of that category "Rex." The former points to an abstract, general item, for which there can be numerous instances, whereas the latter refers to a particular instance of an object located in space-time that possesses a specific identity.

As pointed out by Naikar et al., (2005) and by Kirwan and Ainsworth (1992), WDA must also be starkly contrasted with task analysis (TA). WDA focuses on analyzing and defining certain boundary conditions or constraints on a work system, whereas TA is concerned with describing the tasks and trajectories associated with the particular behaviors of workers (Kirwan & Ainsworth, 1992). Although WDA and TA certainly share points in common, it is important to distinguish between them, because, as will be discussed later in this piece, the formal structure of a work domain is metaphysically distinct from the formal structure of tasks and behaviors carried out within that domain. A work domain is largely composed of spatial items (e.g., spatial regions, objects, spatial relations, etc.). Tasks and behaviors, conversely, are similar to processes, and, therefore, are temporal in nature. Although there is a definite link between spatial items within a given domain and the tasks or behaviors associated with them, it is important not to confuse or conflate the two, because, as will be shown in the following section, objects and processes possess a distinct ontological nature and, therefore, possess distinct ontological characteristics in terms of their part structure.

What Is Ontology?

Ontology, in the purely philosophical sense, is the branch of metaphysics that examines the nature of being (i.e., existence). In this sense, ontology focuses on describing the most basic categories of things such as space, time, matter, process, etc., and how these items relate to one another, forming the fabric of reality (Casati & Varzi, 1999; Simons, 1987; Little, 2003; Husserl, 1900-01; Johansson, 1989). In contemporary information technology applications, ontology provides a means for capturing and modeling categories of entities, their attributes or properties, processes, and relations for a given domain of interest. This is useful in sharing information between computational systems, where integration is often a challenge. For example, ontologies can be used to capture:

- The formal structure of physical items such as people, dwellings, machinery, and geographic domains
- The formal structure of nonphysical items such as threats, commands, doctrines, and roles
- Relations between individual objects and their attributes
- Relations between groups of objects and their attributes
- Relations between individual objects and their aggregations
- Relations between psychological processes and behaviors

- Relations between complex spatial and temporal phenomena
- Relations between causally associated phenomena

Ontologies provide information systems with items such as common vocabularies, formalized data structures, and improved semantic content, which allow an improved understanding of the items found in a given domain, ultimately making data integration easier and more effective (Smith, 2001; Welty & Smith, 2001; Bennett & Feldbaum, 2006). Many current approaches to ontology development, such as the one described in this section, combine the traditional philosophical approach and the information system approach, yielding computationally tractable ontology tools, which are designed in accordance with certain underlying metaphysical principles (Smith, 2002; Sowa, 2000; Gomez-Perez et al., 2004; Little, 2003; Smith, 2001). The following text represents one specific methodology for constructing large ontologies for use in numerous types of applications (Smith, 2002; Smith & Grenon, 2004; Little, 2003; Smith, 2001; Little et al., 2005; Grenon, 2003a; Grenon, 2003b; Grenon & Smith, 2003; Little & Rogova, 2005). The methodology described here is based on a metaphysical approach to ontology construction, which complements a variety of computational approaches used in engineering practices. Though this methodology differs from some other attempts at ontology construction in the literature (see DOLCE Ontology Site; Standard Upper Ontology Working Group), it is designed to provide a useful upper-level (i.e., abstract) segment for existing systems. The methodology for ontology construction described here (modified from: Little, 2003; Little et al., 2005; Little & Rogova, 2005; Little & Rogova, 2006; Little & Rogova, 2007) includes the following steps:

1. Perform term extraction from numerous sources (both free text and structured data).
2. Develop a sufficiently large and representative lexicon of terms to represent all items of interest for that domain (determined to a large extent by domain subject matter experts (SMEs)).
3. Utilize a metaphysically grounded upper-level set of categories (an upper ontology such as Basic Formal Ontology (BFO), SUMO, DOLCE, or others).
4. Construct a sufficiently large set of domain-specific (lower-level) categories based on the domain-specific items gained from point 2.
5. Capture formal relations between asserted terms or categories.
6. Determine a sufficient logical framework (Description Logic, Common Logic, F-Logic, First Order Logic) capable of capturing all items (and relations) in the domain of interest.
7. Design rules and queries to provide adequate reasoning capabilities for extending the ontology.
8. Develop testing methodologies for evaluating the ontology.

A sophisticated ontology provides a comprehensive amount of a priori information about the world by decomposing reality into structured sets of categories that are relationally linked to one another. The categories should correspond to most abstract, somewhat abstract, and least abstract understandings of reality. In this sense, the information within an ontology must be both rationally as well as empirically driven. The upper levels of an ontology are *rational constructs*, meaning they are designed and tested in terms of logically grounded formal relations, which are the product of sound philosophical reasoning. The lower (i.e., domain-specific) levels of an ontology are *empirical constructs*, meaning they contain those materially grounded ontological items that capture and utilize the specific knowledge of SMEs (Little, 2003).

Upper ontological levels model the *most basic types* of objects, object attributes, locations, processes, and relations, whereas lower ontological levels model *the most specific types (and instances)* of those items. For example, categories in an upper ontology can include spatial region, substance, property, and process. There are various theoretical approaches to upper ontology design, for example, BFO, DOLCE, and SUMO (Semy, Pulvermacher, & Obrst, 2004; DOLCE Ontology Site; Standard Upper Ontology Working Group; BFO Home Page), all of which carry with them certain assumed metaphysical underpinnings about the structure of reality. Categories in the lower levels of an ontology can include country, human being, or chemical compound, which, in turn, can contain singular instances of items: *California, 1600 Pennsylvania Ave., Cesium-137, DB_Object_ID S000000296,* or *Red Cross Level 3 of Disaster Relief Operation.* Items in the upper ontology are hierarchically arranged as acyclical tree structures that provide a general framework for all items in the lower levels. Thus, choosing the appropriate upper ontological structure for a given domain or set of applications will affect the quality and effectiveness of the ontology as a whole. Upper ontologies should be chosen on the basis of:

■ Metaphysical (theoretical) positions concerning identity, parthood, universals versus particulars, realism versus conceptualism, and so on
■ User needs
■ Size and scope of the ontology
■ The complexity of the domain in question (objects that change over time or contain complex n-ary relations)
■ Tasks to be performed by the ontology (e.g., querying, rule formation, reasoning)
■ Computational constraints (e.g., runtimes)
■ Abilities to ingest and interoperate with prebuilt subontologies

One of the most powerful features of ontologies is that the lower levels of the ontology can be continuously populated with new and increasingly specific data, because the relational ordering of the domain-specific content is determined by the more general categories represented in the ontology's upper levels. Thus, the levels of an ontology (i.e., upper, middle, and lower) allow the ontologists to represent a given domain in varying degrees of granularity. The granularity of an ontology can vary from the most broad (macro) range to the most minute (micro) range, depending on the scope of investigation under consideration. Ontologies that provide variable granularities ultimately provide both a broader and a deeper understanding of the domain in question, because one can the utilize distinct, but complementary, perspectives on a given item (or set of items; Bittner & Smith, 2001).

The research presented in this piece utilizes the BFO upper ontology (Smith, 2002; Grenon, 2003a; Grenon, 2003b; Grenon & Smith, 2003; BFO Home Page). BFO is a metaphysically driven approach to upper ontology design, which allows for much flexibility and creativity in the design of upper ontologies. The BFO categorical structure is a realist approach to ontology that contains a relatively small, but powerful, set of metaphysical categories that can be easily manipulated and expanded to capture all types of entities, attributes, and processes. BFO exists as two orthogonally related subontologies: SNAP and SPAN (Smith, 2002; Grenon, 2003a; Grenon, 2003b; Grenon & Smith, 2003). SNAP can be understood as providing an ontology that is similar to a "snapshot" of reality, whereas SPAN can be understood as providing an ontology that can capture spans of time. SNAP BFO is an ontology of spatially arranged items that includes objects, spatial regions, object attributes, concepts, agents, etc., all of which can exist in their entirety at a given instant of time. There are no temporal items in SNAP (i.e., no events or processes), because temporal items, by their very nature, unfold over time and, therefore, do not exist in their entirety at any given place.

SPAN BFO, on the other hand, is a purely temporal ontology that contains only events, processes, temporal successions, and sequences, which unfold over (i.e., extend throughout) time. The SPAN portion of the BFO is used to capture not the spatial items, but their temporal successions. SNAP and SPAN are linked to one another via various sorts of more complex (i.e., transontological) relations that represent the ways in which objects can change or unfold over time (see Figure 12.1).

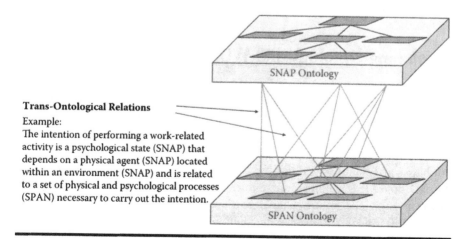

Trans-Ontological Relations
Example:
The intention of performing a work-related activity is a psychological state (SNAP) that depends on a physical agent (SNAP) located within an environment (SNAP) and is related to a set of physical and psychological processes (SPAN) necessary to carry out the intention.

Figure 12.1   Transontological relations between SNAP and SPAN elements.

For example, Object A (e.g., a human being, a nation, an army) can be said to have an identity that endures over time and, thus, is the same at time t_1 as it is at time t_2, even though Object A may have gained or lost parts between those two points in time. One would not wish to say that A is an entirely new and different object at each successive phase because then problems of identity arise, making it impossible to track the object over time. Physical matter decays steadily over time and organisms continuously lose and subsequently regain new cells over time, for example. Yet, common sense tells us that in spite of these kinds of subtle changes, such objects are nonetheless able to maintain their identity over time.

SNAP and SPAN allow us to avoid certain metaphysical problems of identity pertaining to dynamic part–whole relations. Because of issues surrounding identity and parthood, it is important not to confuse an item's spatial parts with its temporal parts. For example, John's hand is a part of John. It is a SNAP item that one can consider atemporally, meaning that all of the parts of the hand and the rest of John's body can be modeled in one go, as a static representation. However, in a different sense, John is a living being who encompasses a life span, which possesses temporal parts (e.g., events such as waving his hand, clenching it into a fist, etc.). The activities of John's life span occur over time, and thus their parts (as temporal segments of durations and processes) are spread over temporal spans, meaning that one cannot capture all the part relations of life span processes in one go, as they can with the spatial parts of the hand in relation to the body. The parts of processes and activities associated with a life span are in a constantly dynamic state of becoming (i.e., coming into being) and dissolution (i.e., fading out of being). According to the metaphysical stance assumed by BFO, the processes and activities that John performs are not SNAP parts of John per se, but rather are SPAN parts of the processes that John undergoes. John and his life span are not identical things, because many of John's spatial parts (i.e., his corpse) can exist for a long period even after the termination of his life span. Conversely, once John's life span ends, so do all the SPAN processes associated with it.

Combining Ontology and WDA

BFO leverages much of its theoretical power from the theory of *realist perspectivalism* (Smith & Grenon, 2004). Realist perspectivalism combines *realism*: the theory that the physical world exists mind-independently and that humans can have direct (and veridical) perceptions of it, with *perspectivalism*: the theory that any single portion of reality can be viewed from various ontological perspectives, many of which can have equal claims to veridicality. For example, one can ontologically examine the same object (e.g., a home) from numerous granular perspectives (e.g., the whole structure, individual rooms, shared portions of various rooms such as adjacent walls and heating or cooling systems, building materials used in the home's construction, the molecular structure of the home's materials, etc.). Depending on the level of granularity from which one is examining it, different veridical perspectives are afforded, some of which are incompatible when considered at different levels of granularity or from different perspectives. In other words, truths about the home's structure may appear different, depending on whether one is discussing the home's building materials versus the overall layout. Realist perspectivalism, therefore, provides the notion of context to one's ontological categorization.

This notion fits well with the overall framework for WDA, as WDA is often used to understand the different kinds of purposes associated with the same object or domain. Naikar et al. (2005) point out that

> A home can be classified as primarily causal if the focus of the WDA
> is on modeling the technical system that provides shelter. Alternatively,
> the same home can be classified as primarily intentional if the focus of
> the WDA is on modeling the social system that promotes the wellbeing
> of inhabitants (p. 7).

The Abstraction–Decomposition Space (ADS)

The ADS is a two-dimensional orthogonal model, which is used by cognitive systems engineers to map both the purposive and physical contexts of workers in a given domain of interest. The two orthogonal dimensions of the ADS represent means–ends and part–whole relations (vertical relations and horizontal relations, respectively). When combined with the orthogonal approach of SNAP and SPAN, it becomes clearer how these two approaches fit together (see Figure 12.2). The top

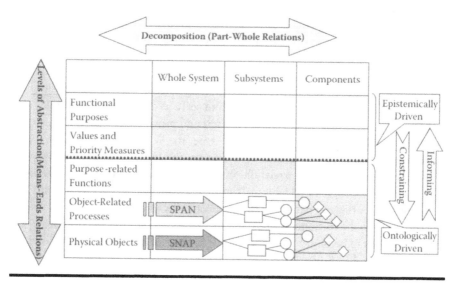

Figure 12.2   Abstraction/decomposition space (ADS) combined with SNAP/ SPAN ontology.

two levels of the ADS (i.e., functional purposes down to purpose-related functions) represent epistemically driven (i.e., cognitive, mind-dependent) factors for WDA, that is, these items serve as inputs to the ontology as overall goals of the system and are provided by SMEs to perform a given set of tasks. Conversely, the bottom three levels of the ADS (i.e., purpose-related functions, object-related processes, and physical objects) contain the items found within the ontology itself. It is at these two layers of the ADS that ontologies become important for WDA, because they provide a sufficiently granular decomposition of all items of interest within the work domain.

The mid-level of purpose-related functions represents the area where the orthogonal ontologies of SNAP and SPAN interpenetrate each other, meaning that this level of the ADS contains basic categories of entities and processes together. As an example, consider the following decomposition of a certain general function of "housework." At the whole system level, purpose-related functions will entail things such as housework, which can contain both SNAP and SPAN items. The category "housework" can be decomposed into certain subcategories such as "cleaning," "arranging," "mopping," "storage," etc. Decomposing one element of housework such as cleaning can include "cleaning the kitchen" or another such subelement of the house, which in turn, at the contents and components level, can entail "scrubbing the kitchen sink." All of these functions specify certain spatial and temporal decompositions (at the category level) when moving from left to right on the table in Figure 12.2.

Ontologies and ADS models can be thought of as interacting via a two-way set of reciprocal relations: *informing* and *constraining* (Little, 2003). In this sense, ontologies inform the epistemic levels of the ADS about the formal structure of numerous kinds of underlying objects, attributes, processes, and relationships. Conversely, the epistemic levels of the ADS constrain the ontology by focusing it on those specific WDA domains of interest to a particular user or SME. This reciprocal relationship is important because it allows the ontology to serve as an input to the ADS in terms of content (particularly in terms of part–whole decomposition of items). At the same time, it allows the ADS to serve as an input to the ontology to narrow its focus to a specific domain of interest and provide empirically driven data about the type of work to be done. In this sense, the WDA as a whole is provided an improved formal organizational structure (in the form of comprehensive and consistent terminologies whose relations can be formally defined), while at the same time, the ontology is provided with real-world, domain-specific data, which fills the lower-level categories and particular instances of items within the domain. A proper synergy between the epistemic and ontological layers also ensures that ontologies do not always need to be built to a massive scale in order to capture an entire domain, instead they can be effectively scoped to fit a certain problem space of interest for some end user with a certain set of tasks in mind.

As is the case in other representations of the ADS (Naikar et al., 2005), the vertical nodes of the ADS are linked casually through means–ends relationships. This can also be ontologically captured (see Figure 12.3), though the ontology utilizes many more kinds of relations than simply causality as pointed out by Little and Rogova (2005, 2007). Moving up through the hierarchy from physical systems to goals reveals the reasons for the existence of the system (its ends), whereas moving down from goals to physical systems reveals the means by which the overall goal-oriented purposes can be achieved. Little and Rogova (2005, 2007) have provided a taxonomy of relations, which includes causal relations within it (as a subset of dependence relations). The utilization of causal relations is one distinct point where the abstraction hierarchy (AH) and ontology overlap, and where it can benefit from a more robust ontological framework that is able to capture and more clearly define causal relations of interest, as well as provide information pertinent to understanding the *relata* (i.e., the basic entities) involved in those relations.

Structural Relations versus Task-Centric Relations

The means-ends relations in the ADS are, according to Vicente (1999) and Naikar et al. (2005), structural relations and not task-centric relations, meaning that they describe the underlying structure of those purposive and physical contexts within which certain actions or tasks can occur. These structural relations influence and shape the activities of workers by either describing, or prescribing, certain *affordances*, whereas task-centric relations describe those specific tasks or activities needed to be performed for certain work-related goals. Affordances can be defined as those perceived or actual properties of items (or an environment) that provide perceptual or cognitive determinations as to how they can be utilized (Shaw & Bransford, 1977; Gibson, 1979; Norman, 1988). The general theory of affordances is most commonly associated with the ecological psychologist J.J. Gibson, and shares a strong tie with the ontological theory of realist perspectivalism discussed earlier. Gibson (1977) states that

> The affordances of the environment are what it offers animals, what it provides or furnishes, for good or ill. [...] the objects of the environment afford activities like manipulation and tool using. The substances of the environment, some of them, afford eating and drinking. The events of the environment afford being frozen, as in a blizzard, or burned, as in a forest fire. The other animals of the environment afford, above all, a rich and complex set of interactions, sexual, predatory, nurturing, fighting, play, cooperation, and communicating. What other persons afford for man comprise the whole realm of social significance. We pay the closest attention to the optical information that specifies what the other person is, what he invites, what he threatens, and what he does (p. 68).

The theory of affordances is useful for WDA because the structure of items within a work environment (either cognitive or physical) affords the performance of certain activities, more or less easily, and restricts or hinders certain other activities, which may not be easily performed within that environment. Ontologies can provide further insight into the environment's affordances by providing a logical decomposition of that environment's objects, their attributes or properties, and associated processes. This, in turn, can lend an improved understanding of the kinds of design parameters associated with the items that make WDA possible.

Donald Norman provides a case for utilizing the theory of affordances for engineering practices. Borrowing from Gibson's general theory (Gibson, 1977; Gibson, 1979), Norman (1988) provides a cognitive-based application for engineering practices, which he terms *the design of everyday things*. It is worth noting that Norman's work bears a striking resemblance to the work of Vicente (1999) and Rasmussen et al. (1994) who also discuss the need for cognitive design principles in contemporary engineering practices. Norman (1988) describes the following situation:

> In one case, the reinforced glass used to panel shelters (for railroad passengers) erected by British Rail was smashed by vandals as fast as it was renewed. When the reinforced glass was replaced by plywood boarding, however, little further damage occurred, although no extra force would have been required to produce it. Thus British Rail managed to elevate the desire for defacement to those who could write, albeit in somewhat limited terms. [...] Glass is for seeing through, and for breaking. Wood is normally used for solidity, opacity, support, or carving. Flat, porous, smooth surfaces are for writing on. So wood is also for writing on. Hence the problem for British Rail: when the shelters had glass, vandals smashed it; when they had plywood, vandals wrote on and carved it. The planners were trapped by the affordances of their materials (p. 9).

As both Norman and Gibson point out, the structural composition of an item (and its relations to other items and to cognitive agents who intend to use it) points to the need for understanding not only what an item is ontologically, but also how that ontological designation (its properties/attributes) leads to applications of WDA. By focusing on structural relations of this sort, ADS allows itself to be married to certain formal ontological approaches. It also treats the underlying relational structure of entities and processes as *categorical items* rather than as lists of specific instances and individuals. In this sense, one can more clearly understand why WDA is not concerned with decomposing specific tasks or items associated with them, but, rather, is concerned with capturing the broader structural relations between means and ends at high levels of abstraction.

Ontologies will be useful for representing those means-ends relations that relate to the purposes and functions of objects. Functional Purposes and Values and Priority Measures (see Figure 12.2) involve intentional (i.e., goal-directed psychological) states, which are the products of epistemic reasoning about a given domain (in abstraction as a set of affordances). Ontologies provide both an a priori, logical structure of a given domain, as well as abilities to reason over that structure, via the use of rules of inference. Reasoners can then utilize asserted facts in the ontology to generate new (i.e., learned) facts about related items and complex states of affairs. Quality ontologies provide a theoretical (yet hopefully veridical) picture of objects, attributes, processes, and relations, upon which, various competing epistemic hypotheses can be laid over and against.

Specific Ontological Relations Useful for ADS

Ontologies can provide improved decomposition of objects and processes in the ADS, because they provide powerful formal machinery for representing the various sorts of relations contained in ADS models that, up to now, have not been sufficiently provided. Examples of these types of formal ontological relations include (but certainly are not limited to):

1. SNAP-to-SNAP and SPAN-to-SPAN relations, which exist internal *to* the SNAP and SPAN subontologies. They include:
   - Genidentity (A=a+b and A=A): Each object possesses a unique identity, even though it can be composed of various parts. For example, a nation is considered to be identical over periods of time although some of its parts, such as bodies of water and coastlines, change over time.
   - Transgranular part–whole relations: There are *part-of* relations, which exist between unitary physical objects and the various parts of which they are composed. They also comprise relations between objects and aggregates of those objects. For example, a house has parts such as rooms, which, have other parts such as walls and a floor, which, in turn, have parts such as beams, studs, plaster, etc. (considered as both individual types of things as well as aggregates).
   - Subsumption relations: These are *is-a* relations, which relate the most general specious items to subclasses and instances that are more specific. For example, a particular house is a house (in general), which is a structure, which is an object.
   - Containment relations: These are relations that delimit a spatial relation of one object being located within a spatial region (sometimes within another object). For example, a car can be contained within the garage (a structural object), which is contained within the boundaries of one's property (a spatial region).

2. SNAP-to-SPAN and SPAN-to-SNAP relations, which exist as transontological relations between the subontologies of SNAP and SPAN (see Figure 12.1). These can include the following:
   - Participation: This is a SNAP-to-SPAN relation (a subset of dependence relations) in which a substance affects or takes part in a process. For example, a broom participates in the activity of cleaning.
   - Realization: This is a SNAP-to-SPAN relation in which dependent substances are realized via their temporal onsets, continued behaviors or activities, and cessations. For example, the role of a worker (e.g., a manager) is realized upon his/her accepting a position or set of tasks, and continues to be realized through his/her performance of those tasks.
   - Involvement: This is a SPAN-to-SNAP relation, which is the converse of Participation, in which a process affects a substance or set of substances. For example, the process of cooking involves pots, pans, or a heat source to perform that function.

These kinds of relations present an initial insight into the kinds of intraontological and transontological relations that are of value to ADS modeling. Smith and Grenon (2004) and Little and Rogova (2007) provide other more detailed forms of these types of relations, which draw upon formal ontological distinctions outside the scope of this chapter. The complexities of these kinds of relations depict the level of detail with which future ontologically driven ADS models need to be designed. By attending to the relational sophistication currently being developed in the field of formal ontology, ADS models could be used to provide more powerful representations of work domains that lead to an increase in the sophistication with which CSE as a whole is performed.

Sample Ontologically Driven ADS Model

Using an ontologically informed ADS model, we can show the effectiveness of understanding the kinds of relation types discussed in the previous section. Coupling an ontology with ADS provides an increased capability for representing the decomposition of objects, attributes, processes, and relations, and an improved WDA model overall. Formalizing pertinent spatial and temporal features of a domain of interest allows a more robust methodology for capturing information relevant for designing the work domain, because implicit information contained in the environment (e.g., definitional information and relational information) can be made highly explicit through the ontology. Figure 12.3 presents a miniature (and highly simplified) ontological model of one type of WDA within a home environment: the domain that affords the baking of a casserole.

The model presented in Figure 12.3 shows SNAP elements of the ontology to the left (under Spatial Item [SNAP]), and SPAN elements to the right (under

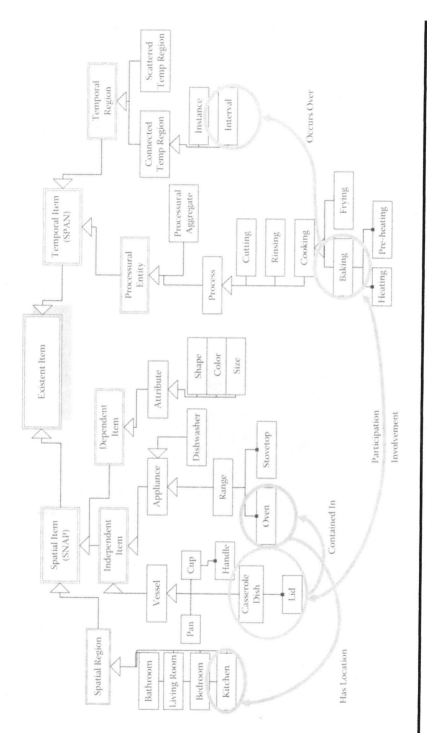

Figure 12.3 Sample of formal ontological relations for baking a casserole.

Temporal Item [SPAN]). The open-headed arrows connecting categories represent subsumption (is-a) relations, whereas the line connectors with the solid balls on the end represent part-relations (e.g., an oven and stovetop are both parts of a range). Certain transontological relations such as Has Location, Participation, Involvement (the converse of Participation), Containment, and Occurs Over are represented by the curved arrows connecting circled areas. This example allows further reasoning: if a casserole dish is contained in an oven and the oven has a location in the kitchen, then the casserole dish also has a location in the kitchen (furthermore, one can also reason that the activity of baking—an activity that involves the oven—would also have its location in the kitchen). This may seem like a trivial form of reasoning, but these kinds of commonsense deductions form the underpinnings for more complex forms of inference, which are often problematic for information systems charged with reasoning about complicated objects, environments, processes, and relations.

Two powerful features of ontologies that provide evidence for their effectiveness in WDA applications are their abilities to be (1) interoperable, meaning that the upper-levels of the ontology allow numerous kinds of distinct and segregated domain-specific information to be coupled and (2) rescalable, meaning that the same ontology model can be utilized for more coarse-grained or fine-grained decompositions of items, depending on needs of the user pertaining to particular sets of tasks or domains of interest. For example, the ontological model mentioned previously (Figure 12.3) could be extended much further to include other important kinds of information such as

- What kinds of food are best prepared in an oven versus a microwave (depending on the affordances of the environment in question, as well as distinct attributes of the food)?
- Who is the agent preparing the food?
- What are the dynamic attributes (e.g., shape, color, size; under Dependent Item) of the items under consideration?
- Where is it to be served, etc.?

These kinds of domain-specific questions provide insight into another important feature of formal ontologies for WDA: their querying capabilities. The ability to query various sorts of information in large-scale data systems is of tremendous importance for a host of applications related to WDA. One may, for example, use such queries as a basis for designing a work domain model in which information must flow between various sources or decision-makers may need to access large amounts of data to perform their jobs. Queries can also be used in real-time scenarios to update both the ontology as well as other connected computational systems, where issues such as resource allocation are pertinent to a task at hand, for example, in emergency or disaster domains (Bisantz, Rogova, & Little, 2004).

Conclusions and Further Research

This chapter has provided an argument for the need to merge ontologies with ADS models to enhance the current capabilities of WDA. By providing some examples of the kinds of formal relation types used within formal ontologies, one can hopefully see both the value and the novelty of this approach for cognitive systems engineers working in WDA domains to more accurately and effectively design ADS models, particularly those areas of ADS responsible for the decomposition of objects, object-related processes, and purpose-related functions. There are methodological benefits to be gained in both disciplines. Engineers working with WDA models are provided with a set of formalized tools for more accurately defining items, processes, and relations in the work domain. On the other hand, ontologists are provided with insights into user needs and domain-specific considerations, necessary for building ontologies that both accurately mirror the real world and provide a genuine use-value to particular work-related domains. This chapter describes one attempt to confront some of the current methodological challenges of ADS design by utilizing formal philosophical categories, which in turn, provide a principled basis for designing ontologically-driven WDA.

References

Bennett, B., & Feldbaum, C. (Eds.). (2006). *Formal ontology for information systems*, in *Frontiers in Artificial Intelligence and Applications*, Vol. 150. IOS Press, Amsterdam.

Bisantz, A.M., & Vicente, K.J. (1994). Making the abstraction hierarchy complete. *International Journal of Human-Computer Studies,* 40, 83–117.

Bisantz, A., Rogova G., & Little, E. (2004). On the integration of cognitive work analysis within information fusion development methodology. *Proc. of the Human Factors and Ergonomics Society Annual Meeting.* New Orleans.

Bittner, T., & Smith, B. (2001). A taxonomy of granular partitions: Ontological distinctions in the geographic domain. Montello, D. (Ed.), *Spatial Information Theory, Lecture Notes in Computer Science,* 2205, 28–43.

BFO Home Page, http://www.ifomis.uni-saarland.de/bfo/

Boury-Brisset, A. C. (2003). Ontology-based approach for information fusion. *ISIF,* 522–529.

Casati, R., & Varzi, A. (1999). *Parts and Places: The Structures of Spatial Representation,* MIT Press.

DOLCE Ontology Site: http://www.loa-cnr.it/index.html

Gibson, J.J. (1977). The theory of affordances. In Shaw, R., & Bransford, J. (Eds.). *Perceiving, Acting, and Knowing: Toward an Ecological Psychology.* New Jersey: Lawrence Erlbaum Publishers.

Gibson, J.J. (1979). *The Ecological Approach to Visual Perception,* Boston: Houghton Mifflin.

Gomez-Perez, A., Fernandez-Lopez, M., & Corcho, O. (2004). *Ontological Engineering: With Examples from the Areas of Knowledge Management, E-commerce, and the Semantic Web,* London: Springer Verlag.

Grenon, P. (2003a). Spatiotemporality in basic formal ontology: SNAP and SPAN upper-level ontology and framework for formalization. *IFOMIS Technical Report Series.* http://ifomis.de.

Grenon, P. (2003b). Knowledge management from the ontological standpoint. *Proceedings of WM 2003 Workshop on Knowledge Management and Philosophy.* April, Luzern Switzerland.

Grenon, P., & Smith, B. (2003). SNAP and SPAN: Towards dynamic spatial ontology. *Spatial Cognition and Computation* (forthcoming).

Husserl, E. (1900-01). *Logische Untersuchungen,* 2 Bde., *Husserliana,* Band XIX, Den Haag: Martinus Nijoff, 1985 edition.

Johansson, I. (1989). *Ontological Investigations: An Inquiry into the Categories of Nature, Man and Society.* New York, Routledge.

Kirwan, B., & Ainsworth, L.K. (Eds.). (1992). *A Guide to Task Analysis.* London: Taylor & Francis.

Kokar, M.M., & Wang, J. (2002). Using ontologies for recognition: An example. *Proceedings of the Fifth International Conference on Information Fusion.* pp. 1324–1343.

Kokar, M. M., Matheus, C.J., Baclawski, K., Letkowski, J. A., Hinman, M., & Salerno, J. (2004). Use cases for ontologies in information fusion. *Proceedings of the Seventh International Conference on Information Fusion.* pp. 415–421.

Little, E. (2003). A proposed methodology for application-based formal ontologies. *Proceedings of the Workshop on Reference Ontologies vs. Application Ontologies.* September 15–18, University of Hamburg, CEUR-WS.org.

Little, E., Eberle, J., & Turino, F. (2008) Utilizing Ontologies for Petrochemical Applications. *Formal Ontologies Meet Industry* (Borgo, S. & Lesmo, L., Eds.), *Frontiers in Artificial Intelligence and Applications* Vol. 174. Amsterdam: IOS Press.

Little, E., & Rogova, G. (2005). Ontology meta-model for building a situational picture of catastrophic events. *Proceedings of the FUSION 2005-8th International Conference on Multisource Information Fusion,* July 25–29, Philadelphia, PA.

Little, E., & Rogova, G. (2006). An ontological analysis of threat and vulnerability. *Proceedings of the FUSION 2006-9th International Conference on Multisource Information Fusion,* July 10–13, Florence, Italy.

Little, E., & Rogova, G. (2008). Designing ontologies for higher-level fusion. *Journal of Information Fusion,* special issue on *Situation Awareness.* Elsevier Press (in press).

Little, E., Rogova, G., & Boury-Brisset, A.C. (2005). Theoretical foundations of threat ontology (ThrO) for data fusion applications. TR-2005-269.

Naikar, N., Hopcroft, R., & Moylan, A. (2005). *Work domain analysis: Theoretical concepts and methodology.* Air Operations Division Defence Science and Technology Organization, DSTO-TR-1665.

Norman, D.A. (1988). *The Design of Everyday Things.* Doubleday Publishers.

Nowak, C. (2003). On ontologies for high-level information fusion. *Proceedings of the Sixth International Conference on Information Fusion.* Cairns, Australia.

Rasmussen, J., Pejtersen, A.M., & Goodstein, L.P. (1994). *Cognitive Systems Engineering.* New York: Wiley-Interscience.

Semy, S., Pulvermacher, M., & Obrst, L. (2004). Toward the use of an upper ontology for U.S. government and military domains: An evaluation. *MITRE Technical Report.* 04B0000063, September. Submission to: *Workshop on Information Integration on the Web* (IIWeb-04), in conjunction with VLDB-2004, Toronto, Canada, August 30–September 3.

Shaw, R., & Bransford, J., (Eds.). (1977). *Perceiving, Acting, and Knowing: Toward an Ecological Psychology.* New Jersey: Lawrence Erlbaum Associates.

Simons, P. (1987). *Parts: A Study in Ontology.* Oxford: Oxford University Press.

Smith, B. (2001). Ontology and information systems. http://ontology.buffalo.edu/ontology(PIC).pdf

Smith, B. (2002). *Basic Formal Ontology,* http://ontology.buffalo.edu/bfo/

Smith B., & Grenon, P. (2004). The cornucopia of formal-ontological relations. *Dialectica* 58 (3), 279–296.

Sowa, J.F. (2000). *Knowledge representation: Logical, philosophical, and computational foundations,* Brooks Cole Pub., Pacific Grove, CA.

Standard Upper Ontology Working Group (SUO WG), http://suo.ieee.org/SUO/Ontology-refs.html

Vicente, K.J. (1999). *Cognitive Work Analysis: Toward Safe, Productive, and Healthy Computer Based Work.* Mahwah, NJ: Lawrence Erlbaum Associates.

Vicente, K.J. (2000). Cognitive work analysis: Research and applications. *Proceedings of the Joint 14th Triennial Congress of the International Ergonomics Association/44th Annual Meeting of the Human Factors and Ergonomics Society (HFES/IEA 2000)*: Vol. 1, (p. 193). Santa Monica, CA: HFES.

Welty, C., & Smith, B. (Eds.). (2001). *Formal Ontology and Information Systems,* New York: ACM Press.

Chapter 13

The Theoretical Foundation of Cognitive Work Analysis

Gavan Lintern

Contents

Overview

In this chapter, I outline my view of a foundational theory for Cognitive Work Analysis (CWA). I describe several theoretical perspectives that provide contextual support for the theory and then describe specific theoretical concepts that underlie each phase of analysis. I use this theoretical outline to identify directions for further development. I also observe that this theoretical foundation asserts immutable constraints on human behavior and, in doing so, offers us clear and specific guidance for the design of complex sociotechnical systems that allow workers to complete the design.

Orientation

I have been troubled for some time that the theoretical foundation of CWA has not been well articulated. Although many of its elements are dispersed throughout the major texts (Rasmussen, 1986; Rasmussen Pejtersen & Goodstein, 1994; Vicente, 1999), they have not been gathered into a self-contained statement such as a single chapter or a journal article. I am also troubled by the misinterpretations of CWA, which abound in the literature; for example, Lind (2003) on work domain analysis, Cummings (2006) on the temporal implications of work domain analysis, and Hollnagel and Woods (2005) on the decision ladder (DL) and the abstraction hierarchy (AH). I suggest that these two issues are related. We of the CWA community have done a poor job in explaining its foundation and have thereby left ourselves vulnerable to misdirected critiques.

In this chapter, I outline my view of the foundational theory. I do so with some trepidation because I believe that many in our community will want to reshape what I say. However, if the structure of the chapter and the purpose I outline for it find general acceptance, I will deem the effort successful even if I cannot achieve consensus on the details. To that end, I will eschew a tightly argued formulation in favor of an account that clarifies foundational concepts and how they relate to form a coherent and understandable rationale for the methods and procedures that make up the framework of CWA. My aim here is to clarify rather than prove, and in service of that aim, I will emphasize narrative and illustration in preference to evidence.

There is not much in my theoretical account that is new. Rather, this is an effort to assemble accepted theoretical ideas, dispersed throughout Rasmussen's work primarily but also with reference to the work of others in our community, into an accessible statement.

A Niche for Cognitive Work Analysis

Some of the negativity toward CWA emanates, I believe, from failure to understand what we are trying to do. Most of the techniques of Cognitive Engineering are aimed at identifying and then working on points of leverage, for example, on developing cognitive support tools in the form of such features as decision aids. In contrast, the framework of CWA was developed for a much larger problem: the design of large-scale sociotechnical systems. Despite the value of other cognitive engineering strategies, they deal only with segments of the design problem for a complex sociotechnical system. I do not intend that remark to be pejorative; many design assignments in cognitive engineering require precisely that form of intervention. My specific claim here is that CWA occupies a niche in the design world that is often not appreciated by those who focus on points of leverage or on the development of cognitive support tools.

A long-standing complaint within Human Factors is that we are brought into the design of large-scale systems only after human integration problems have become apparent. It is commonly argued that the expense of correcting these problems could be avoided if we were consulted earlier, possibly during concept development and then throughout the remainder of the design cycle. In the past, I have been skeptical.

Although I was confident that we could have avoided the kinds of common problems that were emerging, it was never clear to me that we would not have introduced other serious issues. We had no comprehensive analytic framework for addressing issues in concept development and then proceeding systematically through the human systems integration issues in the design of a complex sociotechnical system. Only when I became acquainted with the framework of CWA did I begin to build confidence that I could, if invited early into the design process, contribute as an equal and effective partner.

I also wonder if the emphasis within CWA on representation introduces some negativity. Crandall, Klein, and Hoffman (2006, p. 107) note that knowledge elicitation has received more attention than knowledge representation within the general field of Cognitive Task Analysis. I suspect that formal education plays some role in determining enthusiasm for representation. Engineering disciplines use representation extensively and systematically to impart understanding. In contrast, representation is employed less often in psychological science, and then in an improvised and impromptu fashion.

From my own background in psychology, my initial reaction to CWA was that it was *only* about representation (and therefore insubstantial). I adjusted that thought rapidly as I read further but continue to believe that representation is central to the enterprise of CWA. However, I no longer use the pejorative *only* when I offer that view. I have come to believe firmly in the power of a theoretically motivated and well-organized set of representations for assimilating, building, archiving, and transferring knowledge.

Finally, I suggest that the notion of activity independence troubles many people. Both cognitive psychology and systems engineering are process- or activity-oriented disciplines. Cognitive theories are typically framed as a series of processes or activities, and functional analysis in systems engineering typically results in a representation of functional flow rather than of functional structure. In contrast, Gibson (1979) is one in psychology who has taken this notion of activity independence seriously. I recall that his approach troubled me as I worked through the first two chapters of his *Ecological Approach to Visual Perception*. My first thought was that this was *only* about the structure of the world and that there was no psychology in it. Again, note the pejorative *only*. One should not judge Gibson prematurely, which, I suspect, many people do. One has to get through the complete argument to appreciate its elegance.

I had at least assimilated Gibson's argument by the time I encountered Rasmussen's work, and I did not for a moment cast the same aspersion. I do recall thinking that Rasmussen's distinction between structure and process was much like Gibson's. If we were to take the critics seriously, we would have to assume that Gibson and Rasmussen are similar in that they have built a flawed conceptual structure from a fundamental misunderstanding of the nature of the world. I happen to think otherwise—that each, in his own way, immersed in a conceptually challenging and somewhat distracting intellectual culture, somehow came to remarkable insights about the way we need to approach complex human action.

Rasmussen was concerned with how to integrate multiple, diverse technical capabilities with human capability at many levels of organization into a cohesive sociotechnical system. My emphasis in this chapter is on the theoretical foundation that guided that development and that continues to guide us today.

The Nature of Theory

Opinions on what constitutes a theory are diverse. Sometimes I see summaries of structures as derived through the application of taxonomic methods characterized as theory. Sometimes these summaries incorporate relational statements as might be derived through the application of ontological methods. Indeed, an abstraction–decomposition space is developed through use of ontological methods. The Abstraction-Decomposition Space is not, however, a theory, although I will later argue that the way we build one is guided by a theory of reasoning.

In addition, I have occasionally encountered the opinion that Gibson's ecological approach is not a cognitive theory because it does not posit an internal cause–effect mechanism. I take issue with that opinion on two counts. I suspect that the author of a comment like that is demanding a linear action–reaction event, such as a cue striking a billiard ball, where behavior-shaping constraints will not serve. In addition, given that much of what Gibson said implied internal processes (e.g., resonance to information), I take this view to imply that properties or events external to a central nervous system are not relevant to cognitive theory. These views contrast with the foundational assumptions that guided both Gibson and Rasmussen.

Given this sort of uncertainty, it is worth offering an opinion as to what sort of properties a theory for CWA should capture. My dictionary (Houghton Mifflin,

2000) defines a theory as a set of statements or principles devised to explain a group of facts or phenomena, or that which can be used to make predictions about natural phenomena. That will do for the current purposes, although we should be cautious about the implication of the word "explanation." There can be dissension about what constitutes an explanation versus a description but following Wigner (1979) I conclude that scientists, by use of theory, seek to make sense out of regularities they observe in natural phenomena.

Also note the use of the word *devised* in the definition. A theory is not a statement of fact but rather an imaginative construction. The test of a theory is not whether it is true or false but whether it helps us understand the world in useful ways. Despite being an imaginative construction, theories can be powerful. The theory of gravity, for example, is a relatively simple statement that takes account of a diverse set of natural phenomena. Despite its simplicity, it has remarkable power. There is presumably no one reading this book who doubts that gravity will have its way on every location on our planet: we believe this not only for locations we know but also for places we have never been, for places we will never visit, and even for places we have never heard about.

There is a tendency within behavioral science to envy physicists. They study (or at least used to study) observable and regular phenomena. I suggest in this chapter that the envy is unnecessary: CWA is also based on observable and regular behavioral phenomena that can impart considerable power to our analysis and design activities.

Foundational Perspectives

The foundational perspectives I outline in this section did not necessarily guide developments in CWA, but the concepts they have established represent core assumptions.

Self-Organization

Self-organization is a process of attraction and repulsion in which the internal organization of a system, normally an open system, increases in complexity without being guided or managed by an outside source. Self-organizing systems typically (though not always) display emergent properties.

http://en.wikipedia.org/wiki/Self-organization [accessed March 15, 2008].

Self-organizing systems can transition into a different (and sometimes unexpected) form of organization at a point of nonlinearity. As explained by Prigogine and

Stengers (1984), an adjustment of a control parameter can generate critical fluctuations that cast the system into a new energetic mode. In the case, for example, a change in the control parameter is serving to inject more energy into the system, a point is reached at which the system must reorganize to dissipate that energy. The term *dissipative structure* is often used to characterize the new organizational form.

The patterns of locomotion of a horse offer an illustration. As the rider nudges the horse into increasing its speed, the horse will initially increase the rates of limb motion, but at some critical point will transition into a new, more efficient mode (e.g., canter to gallop). As is true of all nonlinear systems, equine locomotion is mostly linear. However, it is at the nonlinear transitions that interesting things occur. It is invariably true that nonlinear systems can appear linear if one takes a restricted view.

Those who promote self-organization as an explanation of cognition typically emphasize the role of local interactions in the development of patterns. They might offer self-organization as a bottom-up, emergent view in contrast to the top-down view of mental imagery as the shaping influence on cognition. Some caution is needed here. Although local forces of attraction and repulsion play an important role in self-organization, the interplay between local and global constraints are also significant.

If we assume that cognitive states emerge from nonlinear interactions (Freeman, 1995), then cognitive emergence owes as much to the functional layout of the environment as it does to the local interactions of individuals with each other and with artifacts. The cognitive architecture determines the way information flows through the system. This architecture encompasses the functional structure of the physical environment, the social organization of the workplace, and the functional structure of individual minds. New cognitive capabilities emerge from activities undertaken within the constraints imposed by the cognitive architecture and are shaped by those architectural constraints.

Theories that posit a mental image, a mental model, or a mental schema as a formative cause of cognition eschew self-organization (e.g., Johnson-Laird, 1983). In contrast, others argue that an understanding of self-organization is central to understanding cognition. Peter Kugler and I have summarized the basic concepts for this latter view in Lintern and Kugler (1991) and Kugler and Lintern (1995).

Engineers abhor nonlinearities, but biology cannot survive without them. In cognitive engineering, we have a subtle problem. We need to conjoin system components that have been designed with linearity as a design goal to other system components (e.g., the human operators) for which nonlinearity is fundamental. The technocentric approach is to enforce linearity on our nonlinear human workforce: to suppress the self-organizing tendencies of human systems. These self-organizing tendencies are, however, critical to system effectiveness (Lintern, 2003). The humancentric approach we seek in cognitive engineering is to work with, even celebrate, and leverage from the self-organizing tendencies to make our systems more effective.

Situated Cognition

The ethnographic research of Hutchins (1995), Jordan (1989), Lave (1988), Lave and Wenger (1991), Saxe (1991), Scribner and Fahrmeier (1982), and Suchman (1987) offers profound insights. It reveals how adept workers can be at cognitively restructuring their work environment in a manner that appears to have at least some of the characteristics of self-organization. Invariably, the work practices that evolve are cognitively economical and robust, typically more so compared with work practices prescribed by management.

I have reviewed a portion of this work for its relevance to aviation (Lintern, 1995). One lesson to be taken from it (for aviation and more generally) is that workers are both physically and cognitively active, reshaping how they think about their work environment as they develop their own work practices. The conceit of managers, and also of many designers, is that they know how work should be accomplished, and they need to guide workers toward the proper procedures. Ethnographic research on situated work practice reveals how shallow this conceit is.

In thinking about this issue, I reflect on developments in Artificial Intelligence. There are an enormous number of computationally based support systems that would seem to offer huge advantages to current practices. Diagnostic systems for medical practitioners can serve to illustrate. These have been under development for decades but are still struggling to find their way into common usage within the medical profession. It is not unusual to hear accusations that medical practitioners are too arrogant to embrace technology that might replace some of their skills. I suspect otherwise: these systems do not mesh well with the cognitive strategies and workflow processes of medical practitioners. From that perspective, it would seem that it is the designers of these systems who are overly arrogant.

More than anything else, research in situated cognition indicates that we need to be very careful when we adjust cognitive strategies or workflow processes. Existing strategies and processes will have evolved over considerable time to be robust and effective. To change them without a full understanding of the potential repercussions is to risk disaster.

Distributed Cognition

Within the work environment of ship navigation in confined waters, Hutchins (1995) reiterated many of the insights to be drawn from situated cognition but added a particularly evocative and succinct description of distributed cognition. Up to that time, distributed cognition was a somewhat fuzzy concept that even experts in the field would debate.

Hutchins argued that a ship navigation team (together with the accompanying navigational artifacts and procedures) is a cognitive system that performs the computations underlying navigation. It is a distributed cognitive system because various elements of the computations are carried out over time and in different locations. The results of early computations are passed to other locations and then integrated for further computations. Hutchins argued that the system has cognitive properties that differ from the cognitive properties of the individuals within the system, and that the cognitive potential of the group depends as much on its social organization as on the cognitive potentials of its members. Thus, the navigational system performs computations that need not necessarily be within the grasp of all (or even any) of its members.

Most, if not all, sociotechnical systems we design will be distributed. Embedded within the theory of distributed cognition are ideas and principles that can help us understand how such systems work and that can offer important ideas about how to design them. In particular, where we could once rely on normal, social processes to promote collaboration between colocated workers, we must now attend more explicitly to the development of technology and processes that promote collaboration between geographically and temporally distributed work units. In that the social processes that promote collaboration are natural and self-organizing, our role is to provide the structures that will allow them to operate effectively in distributed systems.

Cognitive Systems

A cognitive system is one that performs the cognitive work of knowing, understanding, planning, deciding, problem solving, analyzing, synthesizing, assessing, and judging as it is fully integrated with perceiving and acting. A complex sociotechnical system is an entity that does cognitive work and can therefore be considered as a cognitive system.

The claim that a complex sociotechnical system does cognitive work expands the view of what *cognitive* is beyond the individual mind to encompass coordination between people and their use of resources and materials. This view is aligned with the theory of distributed cognition as described earlier and also by Hollan, Hutchins, and Kirsh (2000). A foremost claim of this theory is that distributed cognition is not a special type of cognition but is rather a characterization of fundamental cognitive structures and processes (Hollan et al., 2000). Thus, all cognition is distributed.

Traditionally, we are used to thinking of cognition as an activity of individual minds, but from the perspective of distributed cognition it is a joint activity distributed across the members of a work or social group and the technological artifacts available for support of the work.* Cognition is distributed spatially so that

---

* Research in situated cognition and ecological psychology focuses primarily on individual interactions with environment or artifacts, but their foundational ideas are entirely consistent with this distributed view.

diverse artifacts shape cognitive processes. It is also distributed temporally so that products of earlier cognitive processes shape later ones. Most significantly, cognitive processes of different workers interact so that synergistic cognitive capabilities emerge via the mutual and dynamic self-organization resulting from both spatial and temporal coordination among distributed human agents.

A distributed cognitive system is one that self-organizes dynamically to bring subsystems into functional coordination. Many of the subsystems lie outside individual minds. In distributed cognition, interactions between people as they work with external resources and the ways they use those resources are as important as the processes of individual cognition. Both internal mental activity and external interactions play important roles as do physical resources that reveal relationships and act as reminders. A distributed system that involves many people and diverse artifacts in the performance of cognitive work is therefore properly viewed as a cognitive system.

The theory of distributed cognition forces a shift in how we think about the relationship between minds, social interactions, and physical resources. Interactions between internal and external processes are complex and unfold over different spatial and temporal scales, and neither internal nor external resources assume privileged status.

A cognitive system is a thinking (or intelligent) information system. However, the enhanced intelligence is not generated by the activity of intelligent technological functions as many in the discipline of Artificial Intelligence will want to claim, but emerges from the coordinated collaboration of distributed human agents via their interactions with each other and with functionally heterogeneous technological artifacts. In the sense that collaboration between human agents and their use of technological artifacts is coordinated, effective, robust, and meaningful, the cognitive system is intelligent.

It is sometimes argued that computer-based agents can be employed to reason about the beliefs of human participants in teams. From the perspective I promote here, people reason but technological devices do not. Two people in coordination can possibly reason more effectively than either in isolation, and if they (as a coordinated dyad) avail themselves of the opportunities presented by technological devices that can compute logical relationships, find and organize information, and probably offer a number of as yet unimagined supporting functions, these entities (the two people together with their technological devices) constitute a reasoning system.

Ecological Psychology

The foundational insight of Ecological Psychology is that cognition is tied up in the reciprocity between an organism and its environment. As with Situation Cognition and as stressed particularly by Hutchins (1995) and his colleagues (Hollan et al., 2000), much of cognition occurs in the world rather than in the head. Some critics of ecological psychology have taken this to be a claim that nothing relevant to cognition happens in the head, although I remain mystified as to how any reasonably competent reading of Gibson can lead to that conclusion.

The concept of affordances is a major contribution from Gibson that is relevant to our work. An affordance is a relationship between properties of an organism and matching properties of its environment. It is a relationship between capability and opportunity. In an explanation of the relevance of affordances to interface design (Lintern, 2000), I drew on work by Warren and Whang (1987), who discussed the relationship between shoulder width and aperture width as a passing-through affordance. Indeed, an affordance is always a relationship. Where the dimensions of that relationship can be quantified, it will be a dimensionless ratio.

Thus, an affordance-based fuel gauge would compare the distance that can be traveled with currently available fuel to the distance that needs to be traveled. Depending on which way you construct the ratio, a value of more or less than one will signify that you can or cannot get to your destination. This strategy removes from the operator the computation that is required when fuel and distance are presented separately. Gibson's affordance claim is that this is analogous to the way we operate in the world.

I suspect that many take an ecological display to be one that is pictorial or richly graphical. However, for a display to be ecological, it needs to be more than that. It needs to incorporate within its graphics a depiction of structure of the work environment in ways that support the natural strategies of human cognition or that otherwise reduce the cognitive complexity of the work. At this time, as our understanding of how to design ecological displays is still evolving, it is unclear how we might depict all important properties in terms of affordances even if that is desirable, but representation in terms of affordances does reduce the need for cognitive computations, which is one of our goals for ecological interfaces.

Requisite Variety

Vicente (1999) argues that the complexity of a technological support needs to reflect the complexity of the work and appeals to the law of requisite variety. Ashby (1957, p. 207), in framing this law, argued that *only variety can destroy variety*, here taken to mean that only variety can *control* variety. In other words, a control system must incorporate as much variety as the system it controls. However, from a human-centric perspective, a control system should not incorporate more variety than the system it controls. Thus, the functional scope and granularity of a workspace must match (it must accommodate but not exceed) the operational complexity of the work.

The law of requisite variety warns us against seeking to reduce control complexity by simplifying displayed information. Hollnagel and Woods (2005, p. 85) also warn against this but mischaracterize ecological interface design as a strategy that reduces complexity. Any ecological interface, when properly designed, will give selective information access to the level of complexity required for the anticipated control problem. It does not continuously display all information at the most detailed levels as, for example, does a single-sensor/single-control strategy, but rather displays patterns that can be selectively interrogated to reveal information for the current control task at the required level of detail. It will therefore guard against the possibility that the complexity of the displayed information will exceed the operational complexity of the work.

A pentagon display for a social system governed by human intentions, shown in Figure 13.1, offers a simple illustration. It follows the polar-star example of Dinadis and Vicente (1999) for a technical system governed by causal physical laws. Five dimensions that contribute to the global construct (in this case, social values) are represented by individual spokes of the pentagon. The measures of those dimensions are normalized to show a symmetric figure under normal or desirable conditions. Where a particular measure reflects an abnormal or undesirable condition, the spoke for that measure generates a distortion in the figure. That distortion will be noticed readily and the offending dimension identified. The relevant spoke can then be interrogated (via mouse click) to foreground more diagnostic detail about the issue. By this means, the requisite complexity is avoided until it becomes relevant, and only that portion of the requisite complexity needed for the current

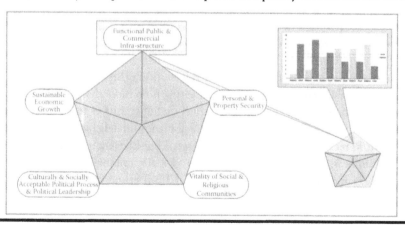

Figure 13.1 **A pentagon representation for a social system governed by human intentions (assessment of progress in building a civil, democratic society). The outset at bottom right shows a distortion for functional public and commercial infrastructure, the details of which can be displayed in a histogram that compares desirable versus actual supply of electricity, gas, telephone, water, gasoline, and public transport.**

situation is displayed. Nevertheless, the entire requisite complexity for the system is available.

In developing complex sociotechnical systems, we need to adhere to this principle of requisite variety. However, we should not allow this principle to encourage more than the essential information variety. At this point in time, I believe that Ecological Interface Design is the only strategy that takes a principled approach to balancing these two constraints.

Work Centering: Whence the Images

Most, if not all, scientific developments emerge from and build on an image that is acquired informally through natural interaction in the world. Theorizing in behavioral science has traditionally derived formative images from some sort of well-known mechanism. Most recently, the digital computer has played a central role, but appeals to formal logic and mathematical relationships have also been influential. I am reminded here of a personal conversation with Gary Klein. He observed that he had entered his early decision research committed to the assumption that he would find evidence of option comparison. Only after confrontation with evidence that suggested otherwise could he reject this concept and develop the notion of recognition-primed decision making.

His work on recognition-primed decision making is now held in high regard, and deservedly so, but there is one aspect of it that attracts a little comment. I suggest that there is a radical move embedded in this evolution. Klein rejected decades, possibly even centuries, of reliance on technological images in favor of a work-centered image, one drawn from the way experienced operators conceptualized their work. Quite independently, it seems, Rasmussen had already made this move, and researchers in situated cognition were actively working through it.

I have heard it said that cognitive engineering is no more than good human factors or good applied cognitive science. I reject that observation and do so primarily because of this move. Human factors is guided predominately by theoretical images derived from technology and logic. In contrast, cognitive engineering is work centered not only in practice but also in theory. We are no longer deriving formative images from mechanism (e.g., computer) but from ethnographic descriptions or analyses of cognitive work.

A Theory of Work Practice

Although each of the perspectives outlined previously can be considered a theory in its own right, and each contributes to how we might understand a theory of work, none constitutes a comprehensive theory of work practice. The theory of work practice needs to characterize the structure within which work is accomplished and the processes with which it is accomplished. Here I argue that the structure is best viewed as functional structure, as represented in an abstraction–decomposition space, and that the work processes can be characterized in terms of work context (such as work situations and work problems), decisions, strategies, social organization (processes of management and collaboration), and levels of cognitive control.

As all will recognize, this is a parochial view from the perspective of CWA. I am not adamant that this view is comprehensive or that it provides the best account for understanding complex sociotechnical systems. I do claim that it constitutes the foundation of CWA as we currently practice it. If laid out in this manner, the theory may stimulate clearer recognition of its implications and thereby encourage refinement, development, and possibly redirection. I would regard any or all of these factors as progress rather than as evidence that our work is based on a flawed theory.

The Functional Structure of Work

It would seem axiomatic that the resolution of a problem would require the problem solver to be cognizant of the purpose of the system in which the problem is observed, the values and priorities that need to be considered within use of the system, the physical resources available for problem resolution, and the uses to which the available resources can be put (i.e., their functionality). In a complex knowledge domain, it is likely that the problem solver would need to decompose some of these elements to fully understand how they contribute to resolution of the problem. The theory that lies behind work domain analysis, the first phase of CWA, is based on the assumption that domain experts reason in this manner when they are resolving knowledge-based problems within their domain of expertise.

The statement earlier, "the uses to which the available resources can be put," implies a means–ends relationship between physical resources and functionality. A physical device such as a sensor provides functional capability (see Box 13.1). Similarly, functional capabilities support realization of values and priorities and, in their turn, values and priorities support realization of the system purpose. Furthermore, the usual approach to work domain analysis as inspired by Rasmussen is to distinguish functional capability as physical functions (directly realized through activation of physical resources) and general functions (more abstract functions, linked indirectly to physical resources via physical functions, established to support realization of system purpose). This structural scheme is shown in Figure 13.2.

Reasoning by reference to the form of functional structure shown in Figure 13.2 is conceptualized as a navigational trajectory through an abstraction–decomposition space. That trajectory is not characteristically systematic. An expert might start anywhere in the abstraction–decomposition space and might wander through it opportunistically, collecting information as the need becomes apparent. The same subject matter expert, on the next exposure to a similar problem, might employ

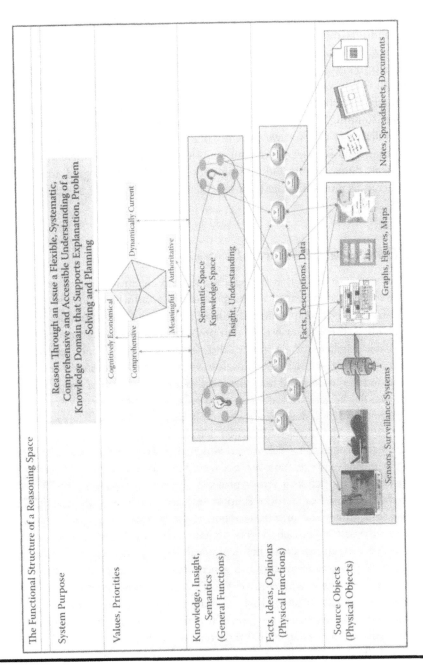

Figure 13.2 The general form of an abstraction–decomposition space.

Box 13.1 Resource-function means–ends relations

A sensor offers the capability of event detection. Thus, the sensor is the means that enables event detection (the end). Note that this is neither a statement of causality nor of sufficiency. A sensor does not cause but rather allows the detection of an event. Neither is it sufficient for the detection of the event (e.g., there must be power available for the sensor). The sensor might not even be essential; one or more other resources, such as direct human observation, might be suitable. However, a sensor does enable the detection of an event.

Furthermore, as shown in Figure 13.2, means–ends mappings may be one-to-one (only one resource required to realize a function), many-to-one (several resources required to realize a function), one-to-many (one resource can support realization of several functions), and many-to-many (several resources support realization of several functions).

---

a different trajectory. All that is required is that the essential information be collected, at least *implicitly*, for the development of an expert solution.

I introduce the concept of *implicit* to take account of the fact that experts might remember information from previous experiences and, in some circumstances, might not even be aware of how that information influences their reasoning. Additionally, information about purposes, values, and priorities may have become entrenched by indoctrination or training. Experts may have tuned their reasoning strategies to take account of that information without necessarily being cognizant of it during problem-solving events. In such circumstances, an expert is unlikely to visit, at least explicitly, all essential nodes of the abstraction–decomposition space.

Figure 13.2 serves two purposes in this account: it illustrates the general form of an abstraction–decomposition space and also the general structure of this form of knowledge-based reasoning. Any information space or display structured on the basis of an abstraction–decomposition space can be thought of as a reasoning space, but here I have chosen to employ Figure 13.2 as a means of explicitly depicting the structure of this form of reasoning.

As shown at the first level of the figure, the system purpose is to reason through an issue. As shown at the level of values and priorities, the reasoning process and its outcome should conform to specific criteria, a notional subset of which is shown in Figure 13.2. The process should be cognitively economical. The outcome should be comprehensive, authoritative, and meaningful. Additionally, it should be based on up-to-date (dynamically current) information.

The information to be consulted for knowledge-based reasoning will be provided by source knowledge objects. These might be various types of sensors as well as documents and graphics. Knowledge objects contain or deliver facts, ideas, and opinions that can be linked through means–ends relations to semantic networks at the level of the semantic or knowledge space. Although different semantic networks can generate different types of understanding, the two shown in Figure 13.2 depict the generation of insight and of interesting questions. As depicted in the figure, some of the same information can contribute to both, but each will draw on a unique constellation of ideas.

Note that this is an activity-independent representation. An abstraction–decomposition space can be likened in this regard to a map, which is a familiar type of activity-independent representation that supports resolution of a straightforward problem set (e.g., navigation through an unfamiliar area). A map supports activity, but none of its elements describe activity. Similarly, an abstraction–decomposition space provides information that can support activity, but it does not contain activity descriptions. The theory on which work domain analysis is based describes the functional structure for the support of reasoning but does not describe the reasoning process itself.

One of the major concerns in reasoning through a complex problem or knowledge-intensive issue is that there can be subtle interactions between seemingly independent functions. For example, a military plan for resolution of an issue might be developed on the basis of availability of certain resources that, if made available, would compromise another critical mission. Expert planners will take such interdependencies into account. An abstraction–decomposition space represents such interdependencies explicitly by use of means–ends links that cross-reference different functional areas.

In summary, the claim is that there is a natural structure to expert reasoning, best described by an abstraction–decomposition space that links the different levels of functional abstraction through means–ends relations and reveals functions and physical resources at different levels of decomposition.

Work Process

Process descriptions of work characterize how work is accomplished within the available functional structure.

Work Situations and Work Problems

Necessary accomplishments of work (problems to be resolved or tasks to be accomplished) are determined by situation or phase. A contextual activity matrix (Naikar, Moylan, & Pearce, 2006) is a useful representation for summarizing work situations and work problems. For reasoning through an issue, there may be phases of formulation, development, and refinement of an argument (Figure 13.3). Several work problems may be involved (see Figure 13.3), each of which has both a normal and a potential span of action.

This form of characterizing work in terms of situations and problems can be taken as a subsumption claim in the style outlined by Clancey, Sachs, Sierhuis, and van Hoof (1998). Subsumption is a hierarchical structure in which activities at a subordinate

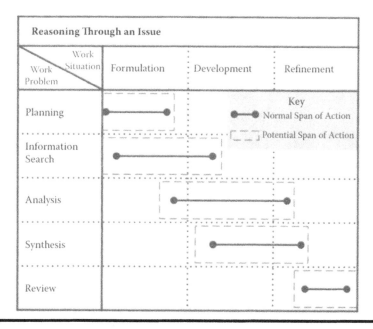

Figure 13.3   Notional contextual activity matrix for reasoning.

level are subsumed under a superordinate activity. It depicts workflow as a composition (performance) versus a causal sequence. This view assumes that human activity is subsumed within and shaped by context, and that most activity is shaped by loosely coupled constraints between several levels of hierarchically nested contexts (Figure 13.4).

The relationship between the superordinate and subordinate activities is one of supervisory management; the superordinate activity initiates, monitors, and terminates the subordinate activity, but beyond that, the subordinate node is autonomous. This is a management rather than a control scheme. The subsumption implication of continuing a work problem into a different work situation is that

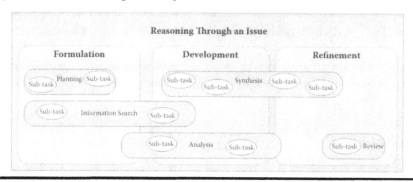

Figure 13.4   A subsumption diagram of work situations and work problems for reasoning.

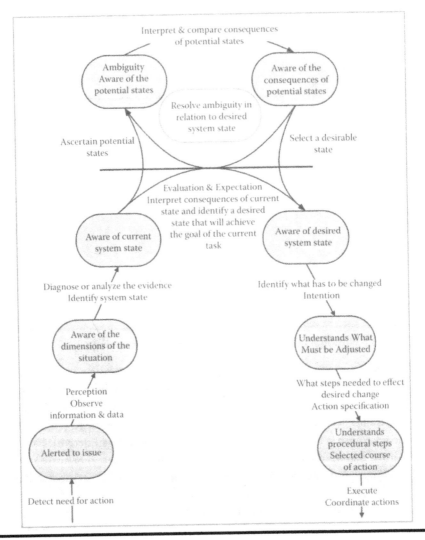

Figure 13.5 A decision ladder inspired by, but modified from, Rasmussen (1986).

it may constitute the same activity in many respects but is being executed from a different frame of reference. Such properties as intensity and detail may be shaped differently by the different frames of reference.

Control Tasks

Tasks are accomplished, problems resolved, and decisions made via transitions between cognitive states. Those transitions are induced by cognitive processes. The decision ladder is the representational form for depicting those cognitive states and cognitive processes. Figure 13.5 depicts the cognitive states and cognitive processes

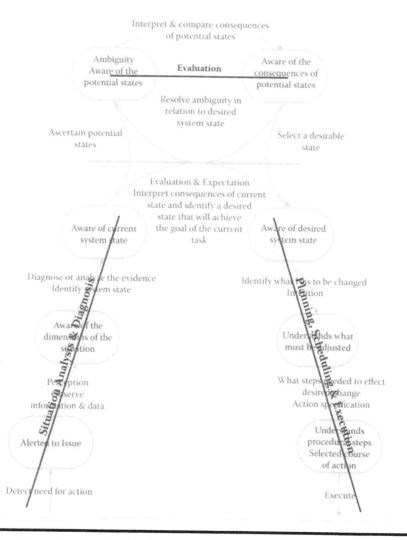

Figure 13.6   The decision ladder's three main phases.

that might be active in the various work problems for reasoning through an issue. States of knowledge are depicted as standardized nodes (ovals) between cognitive processes (directed, labeled arrows). As depicted in Figure 13.6, a task has three phases: analysis, evaluation, and planning.

As noted by Rasmussen (1986, p. 6), breakdown of the decision task may result in different elementary phases, depending upon the context in which the control decisions have to be made, in particular whether it is control of a physical system governed by causal physical laws or management of a social system governed by human intentions. The elementary phases depicted by Rasmussen (1986) are for a physical system governed by causal physical laws, whereas those I have generated for Figure 13.5 are for a system governed by human intentions.

Some take the decision ladder representation as implying a cognitive decision theory that assumes a linear, canonical sequence of information processing, starting from detection of a need for action, progressing through decision processes of options comparison and selection, and finishing with execution. Hollnagel and Woods (2005, p. 63) have apparently converged on that interpretation. Despite their claim, this element of the theory of CWA does not *assume an internal representation of a characteristic sequence of actions as a basis for the observed sequences of actions* (Hollnagel & Woods, 2005, p 63).* Rather, it assumes a set of characteristic cognitive states and processes that can contribute to the accomplishment of a task or the resolution of a problem.

Furthermore, the decision ladder is not a theory or model but rather a template onto which the cognitive states and cognitive processes identified in the analysis of work are mapped. Rather than specifying that these cognitive states and processes must be traversed in a canonical sequence, or even that they are always activated, the theory on which the decision ladder is based merely states that these cognitive states are potentially available and identifies the classes of cognitive process that are needed to change from one state to another.

In principle, any of the available cognitive states can be accessed from any other cognitive state. The CWA literature refers to *shortcuts*, a possibly unfortunate choice of terms because it does imply precedence for transitions of a canonical form. Here, I dispense with that term and also the associated terms, *leap* and *shunt*, and talk only of state transitions, thereby avoiding the implication of a canonical sequence of cognitive states and cognitive processes.

Additionally, within the literature, the form of a shortcut referred to as a leap describes a direct association between two cognitive states. There is an implication of a state transition with no intervening process. I find that implication disconcerting (in the physical world, at least, state transitions require an intervening process), and I prefer to characterize the type of cognitive event that appears to be process free as one in which the process is implicit. This strategy aligns the theoretical argument underlying the decision ladder with a body of research and theory on human expertise that distinguishes implicit from explicit knowledge.

---

* As evidence for their interpretation, Hollnagel and Woods (2005) quote Rasmussen (1986, p. 7): *Rational, causal reasoning connects the "states of knowledge" in the basic sequence.* Whether this can be taken as a claim for *an internal representation of a characteristic sequence of actions* is arguable, but that interpretation is at least incompatible with other arguments by Rasmussen (1986) and also with arguments by Vicente (1999) and Naikar et al. (2006).

Strategies

The decision ladder names cognitive processes but says nothing else about them. For example, a process identified in the left leg of Figure 13.5 is "observe information and data." This label does not, however, suggest how the observational process might be executed. As depicted in Figure 13.7, the execution strategy might employ systematic search based on a planned search strategy, or possibly, opportunistic search to look for relevant information, or even opportunistic search to look for information that contains preconceptualized features.

Within CWA, a strategy is defined as a category of cognitive task procedures that transforms one cognitive state into another. A strategy description is therefore a description of process rather than merely an identification of process. It constitutes a description of the way in which one cognitive state can be transformed into another. Typically, diverse strategies will be available to effect a transition between two specific cognitive states. Furthermore, a worker may shift unpredictably and opportunistically between available strategies during execution of a cognitive process aimed at inducing a cognitive state transition.

Social Organization and Cooperation

A complex, sociotechnical system is distributed and heterogeneous, comprising diverse human and technological functions. A distributed and heterogeneous system remains coordinated in part by direction or instruction and in part by processes of self-organization.

The unschooled view has a large enterprise being organized by management directives. In fact, this style of management can be ineffective. It will consist either of a hierarchy of tightly coupled management layers that result in micromanagement and instability, or else it will be a loosely coupled hierarchy in which the activities of management and operations have little mutual relevance or influence. Characteristically, in either type, management is disengaged from operations, and has a somewhat fanciful view of how operations unfold.

As observed by Weick and Sutcliffe (2001), effective management is mindful in that it pays attention to the complexity and subtlety of operational requirements. It takes account of how operations actually unfold versus how they might be thought to unfold. One particular role of management in this scheme is to set a context that will permit operational staff to develop effective and robust work processes in pursuit of the organization's goals. Mindful managers promote a permissive culture in which operational personnel are encouraged to provide meaningful information about operational complexity and to offer information that can guide development of effective management interventions. It could be said that the directives flow from operations to management rather than, as often stated, from management to operations (e.g., Leveson et al. 2006).

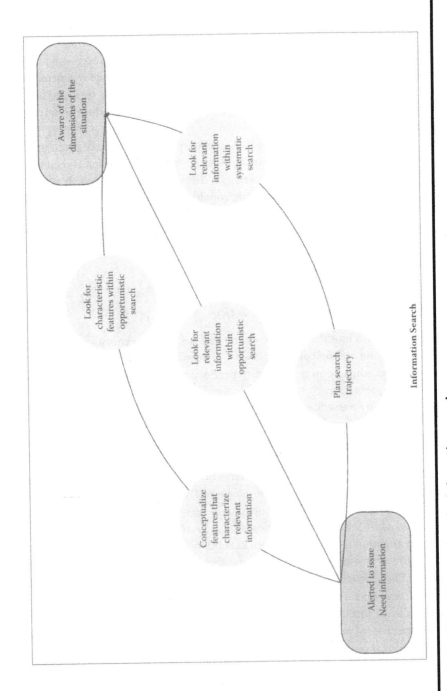

Figure 13.7 Potential strategies for information search.

The proposal I develop here is not, however, to allow operational staff to manage the system. Elsewhere, I have argued that workers are very good at optimizing local practices but, in doing so, can compromise global system integrity (Lintern, 2003). In his analysis of the destruction of two U.S. Army Black Hawk helicopters over Northern Iraq by two USAF F-15s during Operation Provide Comfort, Snook (2000) offers an evocative illustration. Procedures intended to enhance safety and effectiveness had evolved independently over time within the different organizational entities of this operation. The evolved procedures, although locally effective and locally robust, were not coordinated across the organization. The accidental destruction of the Black Hawk helicopters was precipitated largely by the failure of operational staff to understand how the members of other units were organizing their work. More generally, operational personnel should be permitted to develop a robust and detailed appreciation of local constraints on work practice, but management must retain the authority to oversee and maintain the global integrity of the system.

Information is the glue that binds disparate, independent entities into a purposeful collaborative system. Work context is established via management advisories about values and priorities, for example, the daily publication of Commander's Intent as is common in military operations. Doctrinal training also serves in this regard. In today's complex and distributed work environments, work tasks are typically the responsibility of many individuals who also rely heavily on support from technological subsystems. Coordination in such a system requires meaningful and timely communication between interacting workers and also mutual coordination of support functions made available by technological subsystems. Additionally, the character of work practice, work situations, and work problems must be open to the view of management.

All of these communications, both vertical and lateral, must critically promote mindfulness, the apprehension of significant meaning (Weick & Sutcliffe, 2001), as they promote both local productivity and global integrity. So much of today's organizational communication is mindless: meetings devoted to self-promotion and meaningless facts, mind-numbing PowerPoint briefings, and committee reports replete with jargon, acronyms, redundancy, and inconsistency.

One implication of this is that a mindful system is extensively interconnected, although it is not, as some may have it, completely interconnected. In many circumstances, communication can be unnecessary and can even obstruct or compromise work processes. Mindfulness, when applied to the redesign of a system, rationalizes interconnectivity (Klinger & Klein, 1999) so that useful communication is promoted and superfluous communication eliminated.

As becomes evident from the study of self-organizing systems, there is a potential concern with an extensively interconnected system. An adjustment in one part of the system can impact other parts; any design intervention within an interconnected system can have effects beyond those explicitly intended. Sometimes these effects can be undesirable. The solution here is not to limit connectivity but to maintain a view of the system as an integration of local and global processes coordinated systematically via mindful collaboration.

Worker Competency

Workers interact with the system by use of skills, rules, or knowledge (Figure 13.8). Skill-based behavior results when there is no conscious processing between perception and action. An example is that of lane following behavior by the driver of an automobile, where positioning in the approximate center of the lane is guided by the implicit perceptual symmetry of the lane markings on either side of the automobile. In general, skill-based behavior is guided by perceptual information in the form of space–time patterns that have become ingrained by extensive experience or practice.

Rule-based behavior is guided by sets of procedural instructions that specify a sequence of actions, some of which may be conditional, leading to branches or halts in the sequence. Such rules are typically constructed in advance of the required behavior and do not demand reasoning at the time of execution. The decision to advance, to branch, or to halt is typically determined on the basis of perceptually referenced rules.

Knowledge-based behavior is grounded in conscious and explicit reasoning. It is the foundation for deciding, planning, and problem solving. It relies on access to and careful consideration of meaningful and diverse elements of information.

People prefer to operate at the skill-based level of cognitive control but will revert to the rule-based level when encouraged to do so through training or

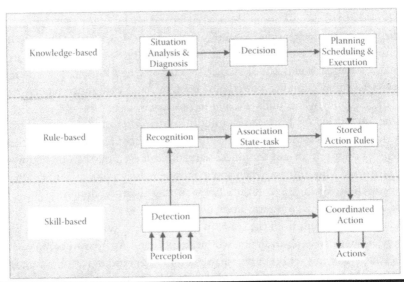

Figure 13.8 Work activity: skill-, rule-, and knowledge-based.

experience, or when an appropriate rule set is readily available and it is apparent that skill-based control will be difficult to execute. We undertake knowledge-based behavior when the problem we face is unfamiliar or possibly of a familiar type but contains unfamiliar details. It could be a problem we know well in an abstract or general sense (e.g., the purchase of a home) but that poses details that need to be assessed and evaluated.

Work constraints and work experience influence the level of cognitive control that is activated. Much activity is an opportunistic mix of skill-, rule-, and knowledge-based control (again, the purchase of a home offers an applicable illustration), the balance of which will tip toward more skill-based behavior as we gain experience. This can be desirable where skillful behavior is more efficient or more effective than knowledge- or rule-based behavior, but can be undesirable in highly regulated industries such as aviation or power generation where some prescribed procedures contain critical steps that cannot be omitted.

Opportunities for Theoretical Development

The Functional Structure of Work

> There is a significant body of empirical research from a number of quite diverse domains showing that problem solving protocols can be mapped onto an abstraction hierarchy representation.
>
> **Vicente, Christoffersen, and Pereklita (1995, p. 530)**

This quote by Vicente et al. (1995) notwithstanding, the relevance of the abstraction hierarchy (more generally, the abstraction–decomposition space) to the design of sociotechnical systems and its relevance to problem solving remain controversial. The nature of the abstraction–decomposition space and of structural means–ends relations is widely misunderstood, and this theoretical perspective on reasoning is largely ignored. I venture to say that beyond the devotees of CWA, there are relatively few who understand the abstraction–decomposition space or accept it as a structural representation of expert reasoning.

In contrast to Vicente et al. (1995), I regard the evidence that experts solve problems by navigating through a dimensional space of functional abstraction and functional decomposition as meager. Although I have not yet been able to access all the evidence cited by Rasmussen (1986), Vicente et al. (1995), and Vicente and Rasmussen (1992), it all appears to rest heavily on the observation that problem-solving protocols can be mapped onto an abstraction–decomposition space (Figure 13.9). I view that as indirect evidence; problem-solving protocols could likely be mapped onto other representational structures such as argument maps (Facione & Facione,

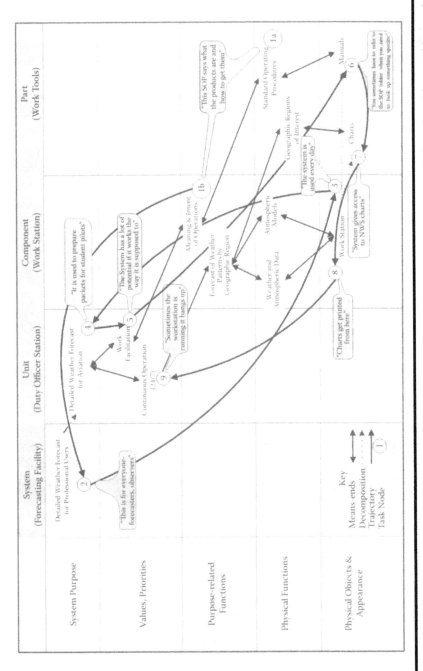

Figure 13.9 An abstraction–decomposition space for a fragment of a weather-forecasting work domain with a problem-solving trajectory overlay (statements in callouts are quotes from a forecaster).

2007) and concept maps (Crandall et al., 2006). Some sort of comparative evaluation of the different protocol-structure mappings seems warranted.

Additionally, the claim emerges from what appears to be a somewhat selective reading. For example, although Dunker (1945, p. 4) notes that the *functional value of a solution is indispensable for the understanding of its being a solution* and follows that comment closely with a diagram somewhat similar to an abstraction–decomposition space, he devotes less than a page to this argument within a monograph of over 100 pages. There is substantive literature on reasoning, problem solving, and expertise, yet any notion of an abstraction–decomposition space together with structural means–ends relations is difficult to find. Most often discussions focus on the use of heuristics (e.g., Simon, 1981). I find it disquieting that our abstraction–decomposition space has raised barely a ripple of interest in this rather substantive body of research.

Sarcedoti (1974) is one who, in proposing a hierarchy of abstraction spaces as a means of reducing combinatorial complexity for planning, appears at first to appreciate the contribution of an abstraction analysis. However, having outlined this proposition, he then proceeds to treat abstraction in terms of decomposition. Ayn Rand also relied on abstraction hierarchies as a basis for rational argument within her Objectivist Epistemology (Rand, 1979/1990), but her illustrations refer to classification hierarchies (Lintern, 2006).

I view the evidence for use of structural means–ends relations to be equally meager. Literature searches with *means–ends* (or its variations) as a keyword find little that is relevant. The most dominant body of research has emerged out of the claim by Simon (1981, p. 223) that *human problem solving is basically a form of a means–ends analysis that aims at discovering a process description of the path that leads to a design goal*. It is however clear from his description of means-ends analysis (Ibid., p. 141–142) that Simon conceives of the links as activity means–ends relationships, as apparently do all others who have been guided by his conceptualization.

An abstraction–decomposition space can be viewed as a form of cognitive map as can, for example, a concept map. There is a body of literature on this topic, apparently stimulated in large part by Tolman (1948) and continuing until at least quite recently (Kitchin, 1994). It would seem prudent to review this body of literature for its relevance to our use of the abstraction–decomposition space and of any insights it might offer. I have yet to undertake that review but continue to think of it as an important project for the near future.

Given my critique, readers may wonder why I remain committed to the abstraction–decomposition space and means–ends relations. I do so primarily because it makes intuitive sense to me. From my ecological perspective, I must be able to describe the functional structure of the environment (the workplace) in some manner. As far as I can tell, Rasmussen was the first to develop a systematic and comprehensive representation of functional structure, and as yet there does not appear to be a plausible competitor. I would be happy to offer a better alternative if I could think of one, but to this point in time, I have not. Colloquially put, the abstraction–decomposition space is the only game in town.

It is nevertheless disconcerting that the evidence is not stronger, and I would like to see us make some progress here. Work is ongoing to show that systems based on the abstraction–decomposition space work well, but applied experimental and design work is inevitably less focused than is desirable to fully establish a hypothesis. We can always question whether a design works well precisely and specifically because of its functional structure or for some other reason.

I would like to see foundational research directed at this issue. Could diverse reasoning and problem-solving narratives be mapped onto different types of structures such as argument maps and concept maps as well as abstraction–decomposition spaces? Does the abstraction–decomposition form emerge as the most suitable for capturing the essence of problem-solving for reasoning? Does it emerge as the most suitable for design?

Additionally, I would like to see experimentation in which information is selectively removed from workspaces designed on the basis of an abstraction–decomposition space. Vicente et al., (1995) have undertaken a test that converges on this but, in my opinion, does not quite get there. Although there is evidence from this research that the reasoning strategies of better performing subjects appear to traverse the abstraction–decomposition space more effectively, the advantage in performance is not tied to any specific information suggested by abstraction–decomposition space.

My preferred investigative strategy is to identify essential information by analysis and then to distort or remove information elements selectively in subsequent empirical tests. I have had success with this strategy in research aimed at identifying information used by aircraft pilots for visually guided landings (Lintern & Liu, 1991; Lintern & Walker, 1991). Because specific hypotheses were linked to the removal or distortion of specific information, the role of particular information elements was demonstrated conclusively.

This strategy could provide worthwhile evidence for means–ends reasoning through an abstraction–decomposition structure. The aim would be to demonstrate that deletions or distortions of information create biases or errors in reasoning or problem-solving performance, as would be suggested by abstraction–decomposition analysis. I recognize that a complex sociotechnical system presents a more diverse and complicated information space than does visual landing of an aircraft, but I nevertheless anticipate that this sort of experiment could provide persuasive evidence for abstraction–decomposition as a basis for reasoning and problem solving.

The Problem of Scale

> If we tried building ships, palaces, or temples of enormous size, yards, beams, and bolts would cease to hold together.

Attributed to Galileo by D'Arcy Thompson (1942/1968, p. 27)

The state should be allowed to grow only so far as it can increase in size without loss of unity

Attributed to Plato by J.R. Bellerby (1945, p. 90)

Problems of scale are well recognized in the architectural and engineering disciplines. As suggested by the quote attributed to Galileo, it would not do to geometrically rescale the physical parameters of a successful bridge design to satisfy the requirement for a larger span. The problem rests in the fact that structural strength scales at a lower order than geometric dimensions. More generally, there are diverse properties in any system that scale at different orders. This statement of scaling relationships is often referred to as the principle of similitude or the Principle of Dynamical Similarity (Thompson, 1942/1968).

One of the earlier discussions is by Buckingham (1914), who focused on the similitude to be observed in physical systems such as aerodynamically similar aircraft propellers of different sizes. This principle is also involved in the study of allometric relations in living organisms, whereby the identity of functions across considerable variations in size remains possible only through systematic variations in form (Thompson, 1942/1968), and in the study of invariant relationships between features that change with development, growth, or natural variation (Kugler & Turvey, 1987).

I have yet to notice a discussion of scale within cognitive engineering. The apparent attitude is that we are designing systems for specific contexts or at least for systems of characteristic size. Nevertheless, we generalize our arguments across work contexts including those that encompass a wide range of sizes. I propose that the scaling issue is important, and I suggest that an approach to it will start with consideration of the principle of similitude.

Communication overhead offers a cogent illustration of this principle. The requirement for communication between workers increases coordinative challenges and cognitive load. Efficiencies will be realized when the complexity and frequency of communications undertaken by human agents are rationalized.

In today's often-distributed work environments, there are multiple options for maintaining connectivity, and every member of the organization can, in principle, communicate with every other member regardless of differences in status or geographic location. That can work well in small organizations, but the number of interconnections scales at $n*(n-1)/2$, where n represents the number of staff members. Table 13.1 shows the growth of interconnections with growth in staffing for selected numbers of staff members. For an organization of even modest size, the number of connections required for full interconnectivity becomes unwieldy at least in human terms. Of course, no one proposes full interconnectivity for large organizations, but, if left to work this out on their own, many in management will strain against the limits of human capability, and some will even go well beyond it.

Table 13.1 Scaling of interconnections against number of staff members for a fully interconnected system

Staff members n	Interconnections $n(n-1)/2$	Interconnections (team size = 5)
5	10	10 + 0
10	45	20 + 1
15	105	30 + 3
25	300	50 + 10
100	4950	200 + 190

Note: Column 3 assumes a single connection between each unique pair of teams.

The issue I address here relates to lateral connectivity structures that will support essential work coordination and collaboration between workers, and vertical connectivity structures that will support essential manager–worker coordination. The methods of CWA can be directed at identifying organizational structures for support of both lateral and vertical connectivity, as Naikar, Pearce, Drumm, and Sanderson (2003) have done for small command-and-control teams. However, despite the success of that work, the principle of similitude suggests that it would be unwise to scale the patterns of team structure identified by Naikar et al. (2003) up to substantially larger command-and-control teams.

My thinking on this issue has been inspired not only by the principle of similitude but also by Simon's (1981) discussion of modularity (or subassemblies, by his account). Extending Simon's argument, any problem in a fully interconnected system can have repercussions throughout, whereas in a modular system, its effects can be localized to the module in which it occurs. Within a work system, work packages developed by small, well-integrated teams and designed as far as possible to be modular in the sense that their reliance on other system products are minimized, can reduce the need for communication connectivity. The third column of Table 13.1 illustrates how the number of interconnections is rationalized for a team size of five; the calculations assume full intra- and inter-team connectivity. The first number in the column represents the number of connections required per team (always 10 for 5-member teams) multiplied by the number of teams, and the second number represents the number of connections required for a single connection between each unique pair of teams.

In formulating these ideas, I ponder on issues such as

- How many peers, managers, or subordinates can someone interact with effectively?
- How is this number influenced by geographic distribution and temporal displacement of work activities?

- How is this number influenced by the nature of the work?
- How is this number influenced by the capabilities of available communication systems?

As far as I can ascertain, there is no body of scientific research or theory that addresses these issues directly.

Note that the scheme as outlined deals specifically with operations (workers communicating with each other about workflow and work products, and workers and managers communicating with each other about the conduct of work). Any sort of scheme that rationalizes connectivity in relation to workflow and work products must also take account of global constraints such as those discussed earlier in relation to communication of purpose and values possibly via use of announcements or indoctrination.

Most generally, I suggest that this problem of rescaling sociotechnical systems is pervasive and that no one understands how to do it effectively. There are many, I believe, who recognize that an effective organizational structure of a small system will not do for a large system, and that system growth must be accompanied by periodic restructuring, but the manner in which it is possible to progress seamlessly through that periodic restructuring remains a mystery. I propose that we may begin to unravel that mystery by serious consideration of the principle of similitude.

Reprise

Earlier in this chapter I suggested that we do not need to envy physicists for the regularity and predictability of the phenomena they study. The general belief that human behavior is unpredictable has emerged, I suspect, from attempts to predict the wrong sorts of details. It remains true that we often cannot predict how even those we know intimately will respond to a suggestion or what they might order at a restaurant. On the other hand, we do know conclusively that their cognitive capabilities are shaped by being situated in their environment, that they can become functional entities of distributed cognitive systems, that their behavior is constrained by the available affordances, and that they exhibit the characteristics of self-organizing, nonlinear systems. The theoretical concepts underlying each of the analyses within the framework of CWA are possibly less well established but, to the extent we have confidence in them, they also imply immutable constraints on human behavior.

These ideas offer little comfort for those who wish to predict and control behavioral patterns by proceduralizing work environments to considerable detail. On the other hand, for those who wish to design a collaborative, distributed cognitive system and are content with allowing workers to complete the design, the behavioral regularities that emerge from these constraints offer important design guidance. As for the law of gravity, these constraints are immutable. They have shaped and will shape the behavior of all people who have ever lived, of all who are living today, and of all who will ever live.

By laying out the theoretical foundation for CWA together with the theory of work practice specific to the analyses contained within its framework, I have sought to satisfy two goals. The first is to promote revaluation within our own community. Is the evidence we use to support our faith in these forms of analysis adequate? Although we may not need better evidence for ourselves, we leave ourselves in a vulnerable position when seeking to persuade others of the sense of this if that evidence is not robust. The second is to help those who are not practitioners of CWA understand the rationale for our commitment to the forms of analysis incorporated into the framework.

My own view is that some of the empirical support for what we do is meager. In addition, I also regard the conceptual ideas behind some of our analytic forms as thin. For example, I suspect there is more that we could say about levels of cognitive control and about social organization and cooperation to strengthen the analyses. Might we follow the guidance of both Rasmussen (1986) and Vicente (1999) in seeking ideas from diverse scientific areas to enrich our framework? At this time, my own thoughts turn to the literature on Macrocognitive Functions (Crandall et al., 2006) and Joint Cognitive Systems (Woods & Hollnagel, 2006), but one of my aims for this chapter is to encourage others to follow their own inclination in exploring ways of strengthening and enriching the theory and practice of CWA.

Nevertheless, CWA has already attained the status of a mature analytic framework and now occupies a special niche within cognitive systems engineering. It addresses systems design problems with far more breadth than any other approach and shows particular strength in the design of the types of large-scale sociotechnical systems that promote enterprise transformation (Rouse, 2005a,b), those first-of-a-kind systems that are complex and distributed. It offers a comprehensive analytic framework for addressing issues in concept development, and then, proceeding systematically through the human systems integration issues. It thereby provides the capability for a cognitive engineer who is invited early into the design of a complex sociotechnical system to contribute as an equal and effective partner. Although we need not dismiss tools of cognitive engineering that target leverage points, cognitive support tools, or other specific system capabilities, we should recognize the strengths of CWA so that we apply it to the design problems for which it is suited.

Acknowledgments

I should like to express my appreciation to Helen Kieboom of Tenix Corporation and to Robin Hutchinson of Monash University for their efforts in reviewing an earlier version of this chapter.

References

Ashby, W.R. (1957). *An Introduction to Cybernetics*. London, Chapman & Hall Ltd.

Bellerby, J.R. (1945). Review of the future of economic society by Roy Glenday. *The Economic Journal*, Vol. 55, No. 217, pp. 89–92.

Buckingham, E. (1914). On physically similar systems: Illustrations of the use of dimensionless Equations. *The Physical Review, Vol IV*, Series II, 345–376.

Clancey, W.J., Sachs, P., Sierhuis, M., & van Hoof, R. (1998) Brahms: simulating practice for work systems design. *Int. J. Human–Computer Studies*, 49, 831–865.

Crandall, B., Klein, G., & Hoffman, R. (2006). *Working Minds: A Practitioner's Guide to Cognitive Task Analysis*. Cambridge, MA: MIT Press.

Cummings, M.L. (2006). Can CWA inform the design of networked intelligent systems? *Moving Autonomy Forward Conference 2006*, De Vere Hotel Belton Woods. Lincoln, UK: Muretex, pp. 136–143.

Dinadis, N. & Vicente, K.J., 1999. Designing functional visualizations for aircraft system status displays. *International Journal of Aviation Psychology* 9, 241–269.

Duncker, Karl (1945). *On Problem Solving*. Psychological Monographs, 58 (5).

Facione, P.A., & Facione, N.C. (2007). *Thinking and Reasoning in Human Decision Making: The Method of Argument and Heuristic Analysis*. The California Academic Press, Millbrae, CA.

Freeman, W.J. (1995). *Societies of Brains: A Study in the Neuroscience of Love and Hate*. Hillsdale, NJ: Lawrence Erlbaum Associates.

Gibson, J.J. (1979). *The Ecological Approach to Visual Perception*. Houghton Mifflin, Boston.

Hollan, J., Hutchins, E., & Kirsh, D. (2000). Distributed cognition: Toward a new foundation for human–computer interaction research. *ACM Transactions on Computer–Human Interaction*, Vol. 7, No. 2, Pages 174–196.

Hollnagel, E., & Woods, D.D. (2005). *Joint Cognitive Systems: Patterns in Cognitive Systems Engineering*. CRC Press, Boca Rotan, FL.

Houghton Mifflin (2000). The American Heritage Dictionary of the English Language, 4th Edition. Houghton Mifflin, Boston.

Hutchins, E. 1995. *Cognition in the Wild*. MIT Press, Cambridge, MA.

Johnson-Laird, P.N. (1983). *Mental Models*. Harvard University Press, Cambridge, MA.

Jordan, B. (1989). Cosmopolitical obstetrics: Some insights from the training of traditional midwives. *Social Science Medicine*, 28(9), 925–944.

Kitchin, R.M. (1994). Cognitive maps: What are they and why study them? *Journal of Environmental Psychology* 14: 1–19.

Klinger, D., & Klein, G. (1999). An Accident Waiting to Happen. *Ergonomics in Design*, 7 (3), 20–25.

Kugler, P.N., & Lintern, G. (1995). Risk management and the evolution of instability in large-scale industrial systems. In P. Hancock, J. Flach, J. Caird, & K. Vicente (Eds.), *Local Applications of the Ecological Approach to Human-Machine Systems* (pp. 416–450). Hillsdale, NJ: Lawrence Erlbaum Associates.

Kugler, P.N., & Turvey, M.T. (1987). *Information, Natural Law, and the Self-Assembly of Rhythmic Movement*. Hillsdale, NJ: Lawrence Erlbaum Associates.

Lave, J. (1988). *Cognition in Practice*. Cambridge University Press, New York.

Lave, J., & Wenger, E. (1991). *Situated Learning: Legitimate Peripheral Participation*. Cambridge University Press, New York.

Leveson, N., Dulac, N., Zipkin, D., Cutcher-Gershenfeld, J., Carroll, J., & Barrett, B. (2006). Engineering resilience into safety-critical systems. In Erik Hollnagel, D. D. Woods and N. Leveson (Eds.). *Resilience Engineering: Concepts and Precepts*. Hampshire, England: Ashgate, Ch 12 (pp. 95–123).

Lind, M (2003). Making sense of the abstraction hierarchy in the power plant domain. *Cognition, Technology & Work*, 5: 67–81.

Lintern, G. (2006). Foundational issues for work domain analysis. *Proceedings of the 50th Human Factors and Ergonomics Society Annual Meeting*. (pp. 432–436). Santa Monica, CA: Human Factors and Ergonomics Society. [CD-ROM].

Lintern, G. (2003). Tyranny in rules, autonomy in maps: Closing the safety management loop. In R. Jensen, (Ed.), *Proceedings of the Twelfth International Symposium on Aviation Psychology*. (pp. 719–724). Dayton, OH: Wright State University. [CD-ROM]

Lintern, G. (2000). An affordance-based perspective on human–machine interface design. *Ecological Psychology*, 12, 65–69.

Lintern, G. (1995). Flight instruction: The challenge from situated cognition. *The International Journal of Aviation Psychology*, 5, 327–350.

Lintern, G., & Liu, Y. (1991). Explicit and implicit horizons for simulated landing approaches. *Human Factors, 33*, 401–417.

Lintern, G., & Kugler, P.N. (1991). Self organization in connectionist models: Associative memory, dissipative structures, and Thermodynamic Law. *Human Movement Science*, 10, 447–483.

Lintern, G., & Walker, M.B. (1991). Scene content and runway breadth effects on simulated landing approaches. *The International Journal of Aviation Psychology*, 1, 117–132.

Naikar, N., Pearce, B., Drumm, D., & Sanderson, P.M. (2003). Designing Teams for first-of-a-kind complex systems using the initial phases of CWA: Case Study. *Human Factors*, 45(2), 202–217.

Naikar, N., Moylan, A., & and Pearce, B. (2006). Analysing activity in complex systems with cognitive work analysis: concepts, guidelines, and case study for control task analysis. *Theoretical Issues in Ergonomics Science*, 7 (4), 371–394

Prigogine, I., & Stengers (1984). *Order out of Chaos*. Bantam Books, New York.

Rasmussen, J. (1986). *Information Processing and Human–Machine Interaction: An Approach to Cognitive Engineering*. North-Holland, New York.

Rasmussen, J., Pejtersen, A.M., & Goodstein, L.P. (1994). *Cognitive Systems Engineering*. Wiley, New York.

Rand, A., (1979/1990). *Introduction to Objectivist Epistemology*, Expanded 2nd Edition. H. Binswanger & L. Peikoff (Eds.). New York: Meridian.

Rouse, W. B. (2005a). A Theory of Enterprise Transformation. *Systems Engineering*, 8(4).

Rouse, W. B. (2005b). Enterprises as systems: Essential challenges and approaches to transformation. *Systems Engineering, 8, (2)*, 138–149.

Sacerdoti, E.D. (1974). Planning in a hierarchy of abstraction spaces. *Artificial Intelligence*, 5, 115–135.

Saxe, (1991). *Culture and Cognitive Development: Studies in Mathematical Understanding*. Hillsdale, NJ: Lawrence Erlbaum Associates.

Scribner, S., & Fahrmeier, E. (1982). Practical and theoretical arithmetic: Some preliminary findings. Industrial Literacy Project, Working Paper No. 3. Graduate Center, CUNY

Simon, H.A. (1981). *The Sciences of the Artificial*, 2nd ed., MIT press, Cambridge, Massachusetts.

Snook, S.A. (2000). *Friendly Fire*. Princeton University Press, Princeton, New Jersey.

Suchman, L.A. (1987). *Plans and Situated Actions: The Problem of Human–Machine Communication*. Cambridge University Press, New York.

Thompson. D.W. (1942/1968). *On Growth and Form* London: Cambridge University Press.

Tolman, E.C. (1948). Cognitive Maps in Rats and Man. *Psychological Review*, 55: 189–208.

Vicente, K.H. (1999). *Cognitive Work Analysis: Towards Safe, Productive, and Healthy Computer-Based Work*. Mahwah, NJ: Lawrence Erlbaum Associates.

Vicente, K.J., Christoffersen, K., & Pereklita, A. (1995). Supporting operator problem solving through ecological interface design. IEEE *Transactions on Systems, Man,* and *Cybernetics, 25,* 529–545.

Vicente, K.J., & Rasmussen, J. (1992). Ecological interface design: Theoretical foundations. *IEEE Transactions on Systems, Man, and Cybernetics*, SMC-22,589–606.

Warren, W.H., & Whang, S. (1987). Visual guidance of walking through apertures: Body-scaled information for affordances. *Journal of Experimental Psychology: Human Perception and Performance*, 13, 371–383.

Weick, Karl E., & Sutcliffe, K.M. (2001). *Managing the Unexpected: Assuring High Performance in an Age of Complexity*. John Wiley, San Francisco.

Wigner, E.P. (1979). *Symmetries and reflections: Scientific Essays of Eugene P. Wigner*. Ox Box Press, Woodbridge, CT.

Woods, D.D., & Hollnagel, E. (2006). *Joint Cognitive Systems: Patterns in Cognitive Systems Engineering*. CRC Press, Boca Raton, FL.

Author Index

Subject Index

Printed and bound by CPI Group (UK) Ltd, Croydon, CR0 4YY

17/10/2024

01775706-0003